THE MACHINE IN AMERICA

THE MACHINE IN AMERICA

A Social History of Technology

Second Edition

❦

CARROLL PURSELL

The Johns Hopkins University Press

Baltimore

Printed in the United States of America on acid-free paper
9 8 7 6 5 4 3 2 1

The Johns Hopkins University Press
2715 North Charles Street
Baltimore, Maryland 21218-4363
www.press.jhu.edu

Library of Congress Cataloging-in-Publication Data

Pursell, Carroll W.
The machine in America : a social history of technology / Carroll Pursell.—2nd ed.
p. cm.
Includes bibliographical references and index.
ISBN-13: 978-0-8018-8578-5 (alk. paper)
ISBN-10: 0-8018-8578-7 (alk. paper)
ISBN-13: 978-0-8018-8579-2 (pbk. : alk. paper)
ISBN-10: 0-8018-8579-5 (pbk. : alk. paper)
1. Technology—Social aspects—United States—History. 2. Industrial revolution—United States. I. Title.
T14.5.P87 2007
303.48′3—dc22 2006052010

A catalog record for this book is available from the British Library.

FOR ANGELA

CONTENTS

PREFACE

THIS BOOK IS A BRIEF introduction to the social history of American technology. It is necessarily selective, and each reader will no doubt fail to discover a favorite machine, process, or inventor. It is social history because the emphasis is on the social role of technology—the way in which it interacts with other aspects of American life—rather than on the internal logic of mechanisms through time. My aim is both to introduce the nonspecialist to the subject and to suggest to specialists some of the connections that I have found interesting over the years.

If looked at through the lens of technology, the periodization of American history might appear very different from what generations of school texts have suggested. My purpose here, however, is to suggest that technology is a part of our general history. Therefore I have been content to march in a rather conventional way, through the colonial period, the Early Republic, the Gilded Age, the Progressive Era, war and depression, and war again, down to the present. Imposed on this traditional chronology are the more particular stories of transplanting a medieval technology to a new setting, then replacing it with a borrowed industrial technology, reforming all this again through the agency of science, and finally perfecting and having to face what Lewis Mumford called the "pentagon of power."

Although the book sacrifices encyclopedic coverage and most technical detail, I have not hesitated to place my own interpretation on events. More than anything else, the stories told here seek to demythologize technology. Just as the emperor's new clothes turned out to be an agreed-upon fiction, so is American technology clothed in a black box more imagined than actual. It is a part of our material condition and is of our own making. Since individual technologies and their networks enhance or undermine the people we want to be and the society in which we want to live, we as citizens must try to understand this mighty force and see it not only for what it is but also for what it does, what it means, and what it might be.

By saying that technology is a part of our material condition and is of our own making, I merely mean to assert that the tools and processes we use are a part of our lives, not simply instruments of our purpose. At the same time, I mean to claim that they do not come from nowhere nor mysteriously out of themselves. The usual way to put this notion is to say that technology is not autonomous. Of course, it is not true that we each have equal power to choose our own technology in every case, nor that our reasons for choosing one over another are always good, nor that we always foresee and desire every consequence of those choices. No aspect of life is so simple or straightforward. I do believe, however, that those with sufficient foresight and power over their own affairs create or choose those technologies they think will preserve and increase that power. We might say that these technologies are socially constructed for certain purposes. The impact they have on any one of us is the result of a complex calculus of class, race, gender, luck, and other similar variables. There is certainly a sense in which we all—perpetrators, victims, beneficiaries, bystanders—collaborate in this social construction. But it is equally certain that we do not all have an equal say in that activity.

While I speak of the people we want to be and the society in which we want to live, I cannot, of course, offer a blueprint that would fit all of those people and all their notions about the good society. We do have in the writings of our social visionaries, from Thomas Jefferson through Mark Twain and Sojourner Truth to Paul Goodman and Lewis Mumford, a consistent theme of hope that America might be a better country than those that have gone before, one characterized by "liberty and justice for all." Such a vision is presumably embodied in our Constitution and Bill of Rights. Technology was hardly mentioned in those documents, but, as the political philosopher Langdon Winner warned, it has had developed around it another, parallel, shadow constitution: one that imposes a logic of hierarchy and inequality, a reification of power, and a signal disregard for liberty and justice. As many of the founding generation feared, a technology not subordinated to our highest political aspirations has become a bulwark of our worst.

❦

The introduction sets the medieval background for European expansion around the world, especially to North America. Both the germs of the idea of progress and the mechanical means for harnessing natural power to human purposes were expanded. The following chapters are divided into five parts. The first, chapters 1 through 3, describes the bringing to America of first a medieval technology and then the new industrial technology that grew up in England during the latter half of the eighteenth century.

The attention in chapter 1 is on the way in which technology was brought from the old world to the new and was there modified to fit the new circumstances of environment and opportunity. Chapter 2 sketches in the British Industrial Revolution and its effect on America, where it replaced the older ways of doing things and undermined idealized political notions of Republican virtue. Chapter 3 shows the effect on transportation, a particularly important aspect of life in a sprawling, sparsely settled, and expanding nation.

Part II, which consists of chapters 4 and 5, tells the story of how in two important areas, manufacturing and agriculture, American technology began to take on a distinctive shape as it adapted and extended the technical base of the Industrial Revolution.

Chapter 4 is about the innovation of an American system of manufactures, its limits, and other changes in industries that owed little to the fabrication of interchangeable parts. In chapter 5, the discussion turns to agriculture, the occupation of most Americans in the early years. Agriculture was significantly mechanized during the nineteenth century, but because of a lack of available power it was not fully industrialized.

Part III, chapters 6 through 8, deals with the way that new American technologies and dynamism reshaped three critical landscapes in their own image. Chapter 6 shows how various technologies both made possible and shaped the rise of American cities in the nineteenth century, and chapter 7 how technology led to the exploitation, and even the definition, of the West.

Technical changes fostered mining, lumbering, farming, and other tasks that were, even at the time, more important than cowboys and sheriffs. Chapter 8 picks up the theme of the export of American technology, at first in the form of new and improved inventions introduced into Europe and then as the imposition of American hegemony on indigenous peoples in two hemispheres.

In part IV I explain how, through the agencies of science, corporations, and government, American technology has created a social hegemony not only over the way we live but over our very evaluation of that life. During a period of some seven decades, the logic of modern technology was worked out in large systems seeking to centralize power and control all significant aspects of national life.

Chapter 9 begins the story of how modern science began to transform the ways, one after another, in which we organize and do things. Exploration, conservation, labor management techniques, civic governance, and industrial research were a few of the important areas in which engineers tried to tame and control the processes of social change. Chapter 10 shows how, between the two world wars, the federal government stepped in to help transfer technological innovation (radios, automobiles, airplanes) to public use, feeding the great burst in sales of consumer durable goods during the 1920s. Chapter 11 investigates the result of that period of technical change, especially the way in which the government attempted, during the depression, to adjust social services so as to revive technological momentum and care for those injured by it. Chapter 12 examines the intense cooperation between science and technology during World War II and the changed expectations and institutions that made the United States the most powerful nation in the world, though one burdened with a military-industrial complex that dangerously compromised its purpose and even its continued existence.

Sometime during the 1970s, technology, along with much of the rest of American society, passed from a modern to what can be called a postmodern configuration. In chapter 13 I attempt to say something meaningful about the ways in which a modernist pattern of technology was challenged and something more "appropriate" and "flexible" was proposed. Chapter 14 deals with the extraordinary expansion of our electronic environment through the spread of computers and the Internet, along with their elaborations, both at home and abroad. Finally, chapter 15 provides a description of the phenomenon of globalization and points out some of the ways in which America contributes to its makeup and workings.

❦

When the plan for this book was first laid out, no one had published a serious history of American technology from colonial times to the present since John W. Oliver's disappointing *History of American Technology,* which appeared in 1956. For many years, those of us who might have done so told ourselves that we as yet knew too little about the details of the story, let alone its larger meaning. Then, perhaps a decade ago, many of us suddenly realized that we now knew too much. Rather than being grateful for new information presented, the reader of today is more likely to complain of things left out and meanings misunderstood, or worse yet, of being smothered in detail. So be it. At a time when the historical profession in general decries the loss of comprehensive narratives and historians of technology are thinking more and more about how to "mainstream" their special interests through both their professional and their public activities, it is proper and inevitable that we should begin to attempt books such as this.

In another sense, however, it is much too late to attempt to write "the" history of American technology. There is not one but many stories to be told. I believe this one to be true, but I know that there are other, different accounts, which are true as well. Nevertheless I hope it is not too optimistic to believe that over the years books like this one will deepen and improve our understanding not just of American technology, but of America itself.

❦

Over thirty years ago Hunter Dupree, relaxing with some of his graduate students, remarked that when the history of American science and technology were properly understood, that knowledge would force us to rewrite American history as a whole. I cannot claim to have made that breakthrough, but that dream has inspired my career. The training and example he has given me over the years have been reinforced by the generosity and decency of Mel Kranzberg, who virtually created the field of the history of technology and gave me my first job in it. In the years since, my doctoral students have been patient with my telling of these stories, and I have learned an enormous amount from them. It gives me great satisfaction to thank them by name: Aaron Alcorn, Molly Berger, Mark Bowles, Betsy Bradley, Gail Cooper, Don Fitzgerald, Mary Ann Hellrigel, David Hochfelder, Susan Horning, Bernie Jim, Russell Jones, Tom Kinney, Arwen Mohun, Peter Neushul, Craig Semsel, Stephanie Smith, Jeffrey Stine, Jim Williams, Neil York, Jeff Yost, and Geoffrey Zylstra.

Finally, I wish I could express how important Bruce Sinclair has been to me over the past four decades. I can't imagine my career without his friendship and example. My children, Becky and Matt, have made my past worthwhile and my future hopeful. And most important, Angela Woollacott, through her love and eagerness to take on intellectual challenges, has made everything I do better than it would otherwise have been.

THE MACHINE IN AMERICA

INTRODUCTION

THE EUROPE THAT FIRST "discovered" then settled America was, according to the medievalist Lynn White, "coming to think of the cosmos as a vast reservoir of energies to be tapped and used according to human intentions."[1] The roots of this attitude went far back in the Judeo-Christian tradition, to the first chapter of Genesis. There it was recorded that God created people in the image of their creator and that they should "have dominion over the fish of the sea, and over the fowl of the air, and over cattle, and over all the earth, and over every creeping thing that creepeth upon the earth." The injunction was added: "Be fruitful, and multiply, and replenish the earth, and subdue it: and have dominion over the fish of the sea, and over the fowl of the air, and over every living thing that moveth upon the earth." The development of both science and technology in the Western world—as well as the tendency to use science itself as a technology—grew in part out of the religious urge to know and worship God and to exercise the dominion that had been given to the human species.

It has often been noted that the ancient Mediterranean world, while being the acknowledged seat of Western civilization, did not produce technological innovations consistent with its advances in other areas of human endeavor. More specifically, mechanical devices consciously adapted to saving labor do not appear to have been aggressively pursued. One common explanation for this failing is that societies based on slave labor did not respect work as such and therefore had little incentive to improve or otherwise "save" it. This lack of interest in labor-saving techniques is also clearly seen in China from the fourteenth century on, when the dominant philosophy shifted away from Confucianism to one that denigrated the material world and thus helped bring a halt to the previous technological progress in the East.

In the West, however, a flexible Christian church gradually developed an attitude toward productive labor, as well as toward devices to enhance it, that proved a fertile seedbed for technological change. The concept of a creator who had literally fashioned the material world and all the creatures therein, the notion of work as a form of praise and prayer, and other ideas growing out of early monastic centers all made it possible to look at technological progress as an acceptable, even praiseworthy aim.

This new attitude was certainly apparent in the comprehensive Plan of St. Gall, drawn up in the early ninth century to guide the construction of a state-of-the-art monastic community center. One outstanding feature of the plan was a water-powered mill to grind the community's grain into flour. Romans knew of the water mill but appear to have made little use of it. The fact that water wheels and mortars and pestles for grinding were included in the Plan of St. Gall strongly suggests that it was the Benedictines who adopted this practical innovation and spread its use throughout Merovigian and Carolingian Europe.

The five hundred years before Leonardo da Vinci, roughly 1000 to 1500, were decisive for the Western attempt to harness the power of nature to the purposes of humankind. Before that time, technical development had been slow and random. This "empirical groping" was now converted into an enthusiastic program, conscious and widespread, designed to make use of the surrounding energy.

By the year 983, for example, water power was being applied to a fulling mill to treat woolen cloth, whereas for centuries earlier it had only been used in mills to grind grain. In 1086 the *Domesday Book,* that great English inventory and census, showed that the three thousand villages of England were served by 5,624 water mills. By 1185 the first windmill had been constructed in Europe. Water, wind, and animal (particularly horse) power remained the basic sources of energy until the nineteenth century. By the time Columbus undertook his voyages of exploration, Europe had access to sources of power more varied than those of any previous culture, and also a wide range of devices for applying that power directly to perform useful work. Even so, fantasy still outran actuality. In the thirteenth century the monk Roger Bacon predicted that in the future, "machines may be made by which the largest ships, with only one man steering them, will be moved faster than if they were filled with rowers; wagons may be built which will move with incredible speed and without the aid of beasts; flying machines can be constructed in which a man . . . may beat the air with wings like a bird . . . machines will make it possible to go to the bottom of seas and rivers."[2]

Despite such predictions, technology was one of the most conservative of human activities. Providing, as it did, the necessary link between human beings and their access to food, shelter, and other basic means of survival, technology was simply too important to tamper with except under profound necessity. The modern saying, "If it ain't broke don't fix it," is only a contemporary formulation of an ancient wisdom. Change was likely to be for the worse, not the better, and for people who lived at the edge of subsistence, innovation was too great a gamble to take.

The England from which so many American colonists set out had not yet, in the early seventeenth century, undergone either the agricultural or industrial revolutions that were soon to set off the modern world from all that had gone before. Great Britain was still an agricultural nation, and not even an intensively cultivated one at that. Medieval methods of farming prevailed, and much potentially usable land was still given over to swamps, fens, wastage, forests, and commons. But of all the technological activities practiced, agriculture was the most basic and widespread, being the primary livelihood of perhaps two-thirds of the population. Aside from the imperial metropolis of London, there were no real cities in the realm, and transportation between towns was virtually limited to horseback or pack-horses for goods. The best way to move bulky goods any distance was by water: down one river, along the coast, and up another river.

Manufacturing establishments were small and scattered. In the larger towns, especially London, guilds closely regulated the numerous traditional crafts. In 1525 there were more than 160 separate crafts practiced in London, but in smaller towns many fewer: 80 in Norwich, the second city of the kingdom; 46 in Sudbury, a town of only twelve hundred inhabitants. Except for woolen cloth, England exported no finished goods during the two centuries before American settlement. The line between industrial and other activities was blurred, as many craftspeople, especially in the countryside and small towns, plied their mysteries only part of the time, giving greater time and energy when possible to agriculture, fishing, or other pursuits.

For most manufactured goods, the people of England either depended upon their own individual skills or sought the importation of goods from abroad. So far advanced were manufactures in other lands that Tudor England tried to lure foreign craftspeople of all types—makers of lace, glass, silk, leather—to leave their homes and come to Britain. The flow of technological know-how across international boundaries is a very old form of commerce, and it has always moved most often and efficiently in the persons of those prepared to practice their skills in some foreign land. Governments then as now encouraged the inflow of skilled technologists but at the same time discouraged their expatriation.

Some manufacturing was done with large machines and on a considerable scale. In the countryside, grain mills, driven by either water or wind power, served a local custom. As early as William the Conqueror's *Domesday Book* in the eleventh century, there were more grain mills than villages in the new realm. Ancient ironworks, some dating back to neolithic times, were still operating. Then in the late 1400s water-driven blast furnaces were introduced from the continent of Europe: three of them were in use by 1530, twenty-six by midcentury, and eighty-six by the mid-1600s. With the growing demand for charcoal for fuel, the forests in those parts of the British Isles where iron ore was worked were nearly denuded. Fuel costs accounted for two-thirds to three-quarters of the expense of producing bar iron. As trees became scarce, increasing resort was made to the Baltic states for both timber and iron. In addition, more mineral coal was mined from the ground, although it could not be used for all purposes.

Industrial processes in which the coal came into direct contact with the material to be worked, as in iron blast furnaces, were beyond the reach of this growing fuel source because of the many impurities in the coal itself. Besides, coal mines (like other mines) quickly became flooded with water from underground springs, and there was as yet no effective way of draining them. Coal was mainly used in small industries that needed to heat liquids (such as beer or textile dyes), in the forges of blacksmiths, and in the home. Because of this domestic use, seventeenth-century London had a serious problem with air pollution, which made it famous for dense and smothering fogs. As the century wore on, the use of coal grew. From the time of Columbus on, European expansion was based, at least in part, on the increase in energy—from wind for the sails of men-of-war to charcoal for the blast furnaces that produced cannon and shot—made available through powerful new technologies.

The productivity, the weight, and the might came none too soon for Europe. In the thirteenth century Mongolian chieftains were still terrorizing the peoples of eastern Europe, and in May of 1453, only thirty-nine years before Columbus sailed, the Turks entered Constantinople, the ancient seat of Byzantium and the eastern Catholic church. As late as 1683, by which time the colony of Virginia had been established continuously for eighty years and Harvard College had graduated many classes of scholars, the Turks were at the gates of Vienna. Someone has taken the trouble to discover that among the books published in France between 1480 and 1609, there were twice as many relating to the Turkish menace as to the newly discovered Americas, a clear indication of the proportionate realities and fantasies of fear to the East and hope to the West.

It was on the high seas that Europe finally became triumphant. On land, Europeans were small in number and split into many continually warring factions. Even after the introduction of gunpowder (cannons were common after 1330), the technical superiority of Europeans in land warfare was small. It was the gunned-ship, developed by Atlantic Europe in the fourteenth and fifteenth centuries, that allowed the West to outflank the East and launch its era of exploration and exploitation. As Carlo Cipolla has written, "religion was the pretext and gold the motive. The technological progress accomplished by Atlantic Europe during the fourteenth and fifteenth centuries provided the means" for conquest.[3] It took another three hundred years for that maritime superiority (and the thin skin of occupation it supported along the coasts of Africa, Asia, and the Americas) to push inland on any scale.

In the Americas, the Europeans were lucky. The Native American peoples were widely scattered and also divided into factions. They were at a technological disadvantage when confronted by the Europeans and were terribly susceptible to their diseases. Even so, the European occupation of vast interior areas (even of the United States) came only as one of the fruits of the Industrial Revolution. The technology brought to North America from 1603 onward was essentially medieval, the product of eight hundred years of remarkable material and psychic development. The problem for the new colonists was to transmit and absorb it: to reproduce tools and methods, and to keep skills at a high level. Only later, when new goals of development and growth became widely accepted, did the need for wide-scale innovation suggest itself.

I

THE TRANSIT OF TECHNOLOGY

I

THE TOOLS BROUGHT OVER

TO LATER GENERATIONS, the pace of technological change in colonial America may have seemed painfully slow. But to the colonists, the medieval technology of field and bench were sophisticated and effective. The fundamental concern of the migrating Europeans was to bring over enough tools to ensure survival in what often seemed a vast and hostile land. They applied traditional and well-learned techniques of farming and house building, cooking, and hunting and modified these techniques only if necessary to meet the new conditions of a new land. So fundamental was this process of technological transfer and modification that it has remained a significant factor even in current technological progress. From the axe to the intercontinental ballistic missile, Americans have readily borrowed technologies from other places or other people and made them serve their own purposes.

Once established in its new setting, however, and modified where necessary to account for new conditions, early American technology changed very slowly. By the mid-eighteenth century, American artisans appeared to work in what newly arrived European observers thought to be quite old-fashioned ways. Only in later times, beginning perhaps in the late eighteenth century, did people begin to expect that invention could be a normal and necessarily positive part of material life. Indeed, the very word *innovation* had negative and even sinister connotations of unsound and dangerously reckless pride.

European technology of the sixteenth and seventeenth centuries, and therefore colonial American technology as well, had at least four distinctive characteristics.

First, it was primarily a handicraft technology. That is not to say machines were completely lacking—indeed, some were both large and ancient—but most tools were hand tools, and for most tasks human beings were the primary source of power, direction, and accuracy.

Second, tools and devices were, whenever possible, made of wood. Houses were usually made of wood, as were plates, shoes, ships, bridges, plows, and a host of other implements. In both the Old World and the New, iron was too valuable and too difficult to work to be used on common things for which wood would do just as well.

Third, most things were made individually, one at a time, by some craftsperson, or by the person who intended to use the item. Guns were made "lock, stock, and barrel," as the saying had it, by a gunsmith, furniture by a cabinetmaker, and so forth. Many common people, particularly those in rural areas, could not afford to have such things made so either did without or made them themselves. One important exception to this pattern lay in the extractive industries, like iron making. Here some division of labor was normal: charcoal burners cooked wood into fuel in the forests, molders cast the liquid metal into shapes, and common laborers did much of the unskilled heavy work like carrying and carting the iron.

Fourth, technological change, when it did take place, was most often a direct result of the experience of the craftspeople. The expectation that such change would arise from scientific knowledge or research was articulated on occasion early in the nineteenth century, but one would be hard-pressed to find examples of this actually happening. A few early Americans, notably Benjamin Franklin and Thomas Jefferson, took the possibility seriously, and the existence of the lightning rod, a most valuable and practical device arising directly out of Franklin's experiments with electricity, proves that it could happen. By the time of the American Revolution, however, science had not yet gained sufficient theoretical grasp of natural phenomena to make such examples numerous.

Spanish, English, Swedish, French, and German settlers all came to the land now claimed by the United States. Of these groups, the first and second put the heaviest imprint on the landscape. The Spanish arrived first and pushed northward, beginning in the 1520s: from Cuba into Florida, the Carolinas, and the lower Mississippi Valley; and from Mexico City through Texas and Arizona to as far north as Minnesota, and up the Pacific Coast to Oregon and Washington. In Florida, they built cities such as St. Augustine (established in 1565), which they fortified against the English. In the Southwest they laid out towns such as Santa Fe (1609), San Antonio (1718), and Los Angeles (1781).

In the Southwest, the Spanish found Native Americans who were primarily hunters and gatherers but who sometimes engaged in agriculture. Indeed, a few of the Indian nations had multistoried dwellings and vast irrigation works. The Spanish themselves were primarily miners and herders of cattle, only occasionally turning to agriculture, usually on a small scale. Beginning in 1769, with the establishment of a mission in San Diego, Father Junípero Serra founded a number of these institutions up the California coast. Here Native Americans were forced to learn and practice the crafts of building, various trades, and agriculture, as well as Christianity. Dams and reservoirs, irrigation ditches, and sometimes gristmills were all part of the missions' technology. The Spanish were never sufficiently numerous during their two-hundred-year tenure, nor were the lands they peopled sufficiently rich, to introduce or evolve the complex technology that developed along the east coast of the continent.

The lands to which most English colonists came proved not impossibly different from their homeland, though New England winters and southern summers seemed remarkably extreme. In the early seventeenth century, however, even the British Isles had not yet gone through either the Agricultural Revolution or the Industrial Revolution. As mentioned earlier, only a small part of England's usable land was cultivated, primarily by medieval methods of farming, and there were no real cities in the kingdom aside from the imperial city of London and several ports like Bristol. A few traditional crafts were practiced in urban areas, while the countryside was dotted with grainmills, often driven by wind.

It was from this rural, agricultural, handicraft society that English colonists came to the eastern coast of North America. In the new country they attempted, as nearly as possible, in technological as well as social arrangements, to duplicate the practice of their former homes. Some tools were brought with them; others were fashioned anew once they had arrived.

Most often, and this continued to be true throughout the next three hundred years, the transfer of technology depended on the immigration of particular people with particular skills. Dispatches from company officials in the southern colonies, royal governors in the middle colonies, and members of the clergy in New England pleaded for tanners and bakers, millers and weavers, sawyers and miners. Once they arrived, these craftspeople went about setting up shop, gathering or making tools and equipment, and arranging for raw materials.

The country to which they came was not an empty wilderness, of course. It had been inhabited for ten thousand years by people the Europeans chose to call Indians, and the apparently "natural" environment in many places had been culturally shaped by generations of hunting and farming. In New England, the perhaps seventy thousand Native Americans lived a life very much attuned to their environment, moving their villages from one type of land to another as seasons changed and different plants and animals became abundant. In the southern part of New England, women provided about 75 percent of the available food through agriculture; the rest came from the hunting and fishing activities of the men. Their technology was ample: snowshoes, traps, bows and arrows, birch-bark canoes for hunting and fishing, hoes, axes, wooden utensils for agriculture and cooking, animal hides and furs for clothing. The mobility of tribes—their adaptation to the change of seasons and resources of the land—helped keep material culture to a minimum, and notions of property were based on contingent use rather than exclusive inherited property rights.

From at least 1524 onward, Native Americans on the coast of northern New England were in fairly frequent contact with Europeans engaged in fishing. With the settlement of the Pilgrims in 1620 and Puritans a decade later, the Europeans taught the local inhabitants as much as they learned from them. Colonists were allowed to train them as assistants in the various crafts, and one settler remarked in the 1630s that the Native Americans "soon learn any mechanical trades, having quick wits, understanding apprehensions, strong memories, with nimble inventions, and a quick hand in using of the axe or hatchet."[1] Puritans in New England, in trying to christianize them, also attempted to give each a "calling" or occupation to go along with the new religion as an essential part of it. In general, the Native Americans proved willing to adopt those European technologies that appeared to enhance their ways of doing things, but in the long run, it was the commercialism of Europeans, their turning of resources into commodities for international trade, rather than their technology, that most disastrously changed the way the Indians lived.

One technology that did greatly affect the Native Americans was the firearm. In 1630 a royal proclamation was issued to prevent colonists from teaching any "Indian" to "make or amend" guns, or "anything belonging to them," but this was an exception. Even it was sometimes violated, and, indeed, traders in Europe conducted a smart business in selling "trade muskets" to them. These were identical to European muskets except that they were made more cheaply and had no bayonet attachments. Gradually during the seventeenth century, Native Americans began to adopt guns for use in warfare. This change, William Byrd of Virginia suggested, made them less dangerous: instead of shooting off several arrows a minute, in silence, they now had trouble getting off one shot, and it made a loud noise and left a tell-tale pall of smoke. Furthermore, some tribes that had traditionally built forts, such as the Narragansett, were now able to do so much faster with the aid of European tools. They also erected blacksmith's forges in the forest, staffed by their own craftspeople, where muskets were repaired and bullets cast from lead. Artillery and bayonets were the only weapons the Europeans had that the Native Americans did not.

The military superiority of the invaders came through training and tactics, as well as a sharper strategic sense. Native Americans preferred to surprise and try to surround the enemy, but they would pull back when they met too much resistance. The British army, not much used in the colonies until 1755, hit upon the tactic of attacking undefended villages and fields. When the warriors were thus forced to mass themselves for defense, the European troops used their disciplined bayonet charge to carry the day. The bayonet, assimilated into European warfare since the 1640s, was a deadly new weapon alien to Native Americans. Gradually, they were driven back to lands not yet coveted by the Europeans. That move and various European diseases —which alone cut the Native American population from seventy thousand in 1600 to twelve thousand seventy-five years later—were unmitigated disasters.

In addition, the Europeans imposed an utterly different relationship between the people and the land. As the historian William Cronon has noted, "the shift from Indian to English dominance in New England saw the replacement of an earlier village system of shifting agriculture and hunter-gatherer activities by an agriculture which raised crops and domesticated animals in household production units that were contained within fixed property boundaries and linked with commercial markets."[2] In 1642, less than a quarter of a century after the coming of the Pilgrims, Miantonomo, a Narragansett sachem, charged: "Our father had plenty of deer and skins, our plains were full of deer, as also our woods, and of turkies, and our coves full of fish and fowl. But these English have gotten our land, they with scythes cut down the grass, and with axes fell the trees; their cows and horses eat the grass, and their hogs spoil our clam banks, and we shall all be starved."[3]

On arriving in America, farmers, and many who had followed other trades at home, discovered that the temperature, seasons, rainfall, soil, and topography of the coastal land were not too different from what they had known at home. In the initial move, therefore, they would merely have to transfer European crops, livestock, tools, and methods of production to their new setting. In each of these areas, however, there was some variation almost immediately. To the traditional crops of wheat, rye, and barley were added Indian corn, tobacco, and other new plants. In place of the familiar cows, sheep, and swine—none of which were immediately available—the colonists came to depend on wild turkeys and deer. Eventually the ever-expanding numbers of domestic animals, especially the oxen and horses needed to pull the plow, wrought great changes in the ecology of New England and set English agriculture off from Indian. Colonists devoted from two to ten times as much land to their animals as they did to tillage.

As far as is known, there were no immediate additions to the traditional tools of farming: the hoes, spades, scythes, reaping hooks, shovels, carts, harrows, and plows. Faced with the curse of too much rather than too little timber, however, the old-style axe was radically improved. One observer in the early nineteenth century described the American axe, as it came to be called, as the instrument, along with steam and corn, that conquered the New World. It was distinctly American, a fact recognized with pride by them and admiration by visitors.

Although details varied from country to country and purpose to purpose, by and large the axe first brought over from Europe had a fairly short handle, a thick wedge, and narrow bite, and the handle was set well back from the cutting edge. Slowly, and by whose agency no one knows, this axe developed into a new type: one with a long, curved, and springy handle set nearer the center of the head, which was broader at the edge than the center and much narrower throughout. Much attention was lavished on the handle, which was often custom-made to fit the size and style of the owner. Technique was also important in the use of the axe.

Testifying before a special committee of Parliament in 1841, an English manufacturer stated that the American axe was, "for a plain article, the most mechanically and the best-constructed little instrument I know; the art being, that a man can fell three trees to one, compared with those which are ordinarily made in England."[4] The great timbered expanse of America put a premium on cutting down forests as quickly as possible. A prominent Ohio pioneer marveled at its excellence: "To this moment it is wonderful to me that so many different things could be done with this simple instrument—that it could be made to perform the function of so many others—and that a single man in a single day could, by its aid alone, destroy so many trees."[5]

It was perhaps the method of farming that changed most radically of all. Indian corn, originating in Mesoamerica and adopted from the Native Americans, was a major innovation early on. Not only was it nutritious and well adapted to the climate of America, but being traditionally planted in hills, it could be put in the ground when the field was still filled with dead trees and not yet plowed. The charming story of Squanto, the friendly "Indian" who taught the Pilgrims to place a fish in each hill of corn, appears to be largely myth. There is no supporting evidence that Native Americans used this method, and it seems unlikely that fish would have been so wasted. As it turned out, Squanto had lived some years on the Iberian Peninsula where this practice was common. He had also lived in London, where he had learned his English.

As the colonists began taking the fertility of the new land for granted, they placed increased emphasis on the production of staple crops for sale—an emphasis often insisted upon by English stockholders and most obvious in the tobacco-growing South. Closely allied with this trend was the rapid abandonment of any form of crop rotation, such as the medieval three-field system of cultivation. Letting fields lie fallow was a luxury the earliest colonists could ill afford. As a result, agricultural practice retrogressed and land became exhausted after years of planting to corn or tobacco. In a country where land was virtually free and appeared to be limitless, soil exhaustion did not seem the problem it might have been under other circumstances.

Long before the American Revolution, however, some farmers, especially those with a different land ethic or those who lived on farms close to cities and large towns, began taking steps to rebuild the soil. The Quaker and German farmers of eastern Pennsylvania reinstated crop rotation, which by 1800 was fairly common in New England. The frontier areas of the Old South and Old Northwest took longer to adopt the method. By 1800 some farmers were interspersing legumes or other soil-building crops between the others and were using fertilizers—marls or lime—in increasing quantities.

In the southern states, tobacco was the main crop, one that had the advantage of feeding a growing habit of many Europeans. Cotton was grown as well, but in an age when linen and wool were still preferred for clothing, it had little market. In addition, it had a severe drawback: removing seeds from the bolls was a tedious and slow process, necessarily done by hand. Cotton had, however, long been grown in the Near East and in India, and in these places gins were successfully at work mechanically pulling the seeds from the fiber. As early as 1725 a gin, described as being from the Levant, was copied in Paris and sent to a plantation near New Orleans for trial. This Indian-type gin cleaned cotton by passing it between smooth rollers, a process that worked well enough on the long-staple fibers of the East but failed on the short-staple cotton of America. Denis Diderot, in his celebrated *Encyclopaedie* of 1762, published a full description of the machine and process and included somewhat idealized plates showing the device.

By the 1740s cotton planters on the French island of Santo Domingo were successfully using a mechanical gin that could clean 50 to 60 pounds of cotton a day, as compared with 1 pound a day cleaned by hand. Such success so near to Louisiana spurred on those who realized that the colony could become one of France's most valuable assets if only cotton could be made an economical crop. Helpfully, by midcentury cotton fabric was becoming more fashionable.

Several avenues of improvement were open to potential innovators, and all were tried. The first such effort concentrated on modifying the eastern gin to clean American cotton. The 1725 Louisiana machine had failed, but such diverse persons in the colony as a wealthy planter, a boatmaker, a physician, an engineer, and a leading Jesuit priest all modified other imported gins, trying to make them do. The colonial records are filled with claims that a successful gin had at last been "invented," but no one made the claim stick. Others, despairing of the success of colonial efforts, commissioned experiments and testing to be carried on in French laboratories and workshops, and Louisiana cotton was sent to Paris to be used in the tests. All failed. Then the experimenters decided to bring the mountain to Mohammed: if the gin could not be modified to process Louisiana cotton, Louisiana cotton would be modified to fit the gin. This, too, failed, as native cotton proved difficult to modify, and imported seed did not thrive. It was not until 1793 that the problem was solved by Eli Whitney.

Agricultural technology was the most important skill in the colonies, but other skills were also abundant, especially in the cities. About 80 percent of the colonists were farmers and 18 percent craftspeople. Although many of the immigrants to the New World brought with them medieval craft skills, except in the few cities along the coast, most of these skills were deteriorating through lack of use. It was widely believed that people from a country where social status and wealth were based on land were quick to abandon any other profession to become farmers in America. It was also true that unless a craftsperson lived in Boston, Providence, New York, Philadelphia, Charleston, or one of the smaller cities like Lancaster or Wilmington, there would hardly be enough customers to allow an individual to practice a craft full time. Even as late as 1700 the crafts were still practiced mainly in the home, for home consumption.

Craftspeople who stayed at their trade struggled hardest in the South. There the inclination of the planters and the necessities of a staple crop agriculture kept cities small in number and in size. One thriving craft was that of cooperage. In 1754 no fewer than 116,231 barrels and hogsheads of rice, tar, turpentine, resin, beef, and pork were shipped from Charleston alone, and 68,000 hogsheads of tobacco and 37,000 barrels of wheat, corn, and flour were shipped from the Chesapeake colonies. Each of these barrels and hogsheads was handmade by a cooper.

Besides coopers, southern plantations required the services of blacksmiths, brickmakers, masons, tanners, millers, spinners and weavers, and often tailors. A large number of Black slaves were taught these crafts, however, and free craftspeople found it impossible to compete with forced labor. George Mason of Virginia, writing in the late 1700s, claimed that his father had slaves who served as carpenters, coopers, sawyers, blacksmiths, tanners, curriers, shoemakers, spinners, weavers, and knitters, and that one was a distiller. Together they made and repaired much of what was used on the plantation in the way of buildings, clothing, and equipment.

The social climate for free labor was better in New England. By 1776 there were 566 towns in the interior regions, in addition to the metropolis of

Boston. Each of these towns had a few craftspeople, and not infrequently the town meeting would decide that the community needed some special skill not yet found there and advertise for someone: a tanner, cooper, blacksmith, or such. Sometimes such people operated not simply as independent entrepreneurs but almost as public utilities. In 1640 the town of Scituate, in Plymouth Colony, offered some 30 acres of land to anyone who would build a gristmill within seven years and keep it going for at least another fourteen.

The middle states, in this as in most respects, presented a mixed picture. On balance, however, the crafts were thriving, with such cities as New York and Philadelphia and large interior towns like Lancaster in Pennsylvania teeming with special needs. By the time of the Revolution, these middle states had become the first mechanical and industrial center of the nation.

Nevertheless, for nearly two centuries the crafts were primarily for domestic use. As a result, domestic needs and utensils have been investigated mainly by antique collectors or museum curators concerned with aesthetic appreciation. In only a few areas is anything known about the exact technology involved. It has been discovered, for example, that the log cabin was introduced into the middle states by Scandinavian colonists and was unknown to the Pilgrims. Gradually, this type of structure, so well suited to the abundant timber of both the Baltic states and the new colonies, came to be widely adopted in frontier regions.

For the inside of the home, nearly everything except the cast-iron cooking pot was made of wood, usually fabricated by members of the family. The standard of living is suggested by the lighting devices. Careful studies have discovered that the technology of lighting used in the colonies was not much better than that found at the beginning of the Christian era. There were lanterns, lamps, candles (eventually) and candlesticks, rush-lights and rush-stands, and pine splints. The lamps were simply saucer-shaped oil containers, no more sophisticated than those used in classical Greece or Rome.

Describing lighting practices in 1630, Francis Higginson wrote from Salem, Massachusetts, that "although New-England have no tallow to make candles of, yet by the abundance of the fish thereof, it can afford oil for lamps. Yea our pine-trees that are most plentiful of all wood doth allow us plenty of them which are useful in a house. And they are such candles as the Indians commonly use, having no other; and they are nothing else but wood of the pine tree cloven in two little splices something thin, which, are so full of the moisture of turpentine and pitch that they burn as clear as a torch."[6]

Unfortunately, pine splints leave behind no physical evidence of their use. A careful study of wills filed in Essex County, Massachusetts, between 1635 and 1681 shows that a large proportion of colonial homes had no more light than that afforded by sun in the day and by the fireplace or pine splints at night. From 1635 to 1664, only 56 percent of the homes had some kind of utensils associated with artificial light, a percentage that remained almost constant at least through 1681. It was hardly a growing market for the crafts.

One domestic appliance of considerable importance that did advance markedly was the stove. Heating and cooking, as well as light, were most often provided by the fireplace. Such contrivances used large quantities of wood, however, were difficult to adjust for heat, and tended to draw heat out of the room and up the chimney rather than warm a room adequately. Benjamin Franklin noted in 1744 that "wood, our common Fewel, which within these 100 Years might be had at every Man's Door, must now be fetch'd near 100 Miles to some Towns, and makes a very considerable Article in the Expence of Families."[7] Colonial Americans were known to be prodigal in their use of fuel: a typical New England household used 30–40 cords of firewood per year. Partly to economize on fuel (Franklin worried that future generations would have to "fetch their Fuel over the *Atlantick*"), and partly to improve health through the better heating of rooms, he devised his famous "Franklin stove," which consisted of cast-iron plates assembled into a box and stood away from the wall.

Franklin asserted, "My common Room, I know, is made twice as warm as it used to be, with a quarter of the Wood I formerly consum'd there." He published, at his own expense, a pamphlet with diagrams and instructions for building the device. Over time such stoves were modified to become the cast-iron ranges universally used during the nineteenth century for domestic cooking.

The colonists' house and the yard area around it were the particular responsibility of the women, who provided the necessary cloth and clothing, tended the garden and raised the poultry, and processed the food for storage or for the family's daily meals. In the nineteenth century the task of cooking became increasingly complex and difficult, since the new iron range that had evolved from Franklin's stove, being more efficient, not only allowed but seemed almost to demand the preparation of more elaborate meals and baking.

Although machines and industry came to the colonies early, agriculture remained the predominant activity for nearly three centuries. Probably the

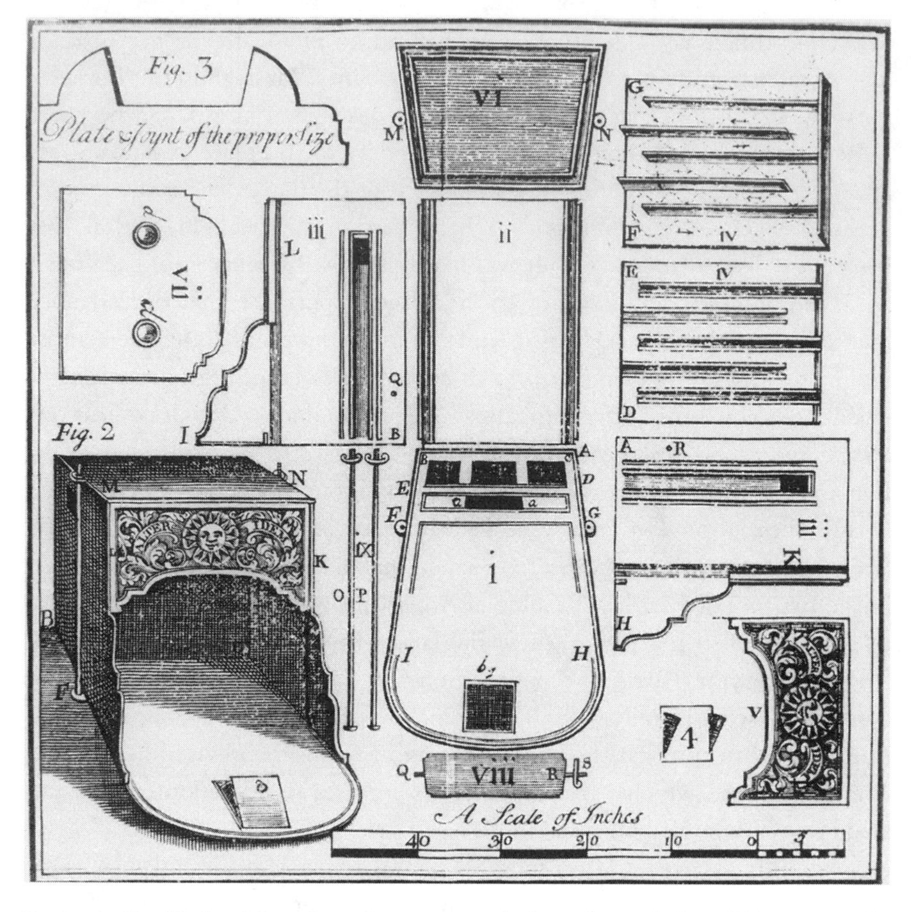

Benjamin Franklin's celebrated cast-iron stove marked an advance in fuel efficiency over the open-hearth fireplace common in colonial homes. Despite urgings to do so, he refused to patent it, hoping that it might find wider adoption. Benjamin Franklin, *An Account of the Newly Invented Pennsylvania Fire-place* (Philadelphia, 1744), fig. 3.

two greatest influences on the nature and scope of colonial industry were the availability of natural resources and of transportation. Because of extremely primitive transportation facilities in the colonies, as in England, large establishments could only be built in sea coast areas, and inland industries were only as large as the local custom and resources could support. It was not until the first quarter of the nineteenth century that transportation other than coastal shipping was able to support interregional trade on any scale.

The natural wealth of the country, rather than any cheap labor or easy capital, was its main attraction, and its earliest industries tended to be extractive. This situation was, of course, clearly in line with and reinforced by British merchantile policy, which decreed that the New World should supply raw materials for manufacture in the Old. Most colonial industries worked directly from local resources and included the making of bricks, pottery, and glass; the extraction of potash and pearl-ash; the extraction of turpentine and other naval stores; the mining and smelting of iron ore; the sawing of lumber; the grinding of grain into flour; and the fulling of cloth.

The manufacture of textiles began early in New England. The first woolen cloth appears to have been made by five professional weavers in Massachusetts in 1638. Of all the steps in making cloth—shearing, scouring,

Agriculture was the common work and culture of early America. This man, with his horse-collar, harness, and plow of medieval design, clearly works in close contact and apparent harmony with nature, represented by the nearby trees and watching bird and squirrel. Taken from a colonial newspaper and reproduced in Elizabeth Carroll Reilly, *A Dictionary of Colonial American Printer's Ornaments and Illustrations* (Worcester, 1975), 292. Courtesy American Antiquarian Society.

carding, combing, spinning, weaving, trimming, dying, and fulling—only the last made use of power other than hand. Fulling, by which the material was compacted and made more uniform, could also be done without machines, the material being soaked in water containing fuller's earth (a fine clay that absorbed oil and helped compact the cloth), laid out on the floor, and either beaten with sticks or trod upon by friends and neighbors invited over for a fulling bee. As early as 1657 enough cloth was being produced to warrant the construction of a fulling mill in Roxbury, Massachusetts, and by 1700 at least seventeen more had been put into operation. The first fulling mill appeared in Virginia in 1692, and there was one in Pennsylvania in 1698. In these mills, large, heavy, water-powered rollers squeezed the material as it passed through a trough of water containing soap or fuller's earth. Alternatively, the cloth was placed in mortars, where it was pounded by water-powered pestels. Separate fulling mills, custom-fulling the cloth brought to them by neighboring weavers, lasted well into the nineteenth century. The colonies long remained a major market for British textiles, with between one-third and two-fifths of their textile production being sent to America as late as 1760.

A more common mill, and one that sprang directly from the abundant needs and resources of the North American continent, was the sawmill. The forest provided most of the resources of the early colonists. Wood was used for every conceivable purpose locally and was an important commodity in foreign trade, in the form of lumber, barrels or barrel staves, or masts and spars. The juices of the pine trees were made into naval stores, the bark of oak provided the tannin to transform hides to leather in tanneries, wood for fuel in home and industry, and the ashes were saved to make potash and soap.

The lumber was first cut by hand in saw pits. One sawyer stood under the log in the pit, pulling down the bottom end of the saw, while another sawyer stood on the log itself. This method persisted in England even after the first sawmill was erected there in 1663. By this late date scores of such mills were already in operation in the colonies. Sawmilling technology appears to have been imported from the continent of Europe, rather than Great Britain. The Dutch had a mill in New York as early as 1623. A mill was also set up in Maine in 1623, and by 1682 there were twenty-four in that area alone. There were sawmills in New York and Massachusetts by 1633, Rhode Island by 1639, Connecticut by 1654, New Hampshire by 1659, and Pennsylvania by 1662. Lumber formed part of the earliest cargoes sent out from

Both the abundance of lumber provided by a sawmill in 1777 and a hint of the environmental consequences are shown in this engraving. Note the presence of both soldiers and Indians, and Fort Anne (N.Y.), which allowed the former to wrest this resource from the latter. Engraving in the American Antiquarian Society, reproduced in Edward P. Hamilton, *The Village Mill in Early New England* (Old Sturbridge Village, 1964), 5. Courtesy American Antiquarian Society.

Virginia, and in 1671 New Hampshire shipped 20,000 tons of boards and sent out ten ships laden with masts.

Aside from getting larger, sawmills evolved little during their first two centuries in America. The basic type consisted of one or more sash saws, which were pulled down by the action of a water wheel and crank and drawn back up by the force of a sappling used as a spring-pole. The log was moved forward by a cog wheel or a weight running over a pully. One adult, perhaps with a young helper, could saw 1,000 feet of pine per day. It was not uncommon for two such saws to be worked together; by 1700 one mill in New Hampshire carried four saws, and there were reports of one in New York that carried twelve saws. By 1769 a mill in the Mohawk Valley of New York carried fourteen saws.

The sash saw benefited from minor improvements over the years, of course. It has not yet been completely abandoned but has been largely displaced by either the circular or band saw. One of the incentives for improvement was the fact that sash saw blades were quite thick and cut perhaps a half inch of sawdust at each pass (the kerf): a gang saw with ten blades would therefore waste 5 inches of wood at each pass. The circular saw appears to have been invented in England about the time of the American Revolution but did not come to America until 1814. For the next thirty years it was used mainly to cut small boards and veneers for furniture because the metallurgy of the blade was such that only small saws could be used. Its main virtue was that it ran at very high speeds.

The continuing search for a better saw was carried on by sawyers of valuable hardwood, since pine and other soft woods were too abundant to worry about. The circular saw turned 312 feet of wood into sawdust for every 1,000 feet cut. If the width of the blade could be cut from 5⁄16 to 1⁄12 inch, the loss would be reduced to 83 feet. The solution turned out to be, in part, better steels through better metallurgical knowledge, and also through adoption of the band saw. This use of a thin, endless band of steel was developed by an American sawyer in the Midwest and patented in 1869.

Another type of mill that was widespread in colonial times was the gristmill, designed to grind grain into flour. Grinding grain between stones was an ancient process. The Native Americans in some areas used stone mortars and pestles to grind corn, acorns, and other foodstuffs. In the early colonial period and well into the nineteenth century in some areas, grain was ground with hand *querns.* These devices consisted of a solid lower stone and a top one with a hole in the middle and were similar, except in scale, to the stones used in watermills.

In the typical water-powered gristmill of colonial times, grain cleaned of dirt and chaff was emptied from bags into a wooden hopper mounted above the eye of the top millstone. Falling between the stones, which were dressed with cutting edges, the grain was ground as it moved toward the edge of the stones. The fresh flour, being moist and warm, was spread out on the mill floor, where it was raked about and eventually shoveled into a revolving, slanted cylinder covered with cloth (the bolter), which separated out the fine from the super-fine flour and the bran. The different grades of flour were then shoveled into sacks or barrels for removal.

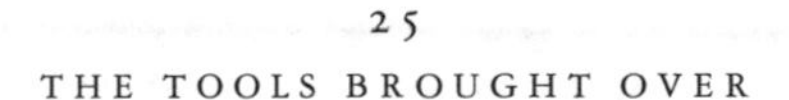

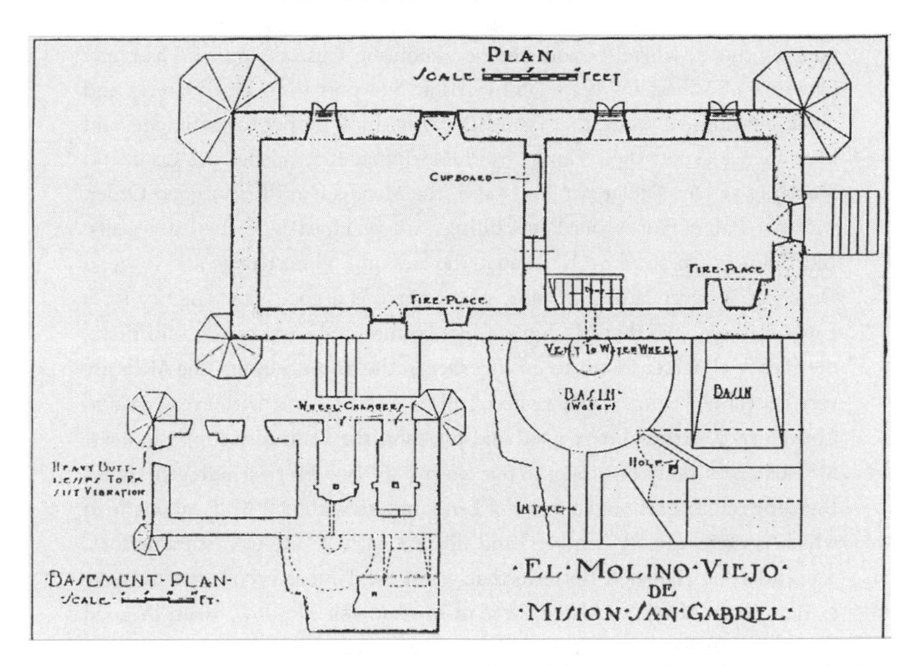

Mills were found in colonies other than those established by the English. About 1810 the first of two gristmills was built, under the supervision of priests, in southern California at the Mission San Gabriel. Still standing in 1994, El Molino Viejo represents the medieval tie between the practical problems associated with religious institutions and the technological means adopted for solving them. Rexford Newcomb, *The Old Mission Churches and Historic Houses of California* (Philadelphia, 1925), 195. Reprinted by permission of HarperCollins.

In the colonial period, the middle states—especially southeastern Pennsylvania, Delaware, and parts of Maryland and Virginia—were known as the "bread colonies." The area compared favorably with the best wheat-growing areas of the world, produced a large cash crop for both foreign and domestic commerce, and was not surpassed until the opening of the Erie Canal in 1825 brought into production the lands of upper New York State. The Wilmington, Delaware, mills, where tidewater met the falls on the Christiana River, were famous for the bounty and quality of their product.

A typical gristmill of the middle states was described in an advertisement in the *Pennsylvania Gazette* in 1765:

> To be sold by the Subscribers . . . a complete Merchant and Saw-mill, situate on Redclay Creek, two miles from Newport Landing, in a good

> Neighborhood, where, besides the neighbouring Custom, may be had, any Quantity of Wheat by Water, delivered at Newport from Duck Creek, and other Landings adjacent, that the Miller may think proper to encourage, and from thence export their Flower to Philadelphia, at Six-pence per Cask, and Scantling at One Dollar per Thousand; the Merchant-mill is in good Order, with two Pair of Stones, one Pair Cullings, where, Hoist-jack, Boulting-geers, and Country Chest all go by Water; the Saw-mill almost new, and in good Order, on a never failing Stream, where there is a good Conveniency for a Paper-mill, or any other Machine of like Nature, to be erected on said Place, the Plenty of Water for all to go together in the driest Seasons; the Mills are very forceable, having about 12 Feet Fall on a short Race, with every natural Conveniency, extraordinary good and pleasant, the Dam easy supported; the Mill-house is built of Stone; about 45 by 26, two Stories high; the other Buildings are Log; about 50 Acres of Land may go with said Mills, about 36 of which is extraordinary Timber Land, loaded with White-oak, suitable for a Saw-mill, one Half rich Bottom, that when the Timber is cut off will make extraordinary Meadow. Also a Piece of good Marsh Meadow, when in good Order, is sufficient to produce 20 Tons of Hay a Year.[8]

The advertisement is rich in the details of colonial American milling. It indicates that local resources and good transportation were paramount needs and that the miller was also a part-time farmer. Waterpower was adaptable for several kinds of mill: saw and flour were then operating, paper was suggested as an addition, and it was in just such mills that some space and power were first given over to cotton spinning early in the American Industrial Revolution. Not all mills took such advantage of waterpower to operate ancillary machinery.

Writing in 1804 and looking back over a thirty-year career as a millwright, Thomas Ellicott recalled that

> when I first began the business, mills were at a low ebb in this country; neither burr-stones [a particularly hard and rough type of stone imported from France], nor rolling screens being used; and but few of the best merchant mills had a fan. Many carried the meal on their backs, and bolted by hand, even for merchant work; and I have frequently heard, that a little before my beginning the business, it had been customary, in many instances, to have the bolting mill some distance from the grinding mill, and there bolted by hand. It was counted extraordinary when they got their bolting to go by water.[9]

Nor were all mills so fortunately situated for merchant work (producing flour for sale rather than grinding on custom). Most were small mills, multiplied a thousand times across the face of the colonies.

Just after the close of the colonial period, this traditional flour-milling technology was radically changed, and America had its first example of a truly revolutionary invention creditable to "Yankee ingenuity": the "automatic" flour mill of Oliver Evans, a Delaware millwright. In 1782 Evans purchased an old mill and began to install the five separate inventions that were to improve so radically the process of flour milling: the hopper-boy, elevator, descender, conveyor, and drill. In the common mill, the grain was carried to the second floor and then released into millstones on the first. The flour was collected in the basement and carried to the third floor, where it was spread to dry and cool. It was then raked down a chute to be boulted on the floor below. This process had several drawbacks: a large amount of hard, manual labor was used, a considerable amount of material was wasted at each step of the operation, and the quality of the flour was never what it should have been since, as Evans put it, "people did not even then like to eat dirt, if they could see it."[10]

Under Evans's plan, these three drawbacks were greatly lessened by mechanical contrivances designed and arranged so that the power of the waterwheel made the entire operation an automatic and uniform process. In Evans's mill, an elevator (with buckets moving on an endless belt inside a closed chute) carried the grain to the second floor and dumped it into the hopper above the millstones. Gravity carried the grain down to the stones on the first floor and the moist, warm flour from the stones to the cellar. Another elevator carried the flour to the third floor, where it emptied into the hopper-boy. This simple device consisted of two long arms set with teeth, which, when it slowly revolved, stirred the flour within an area of the floor enclosed for that purpose. The flour then fell to the boulter below and thence into waiting barrels. Thus the miller, in theory at least, had only to empty the grain sacks and cover the barrels; the rest was automatic.

The elevator and hopper-boy were the most commonly adopted and, indeed, indispensable parts of Evans's improvements, but the other three were sometimes used as well. These consisted of the descender (an endless belt set at a slope on which material could ride downward), the conveyor (a screw made of sheet iron on a wooden core shaft), and the drill (essentially an elevator for moving materials horizontally rather than vertically). In 1785 the

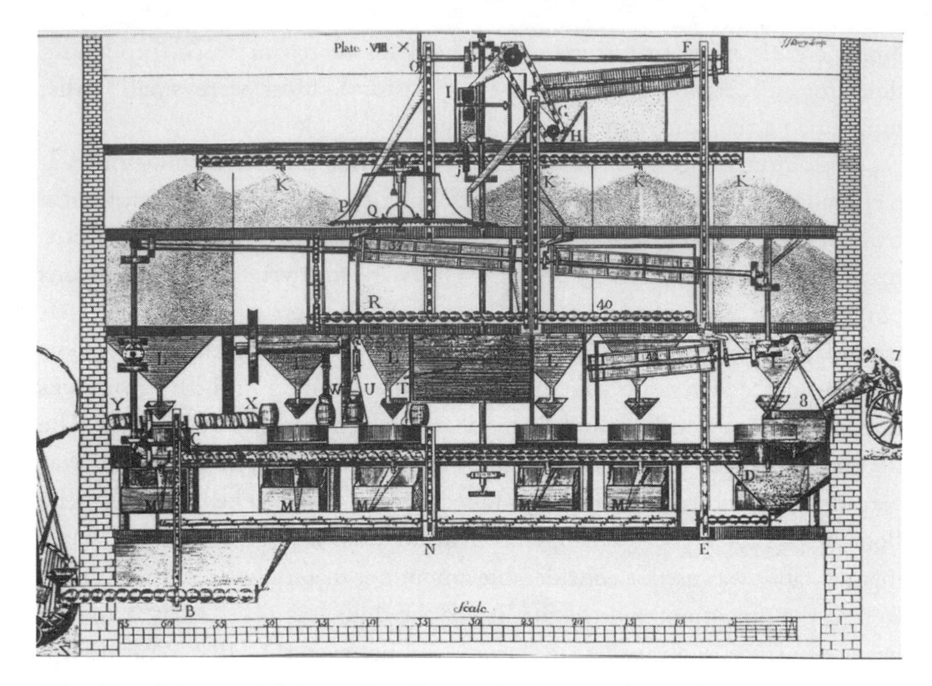

Oliver Evans's "automatic" flour mill as illustrated in the 1795 edition of his popular *The Young Mill-Wright and Miller's Guide* (Philadelphia, 1795), Plate X.

improvements were installed in Evans's own mill, to serve as a showplace for the new process. The result was a classic case of the injustice of which so many inventors complained: his equipment, so simple and superior, became standard within a few years, but even with a patent he was unable to collect royalty fees from all those who used it. It was a matter of succeeding too well. His illustrated book, *The Young Miller and Millwright's Guide,* presented such a cogent argument for the process that many country millers were persuaded to try it. And with the help of Evans's drawings, every country carpenter could install the machinery.

One of the most significant results of Evans's improvements was that the way was now open for large-scale milling operations. In all but the most remote sections of the country, as soon as transportation facilities allowed, merchant mills grinding for the general market came rapidly to replace custom mills grinding local grain for local farmers. There was no significant change in design for nearly a century.

The lack of sufficient power was a problem faced by all colonial mills. There were only four sources available—hand, animal, wind, and water—and their quantities were limited. Most processes could be as well operated by the first two as the last, but in actual practice most millers turned to waterpower to supply their motive force. Under normal conditions, a stream was dammed and a portion of the water was diverted to the side, through a millrace. The race was dug in such a way that it was fairly level and thus was increasingly higher than the descending stream itself. At some point along the way, the mill was built and equipped with a waterwheel. These devices came in three configurations: the undershot (the water passed under the wheel and operated solely by impact), the overshot (the water hit the top of the wheel and added the weight of the water in the buckets to the impact), and the breast wheel (the water struck the center of the wheel). The overshot was the most efficient in using both the kinetic energy of the stream and the weight of the water, but it required a greater "fall" of water, that is, a greater differential between the elevation of the stream and the race. The power was carried by the wheel shaft into the mill and then, through gearing, to the machinery itself.

Watermills developed only about four horsepower, and many of the rivers of the eastern seaboard—such as the Merrimack, the Connecticut, the Hudson, and the Schuylkill—were too large to be dammed and harnessed until well into the nineteenth century. Because the size of a mill was limited by the market that primitive transportation would allow it to serve, however, the lack of power made small difference in most cases, although it did prevent the rise of power-driven manufacturing. Cities, until the mid-nineteenth century, remained largely mercantile in purpose. Colonial mills remained in the countryside, and as new ones were added they had to be located where new sites were available, farther and farther upstream and more distant from urban markets and good transportation.

The manufacture of iron, which by the time of the American Revolution was a well-established and important part of the colonial economy, was different from saw- or gristmilling in several ways. First, it was more limited in range, since iron ore was not as readily available as either good farmland or timber. Second, not all phases of the activity received official encouragement from England, and indeed parts of it were expressly forbidden to the colonists. Third, it was necessarily carried out on a large scale in relative isolation, most often on extensive iron plantations in the countryside. Fourth, iron making consisted of a complex series of steps: mining the ore, cutting trees and burning the charcoal for fuel, smelting the ore in blast furnaces, casting the molten iron into either molds or pigs, converting the pig iron to wrought

W. Hawxhurſt,

STILL carries on the Sterling iron works and gives the beſt encouragement for founders, miners, mine-burners, pounders, and furnace fillers, bank's-men, and ſtock takers, finers of pig, and drawers of bar; ſmiths, and anchor ſmiths, carpenters, colliers, wood cutters, and common labourers: They will be paid ready caſh for their labour, and will be ſupplied with proviſions there, upon the beſt terms.

N. B. Said Hawxhurſt continues to ſell pig, bar iron, and anchors, which he makes of any weight under 3500, and as he has by him a conſiderable quantity of anchors, he would ſell them by the ton, to retailers or exporters at a lower price than the importers from Europe, or the neighbouring colonies; he has alſo cart, waggon and chair tire; which he ſells on the moſt reaſonable terms, for caſh, or Connecticut Proclamation bills. He alſo will take old and caſt by anchors in part of pay for new ones, in proportion as they are in value.

21-

A scene of colonial iron making at a blast furnace. On the right, pack animals bring material (possibly charcoal) up the hill. At the top mining appears to be under way as a worker carries a load (of iron ore) on his head. At the bottom left, two workers shape an anchor. The accompanying text indicates the extent to which division of labor was carried forward in such enterprises. Woodblock print from a colonial Connecticut newspaper.

iron in forges, and finally, rolling, cutting, and slitting the wrought iron into useful dimensions. Some of the wrought iron could also be converted into steel in small furnaces. The earlier steps in this long process—being of an extractive nature and therefore industries approved of under the British mercantilist system—were encouraged. Later stages were more akin to the manufacture of finished goods, which the British hoped to reserve for themselves.

Mechanical progress was encouraged in the colonial period through the granting of patents of monopoly. The colonial practice was patterned on that of England. British kings and queens had for many centuries been in the habit of granting monopolies to favorites or to creditors. These covered the introduction of new machines and industries into the kingdom but also gave exclusive rights to certain trades, markets, and products (such as the right to sell tea in the American colonies).

So open to abuse was this system, that British lawyers argued in 1602 that

> when any man by his own charge and industry, or by his own wit or invention doth bring any new trade into the realm, or any engine tending to the furtherance of a trade that never was used before; and that for the good of the realm; that in such cases the king may grant to him a monopoly-patent for some reasonable time, until the subjects may learn the same, in consideration of the good that he doth bring by his invention to the commonwealth, otherwise not.[11]

The point was not completely won, however, and James I was so prodigal in his granting of monopolies that failed this test that Parliament wrung from him the Statute of Monopolies in 1623. By this act all monopolies were declared to be illegal, and only those were permitted that introduced a new industry either by importation or by invention.

Colonial practice tended to be somewhat permissive, although a 1641 law in Massachusetts declared that "no monopolies shall be granted or allowed amongst us, but of such new Inventions that are profitable to the Countrie, and that for a short time." That same year the Puritan colony issued a patent for a novel way of making salt, a case that was clearly of the type approved of by English law. Such patents joined with other devices in the encouragement of mechanical ingenuity. At times, Virginia even granted a sum of money as a reward for some invention. The results of such practice are not clear, and the effort was perhaps more a result of legal and legislative habit than proven effectiveness.

Each colony issued its own patents under separate pieces of legislation wrung from a sometimes reluctant assembly and unrelated to others granted in that or any other colony. One exception was Pennsylvania, where William Penn's strong moral disapproval of monopolies prevented the issuance of patents for many years. When Sybilla Masters, wife of the governor of Pennsylvania, sought a patent for her improved mill for "Cleaning and Curing the Indian Corn Growing in the several Colonies in America" in 1716,

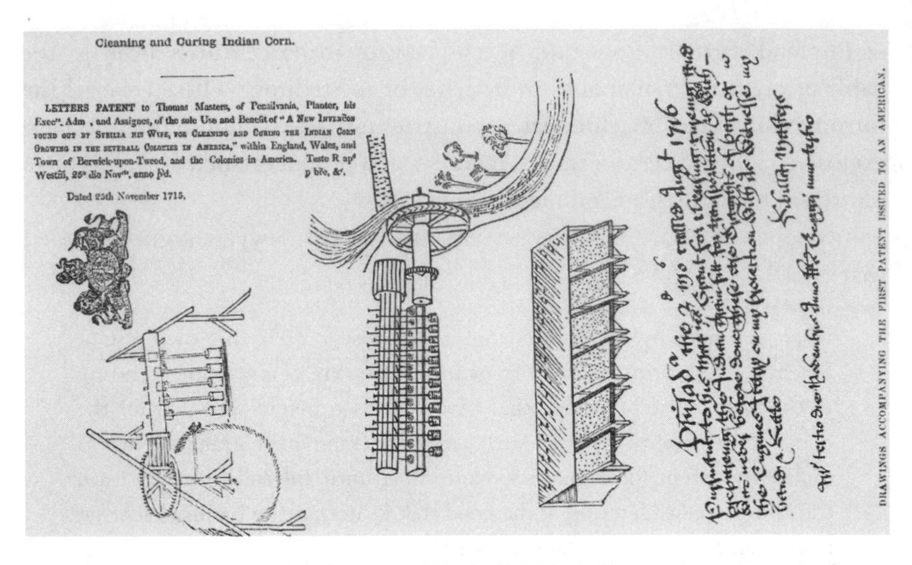

It has been claimed that this patent, issued by the British Parliament in 1715, was the first to be granted by that body to an American. The device, a gristmill to be powered either by water or an animal harnessed to a sweep that would activate a series of mortars and pestles, is not surprising. Note, however, that although the patent was issued to Thomas Masters, a planter of Pennsylvania, it was acknowledged to be for an invention "found out" by his wife, Sybilla. Samuel H. Needles, "The Governor's Mill, and the Globe Mill, Philadelphia," *Pennsylvania Magazine of History and Biography,* 8 (1884), facing 286.

she not only had to go to the trouble of obtaining it from the British Parliament but suffered the indignity of having it issued in her husband's name. So strong was this tradition that when the governor of the colony urged Benjamin Franklin to patent his new stove in 1744, the latter "declined it from a Principle which has ever weigh'd with me on such Occasions, viz. That as we enjoy great Advantages from the Invention of others, we should be glad of an Opportunity to serve others by any Invention of ours, and this we should do freely and generously."[12]

Mostly through borrowing, but partly through innovation, the technologies available in the colonies were sufficiently established and varied by the eve of Independence to form a foundation for a new imperial economy. It was this possibility, of course, that British policy had been designed to prevent. Laws were passed restricting the emigration of skilled workers from Britain, and others forbade the construction of certain mills and manufactories in the colonies. In important areas, however, it was already too late. In the iron industry, for example, by 1776 the colonies were operating more blast furnaces and forges than existed in the British Isles and were already producing one-seventh of the world supply of pig and wrought iron. European technology, augmented by that learned from the Native Americans and enslaved Africans, had taken strong root in the new soil and threatened to bear rich fruit.

After 1763, when relations between the colonies and Great Britain began to deteriorate, the once leisurely pace of technological change began to quicken in America. Calls for economic boycotts and self-sufficiency and eventually the growing prospect of war all put a heavy premium on quick technological progress in the colonies. In England the Industrial Revolution was already gathering force, but in America it was merely a matter of rounding out and perfecting a preindustrial technology to form the basis of a more independent American society.

2

IMPORTING THE INDUSTRIAL REVOLUTION

THE MOST IMPORTANT FACT about the history of early American technology—and perhaps about our entire early history—is that the American Revolution and the Industrial Revolution happened at the same time. In those years of political turmoil between 1763 (when the French and Indian Wars ended) and 1787 (when the Constitution was adopted), James Watt improved the steam engine; James Hargreaves, Richard Arkwright, Edmund Cartwright, Samuel Crompton, and others mechanized the textile industry; the first canal was built in England; the first steamboats were constructed; and the iron industry was revolutionized by the introduction of fossil coal, the puddling process, and rolling mills.

It was during these same years that the American colonies carried out their political revolution against the British Empire and set out to erect an empire of their own. By the 1770s the original colonies had begun to experience tremendous growth both in population and in economic potential. Their premier cities were second only to the imperial seat of London, and already their iron industry was larger than that of England. The population growth, largely the result of natural increase, but of immigration as well, created a tempting market for businesspeople. The attempts of the British Crown to regulate and change the direction of this activity led some bold thinkers to dream of an American empire, free to grow at its own pace and to appropriate its resources for its own benefit.

Once free of the British throne, most of the leaders of the new United States believed that growth—primarily economic growth—was essential to the survival of the nation, which was then still a thin strip of land along the Atlantic Coast. The increase in demand was partly the result of a growing population but also of growing expectations. People were confident that the United States would settle the West, create its own literature, preserve its liberty, increase its foreign trade, encourage a flowering of science, and support a rising standard of living for its citizens. Alexander Hamilton's plans for public credit and trade fostered this belief, as did the more comprehensive imperial program of James Madison and Thomas Jefferson. Indeed, it was the primary commitment of the Jeffersonian Republicans to a program to stimulate economic growth that kept them in power for so many years.

The calculus of success was implicit. To preserve the nation's liberty and ensure prosperity, the economy had to grow. Growth would depend on a population increase, but equally on the progressive exploitation of abundant natural resources. Natural resources, in turn, were defined by the technology available to use them. Thus the nation's well-being and very survival depended on a powerful technological base. In the euphoria of independence and of what seemed to be unimaginable resources to the west, the plan appeared to many to be without cost. The day when the resources might be used up, when the waste generated might begin to poison the environment, and when technological means would threaten to overwhelm social purpose could hardly be imagined.

The problem of stimulating a growing economy with a continuing shortage of labor was solved by the contemporaneous Industrial Revolution in England. The American Revolution safely won, American leaders looked across the Atlantic with shock to discover the changes that had been wrought there in so short a time. When Thomas Jefferson was sent as U.S. minister to France he stopped over in England and got his first glimpse of the technological changes then taking place. "Strange as it may appear," he reported, they "card, spin, and even weave, it is said, by water in the European manufactories."[1] He also saw his first Watt steam engine. He had studied the older engine of Thomas Newcomen in books while a student at William and Mary College, but this was different.

Here was the technological solution to the nation's economic problem: hands could be replaced by machines. Machines were especially effective in industry, and manufactures could therefore be encouraged. To the objection that there were no such machines in America, Hamilton, in his 1791 *Report on Manufactures,* replied: "To procure all such machines as are known in any part of Europe can only require a proper provision and due pains. The knowledge of several of the most important of them is already possessed. The preparation of them here is, in most cases, practicable on nearly equal terms."[2]

Thus Hamilton advocated, and James Madison heartily seconded, what was much later termed the "Transit of Technology." When the American colonies were first settled, the immigrants had brought with them the essentially medieval technology they had known at home. In their absence, a new, industrial technology had been born. It was now necessary to import that as well. The transformation was not easy, but it was done largely through the expatriation of British mechanics, the very route that the Crown had feared and hoped to guard against. The process was most encouraged, and is best studied, in the textile industry.

As late as 1750, the production of textiles was still an ancient art. The process consisted of four basic steps developed at different times: the fiber was prepared by hand carding; the spinning was done on the familiar spinning wheel, which had reached its near-final form in the fifteenth century; the weaving was done on looms not much changed from pre-Christian times; and the fulling mill, introduced by A.D. 1000, was the only mechanical aid to finishing (fulling, dying, trimming). Between 1750 and 1800 all of these basic steps were greatly improved and aided by mechanical power. Furthermore, partly in response to the introduction of power machinery, the textile industry, both in England and later in America, was reorganized during these same years from the predominant cottage or household system to the new factory system.

The improvement of the spinning process began in England in the 1730s when John Wyatt and Lewis Paul developed a hand-powered spinning machine, patented in 1738. The first mill using their machinery was placed in operation in 1740. Until about this time, cotton was not a particularly popular fabric. Imported mainly from India, cotton was exotic and ranked well below wool and linen as common cloths. Then, about midcentury, cotton cloth became something of a fad, which, along with calico printing, helped establish a new cotton textile industry in the Lancashire district of

England. The resulting increase in demand for cotton yarn was too great for the Wyatt and Paul mechanism to handle. Then came that astonishing series of inventions so closely identified with the Industrial Revolution. James Hargreaves is said to have invented his spinning jenny in 1764, but little is known about it before 1767, and it was not patented until 1770. Operated by a large hand crank, this device pulled rovings (long rolls of cotton) into thread and twisted it as it was taken up on a roller.

In 1769 Richard Arkwright patented his water frame, a device not unlike the jenny but operated by waterpower. In 1775 he also patented a carding engine (to replace the hand cards) and the use of rolls in preparing the rovings. With the water frame, four pairs of rollers pulled threads out of the rovings and fed them onto bobbins, giving them a twist for strength. In about 1774 Samuel Crompton began work on his mule, so named because it was a hybrid of the jenny and the water frame. Although these three machines produced the same result, they did so by somewhat different means and so were complementary rather than competitive inventions.

These devices all operated on the same principle: the object was to draw out the cotton fibers into a thread. The problem was to coordinate the size, placement, and speed of the various rollers and bobbins in such a way that the resulting thread would be even in strength and thickness throughout its length. As so often happens, the basic idea was obvious and occurred to many people. The real invention lay in the delicate and sensitive arrangement of the machine's parts. Crude as these early wood and brass machines may have seemed by later standards, they represented great advances, made possible by new clockmaking and lathe-work techniques developed in the preceding half-century. So much was improvement the order of the day that Hargreaves was never able to enforce his patent on the jenny, Arkwright finally had his patent voided by the courts, and Crompton never even bothered to patent his mule. Nevertheless, a mechanized cotton industry was founded, beginning with Arkwright mills established in 1769 and later. In 1788 a pamphlet on the English cotton industry reported 120 spinning mills in operation.

Weaving was mechanized somewhat later than spinning. The first major improvement of the eighteenth century was John Kay's flying shuttle of 1733. It is often said that this development made such an improvement in weaving that a shortage of yarn developed, which led to improvements in spinning and again put pressure on weaving until the introduction of the power loom. This neat formulation is undermined by the fact that Kay's admittedly useful flying shuttle was not adopted for many years, by which time the spinning process had already been mechanized.

Before 1810 almost all weaving was done on hand looms. The flying shuttle increased the speed of weaving narrow cloth (18–30 inches wide) and dispensed with the second weaver normally needed to help weave broadcloth (over 30 inches). Furthermore, the flying shuttle had suggested how power might be applied to the weaving process. Again, the difficulty proved to be exercising the proper control of the necessary motions. The material had to be kept at a uniform length, the shuttle had to move with uniform speed, and the reeds had to fall with uniform pressure. Moreover, when a thread broke the whole loom had to be stopped.

The Reverend Edmund Cartwright patented a power loom in 1785, and two years later a small factory with twenty of his looms was established. The type of loom that eventually became dominant in England was developed by others between 1813 and 1821. Then between 1822 and 1830 Richard Roberts, an English machinemaker, brought the power loom to something like its ultimate dimensions and workings. Between 1813 and 1833 the number of power looms in England jumped from 2,400 to 85,000. Although in 1830 there was still a great deal of room for improvement in all the cotton textile machines and the mechanization of wool production was not yet as perfect as that of cotton, it is safe to say that the industry as a whole had become mechanized.

Needless to say, all of this development was British rather than American. As Hamilton and others pointed out, the American problem was how to get a hold of these devices rather than how to invent them. A jenny was exhibited in Philadelphia on the eve of the Revolution, but that long and difficult war isolated Americans from British improvements for a decade. At the end of the war, American textile manufacturers were totally hand powered and discontinuous. Then within the next twenty-five years they became mechanically powered and continuous.

As late as 1790 the only yarn spun anywhere in the world by waterpower was manufactured in England on Arkwright machines. Then, during the winter of 1790–91, a cotton spinning mill using Arkwright machinery began production in Pawtucket, Rhode Island. This machinery was the handiwork of Samuel Slater, a young spinner from England familiar with the industry there and the first to escape England and use his knowledge to set up a mill in America. He was shortly followed by a number of others.

During the eighteenth century (as before and after) many nations engaged in a great game of "beggar-your-neighbor." England was one of the most successful, having attracted a steady flow over the years of German miners, Italian silkmakers, and Flemish and French craftspeople of various kinds, who plied their old trades in this new setting, very much to the benefit of the British economy. Perhaps because they were themselves so successful, the British were exceedingly jealous of their own capabilities. In 1750 Parliament passed an act that forbade the exportation of silk- or wool-making machinery and the expatriation of workers in those trades. In 1774, in recognition of the recent explosion of the cotton industry, Parliament passed an act prohibiting the export of tools or utensils used in the manufacture of cotton and linen-cotton textiles. Loopholes in previous legislation were plugged in 1781 by a law that attempted to prevent the escape of sketches, plans, models, and other methods of reporting information about the construction of textile machines.

Despite fines of £500 and twelve months of imprisonment for transgressing these laws, Americans interested in manufactures busily scoured the British countryside for workers willing to go to America. Machines, too, were sought. Commenting on the smuggling out of the new Crompton mules, Thomas Digges, a shadowy character from Maryland, pointed out that they were

> of such a size as not to be admissible in the hold of any common ship, and are brought [to Ireland] covered upon the quarter deck, one on each side only. They might be got to America with a little address, and some risque both to the person shipping them and to the ship—The vessel must be *English* and she cannot clear out direct for America, but may clear for Cork or the Isle of Man and so proceed on—They stop no sorts of Machinery coming from Manchester or Liverpool to [Ireland]. . . . But they are so watchful in England as well as here [in Ireland] for any going to America that upon the slightest suspicion they stop and search the Ship.[3]

Here, indeed, was Hamilton's "proper provision and due pains."

It was easier to send people than machines, however, and in April of 1792 Digges claimed to have sent over, "by some art, and very little expense," eighteen to twenty "valuable artists and machine makers in the course of last year." To operate the machines these artisans would build, Digges predicted "a considerable migration this Spring from this place [Belfast], Derry and

Newsy, I dare say not less than 10,000 people, and in every family . . . spinners and often weavers—almost wholly a protestant body, sober, and industrious." He expressed the common fear that although many artisans were crossing the Atlantic, "the ease" with which they obtained Freeholds would make them "quit the Loom for farming." He had high hopes, however, that "in the Circumstances of the Cotton Manufactory, by improved Mill Work and the new invented Machinery for spinning, . . . such high price of labour in America may be in some measures counteracted by these mechanical inventions." There is no way now of knowing how many artisans Digges managed to send to America, but the career of one may stand for that of others.

William Pearce—a "second Archimedes," Digges called him—was apparently of English birth, although his early years are wholly obscure. According to Digges, he "was the inventor of Arkwrights famous Spining and Weaving Machinery at Manchester, but was robbed of his invention by Mr. Arkwright (then a Hair Dresser of Manchester and since made a Baronet from his wealth and consequences)." Also according to Digges, Pearce was "the artist who erected the Mills of Messrs. Cartwrights, which dresses the Wool, spins, and weaves Broad Cloth by force of Water, steam, or by a Horse." About 1790 this able mechanic moved from Doncaster, in Yorkshire, where he had been working for the Cartwrights, to become a partner of Thomas McCabe, a watchmaker of Belfast. The two were "jointly Concerned and Bound to each other in the line of making Machinery, Looms, etc. for the Manufacture of linin, Cotton, Callico, Checks, Thicksets or Corduroys, etc. etc., and for the getting of Pattents or Premiums for their inventions."

Digges set out, successfully, to seduce Pearce away and get him to the United States. Frustrated in his attempts to obtain a premium from the Irish Parliament, Pearce agreed to emigrate. Well aware of the British laws against seducing away artisans, Digges arranged the details with some discretion. Some key parts of the machines, as well as drawings, were sent on separate vessels. An American-owned vessel was selected for Pearce himself, and he was careful to travel under an assumed name. Somehow word leaked out to the authorities, and an express was sent to Dublin to stop him. According to Digges, "a Cutter pursued and search[ed] the Vessel twice for His double Loom and they would have brought him back had He not entered and given a different name—this was done in my sight and within a half hour after I had parted with him sailing out to Sea."

Pearce had letters of introduction from Digges to several leading figures in the United States—George Washington, Thomas Jefferson, Alexander Hamilton, and others. By September of 1791 he was in the employ of the Society for the Establishing of Useful Manufactures, a company led by Hamilton that was seeking to erect an extensive cotton manufactory at the new town of Paterson, New Jersey. At the same time, he established his own cotton manufactory in Philadelphia, the site of the shops in which he built machines for Hamilton. In 1792 his manufactory was visited by George and Martha Washington. "The President," reported the local newspaper, "attentively viewed the Machinery, etc. and saw the business performed in its different branches—which received his warmest approbation."

After a falling out with the Paterson enterprise, Pearce contracted in 1793 to build a mill on Brandywine Creek just above Wilmington, Delaware. Four years later, shortly before it was destroyed by fire, the mill was described by a French traveler as having "the whole of the machinery for carding, spinning, etc . . . constructed on Arkwright's plan."

Not all immigrant English artisans were as useful to the new nation or as successful as Pearce. Henry Wansey, himself a British textile manufacturer, traveled through the United States in 1794, the very year that Pearce began the Delaware mill, and noticed with disgust that prospective manufacturers usually "put the whole business under the care of a chief workman (being ignorant themselves) who has no interest in an economical management of the concern." Passing through Connecticut in May 1794, Wansey recorded that

> one of my companions in the coach, was a Mr. M'Intosh, originally from Bocking, in Essex. He took me in a one-horse chair to see his large manufactory, which had lately been established at a head of water, about three miles from Newhaven. It is patronized by the State, which has already advanced him ten thousand dollars, and engaged to go as far as sixty thousand; they being very anxious to establish the woollen and cotton manufactory in that district. But from what I saw of the undertaking, I am convinced, a great deal of money will be sunk to very little purpose. . . . There were two carding engines finished and at work, but both very much warped and cracked, by the heat and dryness of the rooms, as well as being made of unseasoned wood. Two slubbing and two spinning machines of good and complete workmanship, but the cotton yarn, which was then spinning, was not better than candlewick yarn. He has a

> wheel of thirty feet diameter, and eight feet wide, but I think they will often be in want of water to drive it: the cards were very badly made. He has erected forges there, and is making the heavy wrought and cast iron wheels, brasses, screws, spindles, etc. on the spot, at a very vast expense. The coal for working and smelting is brought from Virginia. A vast number of workmen are employed in this department at a very heavy expense. He has many English workmen engaged at great wages, particularly one from Sir George Young's manufactory at Ottery, in Devonshire, who engages to undertake the spinning worsted by water; a promise I do not think he will ever perform.[4]

Despite such pessimistic appraisals, many of them no doubt warranted, American manufacturers pressed ahead to establish the industry. In 1792 the *Gazette of the United States* reported that "an association in Virginia, another in the territory south of the Ohio, and a company in the western district of South-Carolina, have provided themselves with carding and spinning machinery on the British plan, to manufacture their *native* cotton."[5] Contrasting sharply with Wansey's testimony, Tenche Coxe asserted that "a large proportion of the most successful manufacturers in the United States are persons, who were journeymen, and in a few instances were foremen in the work-shops and manufactories of Europe, who have been skillful, sober and frugal, and having thus saved a little money, have set up for themselves with great advantage in America."[6]

Out of such beginnings an American textile industry began to develop. Through an uncoordinated system of bounties, patents, investment of public funds, industrial espionage, encouragements to emigration, permission to incorporate, and a large amount of self-education from practice, reading technical descriptions, and talking with others, Americans gained sufficient knowledge to construct and sustain a textile industry on modern principles.

Furthermore, the contemporary turmoils of the French Revolution and Napoleonic wars served as a twofold boon to manufacturers: it tended to cut Americans off from foreign sources, thus producing a kind of protected market for home manufactures; and it influenced a large but unknown number of British artisans to seek the peace and patronage of the United States. By 1813, in the midst of the War of 1812, when foreign cloth was completely cut off, *Niles' Weekly Register* reported that there were seventy-six cotton mills within a 30-mile radius of Providence, Rhode Island, operating a total of 51,454 spindles.

At the end of the War of 1812, large quantities of British textiles were dumped on the American market, especially in New York, a practice that caused great distress in the industry. Many mills, overcapitalized in the beginning, went bankrupt and were then reopened under new owners with a lower capitalization and better chance of economic success. By the fall of 1816 E. I. Du Pont's woolen mill in Delaware was in operation, and he found it necessary to advertise in the Philadelphia newspapers for "any person lately arrived from Europe, and well acquainted with some branch in the finishing department of a CLOTH FACTORY."[7] The "late" arrival of such a person would, he hoped, ensure that they were knowledgeable about the latest technological improvements in England. He also wanted "as apprentices, two lads of good disposition, and respectable connections." This, too, was significant, because the shift from using apprentices who had to be taught the entire business and "hands" who could be exploited for their labor alone was already under way.

To the extent that there was unused waterpower within a reasonable transportation distance of their markets, the textile mills were built new from the ground up. In many places, such as Delaware, milling had already been going on for nearly two hundred years, and most prime sites had long since been appropriated for gristmills, sawmills, and other types of mills. One typical advertisement (1822) for a millseat in Delaware contains a host of social data:

> FOR SALE, OR TO RENT, Or a partner would be taken in the concern.
>
> About 7 acres of land on both sides of Red Clay creek, about 4 1/2 miles from Wilmington, and 2 from Newport, New Castle county, whereon is erected a large mill house, 50 feet long and 39 wide, 3 stories high, with an addition on one side of 45 feet long and 25 feet wide, 3 stories high, and at one end an addition of 28 feet long and 24 wide, one and a half story high, with 2 large sheds, and a shear shop. There are 2 water wheels, and water sufficient to drive them at all times (or nearly so) to run five feet stones powerfully. There is likewise a saw mill, and sufficient water (exclusive of what is necessary for the grist mills) to drive it eight or nine months in the year. One water wheel in the mill is now employed in a woolen establishment, which is carried on pretty extensively; the other wheel in the milling business. There is house room (by evacuating the grist mill machinery, etc.) sufficient for two thousand cotton spindles, with all the machinery necessary for them, besides the woolen establishment, and water sufficient for both establishments. There is a large, tight

> dam, a short head and tail race, with twelve feet head and fall, a large stone mansion house and kitchen, 2 stories high, cellared under the whole, a good frame barn, with stabling under it, and six tenements for families to live in. It is a healthy neighborhood and handsomely situated. More land can be had convenient.[8]

Milling was clearly quite flexible, and not yet entirely divorced from farming. Both these circumstances no doubt made accommodation to the new Industrial Revolution easier and less risky.

The eventual future of the American textile industry lay not with such small and rural mills as this, but rather with the new "factories" of New England. Partly the result of the introduction of power looms, but more the result of a combination of machines and new organization, the first true factory was probably the one established by the Boston Manufacturing Company at Waltham, Massachusetts, in 1813. It was the earliest textile mill in the country in which all the processes, including weaving, were power operated. Here, and at Lowell, large factories drew power from specially dug water-power canals and recruited young women (the so-called Lowell Girls) from nearby farms to live in supervised dormitories.

The "Rules and Regulations of the Matteawan Company," published in 1846, gives some flavor of life in these new factories. This establishment, located in a small manufacturing village on Fishkill Creek, near the Hudson River, in New York, consisted of a cotton factory, a machine shop that built machines both for the factory and for sale, and a card factory. Three hundred men, women, and children worked in the cotton factory and lived in 100 company-owned tenements. They lived under rigid rules:

> No person will be admitted into the yard during working hours, except on business, without permission of an agent. At all other times, the watch man will be invested with full control.
>
> The work bell will be rung three minutes, and then tolled five minutes; at the expiration of which, every person is expected to be at their work, and every entrance closed, except through the office, which will at all times be open during the working hours of the factory.
>
> No person employed in the manufacturing departments can be permitted to leave work without the permission from their overseer. All others employed in and about the factory are requested to give notice to the agent or superintendent, if they wish to be absent from their work.

> No spiritous liquors, smoking, or any kind of amusements, will be allowed in the workshops or yards.
>
> Those who take jobs will be considered as overseers of the persons employed by them, and subject to these rules.
>
> Should there exist among any of the persons employed, an idea of oppression on the part of the company, they are requested to make the same known in an honorable manner, that such grievances, if really existing, may be promptly considered.
>
> To convince the enemies of domestic manufactures that such establishments are not "sinks of vice and immorality," but, on the contrary, nurseries of morality, industry, and intelligence, a strictly moral conduct is required of every one. Self-respect, it is presumed, will induce every one to be as constant in attendance on some place of divine worship as circumstances will permit. Intemperance, or any gross impropriety of conduct, will cause the immediate discharge of the individual.
>
> The agent and other members of the company are desirous of cultivating the most friendly feeling with the workmen in the establishment, believing they are to rise or fall together. Therefore, to promote the interest and harmony of all, it is necessary there should be a strict observance of these rules and regulations.[9]

Although the key to the factory's success was the careful integration of all the basic steps in cloth production under one roof, which were run and regulated from a single power source, the separate cardmill at Matteawan is a reminder that the several parts were developed at different times. It was not until Richard Arkwright's machine of 1775 that a practical carding engine was available. This model had not only a cylinder but also a comb and crank to remove the carded fibers and run them through a funnel and between two rollers, which produced a continuous sliver of material ready for spinning.

Some sort of carding engine was possibly shown in America as early as 1770, but even in 1800 most carding was still done by hand. It is not uncommon that as new technologies threaten to displace older ones, the latter, in turn, are improved in some way so that the displacement is never complete or at least is delayed beyond what at first seemed to be reasonable. To help fill the need for hand cards, the American inventor Amos Whittemore invented one of the country's earliest and most ingenious machines.

Born in Cambridge, Massachusetts, in 1759, Whittemore was apprenticed to a local gunsmith and then variously employed. Eventually he became, with his brothers, a partner in the firm of Giles Richards & Co., manufacturers of wool hand cards. Such cards were an important item of commerce in America. When the British Parliament forbade the exportation of textile machines, plans, or models in 1774, the flourishing card trade with the American colonies had been cut off. In response to pressure from British card makers, the following year the authorities excepted wool cards from this general ban. When the new United States passed its first protective tariff in 1789, however, wool cards were one of the few items on which a specific duty was placed.

That same year President George Washington traveled north to Boston and visited the Richards enterprise. "I was informed," he recorded, "about 900 hands of one kind and for one purpose or another [were employed]—all kinds of Cards are made; and there are Machines for executing every part of the work in a new and expeditious man'r, especially in cutting and bending the teeth which is done at one stroke. They have made 63,000 pr. of cards in a year, and can undersell the Imported Cards—nay, Cards of this Manufactory have been smuggled into England."[10] The source of this machinery is not given, but may have been the Delaware inventor Oliver Evans.

A more detailed description of the works was given in 1794:

> Cards of the various kinds used in other manufactures are made in this town [Boston] in large quantities, and with great despatch. The manufacture of cards was begun here before the revolution; but the improvements made in it since, have discouraged, and operated to exclude importations of this article into this commonwealth, and in a great measure into the southern states, they being supplied with a large portion of what is made at the manufacture in this town of Mr. Giles Richards, who was first named in a company that began this business in 1788, by newly invented and improved machines, the effects of American genius. The principal manufactory is at Wind-mill walk, contiguous to the grist-mill at the mill bridge. The card boards are cut by the operation of a wind-mill. One man working at the machine used for cutting and bending the wire, and pricking the leathers, can prepare a sufficient quantity of wires in twelve hours to stick upwards of twenty dozen pair of cards.
>
> One half of the number of men skilful in using these machines, can perform the same work, in the same given time, which can be done by any other method yet discovered.

> Between six and seven thousand dozen have been made annually, and, as hinted above, exported to the southward. Not less than twelve hundred persons, chiefly women and children, have found employment is sticking the cards; and as the manufacture advances in credity, the demand for the cards will probably increase and furnish employment for a much larger number.
>
> This is a very valuable manufacture, not only as it employs women and children, but also a great number of others in the commonwealth, in manufacturing the sheep skins, and making the tacks, &c. Four fifths of the cards manufactured in the commonwealth, are made in the town of Boston. The new inventions in cutting the wires and boards, not only diminish the toil of labour and expedite the work, but also occasion the price of the cards to be reduced.[11]

The machinery used here merely cut and bent the wire and punched holes in the leather cards. The actual setting of the wires was still done by hand by the two thousand or so women and children. This labor, far from being deplored, was seen by many as a public benefit, since it made productive people who presumably would not otherwise have been so. Ironically, in the atmosphere of technological euphoria of the early Republic, both the employment and the "saving" of labor were heralded as public boons.

When Whittemore became interested in the business, he sought to further improve the firm's machinery so that hand labor, specifically that of the women and children, would be eliminated. The mechanical solution apparently occurred to him in a dream, after which he was able to finish the design within three months. In 1797 he patented his new machine and introduced it into his manufactory. Two years later he traveled to England to secure a British patent. His interests there were eventually taken over by Joseph C. Dyer, who established the manufacture of his machines abroad.

> About the beginning of this century Mr. Amos Whitimore [*sic*], of Boston, commenced his experiments for making cards by machinery. His first step was to examine the movements required to form and set card teeth for hand carding. He was for a long time engaged with trial machines, and ultimately succeeded in performing the operations for making and setting the card teeth by movements effected by eccentrics on a driving shaft, viz. (1) feeding the wire, (2) holding it, (3) cutting off and (4) bending the wires into staples, (5) piercing holes in the leather, (6) passing the staples through it, (7) pressing their crowns to the sheet, (8) crooking the teeth to

the knee bend, and (9) advancing the leather to receive the next row of teeth.[12]

Dyer noted that along with a lace-making machine by a Mr. Heathcoat, Whittemore's card-making machine had an even wider effect:

> A new principle of action was adopted to produce and govern the movements for making wire cards, viz., that of eccentric curves revolving on a driving shaft and guiding the motions of the traversing parts of the machines in their due order of succession for making and setting the card teeth as before stated. This application of curvilinear projections or cam pieces has since been extensively employed for giving intricate motions in many other machines invented during the last fifty years. Whence it appears that both Mr. Heathcoat and Mr. Whitimore became pioneers and guides to other able mechanicians in their several labours for the advance of mechanical science.[13]

What Whittemore did was to "program" into the size and arrangement of these cams the information needed by the machine to make the cards. The spread of this technique to other machines to solve similar problems is an excellent example of the way technological innovations sometimes spread from one industry or use to others. In testimony before a parliamentary committee in 1841, Dyer's successor in the machine-making business explained how the device spread from Great Britain to Europe:

> That machine remained for a great number of years only very partially employed, but a Frenchman of the Screeve came over, and got two of those machines from Dyer; that machine is the most complex machine in the whole circle of manufactures, with the exception of the lace machine; now, that machine for card-making, so complex and difficult, having so many different movements, all of them requiring very great nicety in the adjustment, and very great knowledge of mechanical combination, is made by at least 50 different houses in France and Belgium, and can be worked perfectly. They are exporting those machines, and the produce of them, to every country in Europe.[14]

It seems a pity that such an elegant machine should have eventually been rendered obsolete because it produced hand cards that were no longer needed. The machine carding of fibers, however, had long been sought, and in 1793 the brothers Arthur and John Scholfield sailed from Liverpool bound for

America carrying with them an extensive knowledge of textile manufacture, including machine carding. Upon arriving in Boston, they stayed at the home of the American geographer Jebediah Morse, who, being impressed with their mechanical abilities, recommended them to a group of Newburyport merchants who wanted to build a textile mill. Early in 1794 they finished the machinery, including excellent carding machines, all of which, except the looms, were water driven. Then in 1799 they moved to Connecticut, where they established their own woolen mill. While John continued to operate this mill, Arthur, in 1803, began to make their carding machines for sale. He may well have been the first to do so in this country.

Meanwhile improvements continued to be made in England, and each, in turn, was imported. The mule was made self-acting about 1830 by Richard Roberts of Manchester, but because the export of textile machinery was still forbidden by British law, it, too, had to be brought over discreetly. A special act of Congress seeking to facilitate this process was helped along, according to later reports, by a member from Kentucky. The "final passage of the bill was in much doubt" until he expressed his determination that "every thing should be done that could possibly benefit his constituents, [and] strongly advocated the passage of the bill, on the grounds that his State and the whole West were as much interested in the improvement of the breed of mules as the North; and he declared, as his 'firm conviction, that Kentucky could raise more and better mules than any other section of the country.'"[15]

By this time, the country's textile industry was already quite large, ranking as a major industry along with flour grinding, sawmilling, and iron production. By 1840 the cotton industry alone operated 2¼ million spindles. Until the mid-twentieth century, it proved to be a hearty transplant. Furthermore, it became the nursery of two other important developments. One was the rise of the machine-tool industry, which received a strong impetus in the shops maintained or patronized by the textile mills. The other was the knowledge of waterpower and its use that grew out of such careful studies as those of James B. Francis, undertaken at the Lowell mills in the 1840s and 1850s.

Like the textile industry, the manufacture of iron in England, which had changed little since the Middle Ages, began experiencing a series of changes after 1750 that both precipitated and also exemplified the Industrial Revolution. During the next half century (1) mineral coal replaced charcoal as the principal fuel, (2) blast furnaces became larger and more efficient, (3) tilt hammers and forges were replaced by rolling mills for the production of wrought iron, (4) steam replaced water as the source of power for both furnaces and forges, and (5) the puddling furnace replaced the forge fire in the refining of pig iron. As with the new developments in the textile industry, new processes and devices in the British iron industry eventually arrived in America and transformed the iron industry on this side of the Atlantic as well.

The first ironworks in British America were built at Falling Creek, near Jamestown, Virginia, in 1619, but three years later the establishment was destroyed during a general uprising of Native Americans. From this inauspicious beginning grew an industry that would be found in all thirteen colonies, except Georgia, and would include Maine and Vermont, as well. Forges and furnaces were operating on many isolated "iron plantations," some using slaves, and slitting mills and steel furnaces were located in some colonial towns and boroughs.

The iron manufactories spread mainly from north to south and from forest to town, but they were not distributed evenly over the colonies. The early promise of Virginia had not been fulfilled, and appreciable amounts of iron were not produced in that state for a hundred years after the massacre at Falling Creek. The building of the Saugus forge and furnace at Lynn, Massachusetts, in the 1640s marked the beginning of roughly a century of New England dominance of the American iron industry. In 1759 Israel Acrelius, a Swede touring the colonies, remarked that American iron, being both soft and tough, was well-suited for shipbuilding. The large shipbuilding industry of New England needed quantities of this iron and helped make eastern Massachusetts the largest market for American iron products.

About the middle of the eighteenth century, however, the bog-ore regions of Massachusetts began to lose their predominant place in American iron production. Most of the individual ironworks had been little more than oversized blacksmith's forges (bloomeries), and America still looked to England for its supply of iron products. By 1750 the bloomeries and bog ore of Massachusetts were giving way to the furnaces and pit ore of the middle-Atlantic states, particularly Pennsylvania. In 1759 Acrelius reported that Pennsylvania had the most advanced ironworks in the colonies, and that when its production was added to that of New Jersey, the Delaware River region had to be the center of the American iron industry. The typical Pennsylvania iron plantation at this time flourished far out in the countryside, surrounded by forests, which were subsequently cut down and baked into charcoal. Local ores were smelted in stone blast furnaces and the product was cast into salable products (such as kettles or fire-backs) or into pigs for sale to forges. The machinery was run by waterpower, and the large number of specialized and often skilled workers lived in what amounted to a company town, served by a company store.

Although the American iron industry was growing in extent and importance, it was not developing technologically. New processes that were revolutionizing the industry in Great Britain were slow to be adopted in America, and units of production were multiplied rather than enlarged to meet increasing demand. Between 1750 and 1830 American ironworks were generally scattered in area, small in size, and conservative in technique. The pressures of fuel exhaustion and the demands of war that were forcing English ironmasters to use steam engines, rolling mills, puddling furnaces, and mineral coal were not operating to the same degree in America.

After 1830, however, dramatic changes took place in the American industry as well. Shortages of labor and of fuel made improvements necessary, first in the secondary (forges) and then in the primary (furnaces) manufactories. The country's population was growing, manufacturers and builders (especially builders of steam engines) were demanding more iron, public works were multiplying, and railroads were beginning to fan out across the nation, gathering the reaches of a continental empire.

By the time of the Civil War, America's iron production was second in value only to that of flour milling and, like the latter, was moving its center westward from the scene of its colonial beginnings. With the introduction of coal and coke, and ore from the Great Lakes region, the iron production of eastern Pennsylvania became somewhat less important.

In the late eighteenth century iron was used in one of three basic forms: cast iron, wrought iron, and steel. The differences between these, though large in practice, arise from small differences in the amount of carbon they contain. Cast iron has 2½ percent to 3½ percent of carbon, steel has ¼ to 1¼ percent, and wrought iron has essentially none. Because charcoal, the fuel used to smelt and refine iron, is a form of carbon and enters into chemical interaction with the iron itself, percentages were difficult to control. In 1750 only about 5 percent of the iron made was used in the form of castings, and virtually all the rest was wrought iron. Steel, being expensive, was used mainly for specialty items such as blades.

Iron was manufactured in a variety of mills, each designed for a special purpose. In Europe until well into the seventeenth century and in America during the early years of colonization, iron ore was first reduced in a bloomery. In Europe the ore was mined from holes tunneled into the earth, but in the American colonies it could still be dug from open pits and even scraped off the bottom of swamps in the form of bog ore. Bloomery temperatures never rose high enough to melt the iron into a liquid state. The mass of spongy iron produced was called a bloom and was hammered so as to expel the rocky material foreign to the metal.

As the industry progressed, these bloomeries were replaced, as they had been in Europe, by blast furnaces. If built large enough and provided with enough draft, these furnaces could melt the iron. The blast furnace was a tall chimneylike structure with a hearth at the bottom, often built against the side of a hill to facilitate its loading. Measured amounts of iron ore, charcoal for fuel, and some sort of limestone or oyster-shell flux were poured down the stack and set afire. The burning was urged on by a blast of air from tubs (operated by waterpower) and introduced into the hearth through a nozzle called a tuyere. When this charge was melted, the slag was skimmed off the top and the molten iron allowed to run into trenches (sows) dug into the sandy floor, and thence into perpendicular, shorter, attached trenches, called pigs. Alternatively, the iron could be discharged directly into molds, such as for cannon, pots, bells, and so forth. The efficiency of these furnaces was controlled, in part, by the quality of the blast they used. Their size was limited by the fact that the charcoal they used for fuel tended to be crushed if the charge was too large and heavy. These problems were eventually solved by the use of steam engines, the hot blast, and mineral coal (increasingly in the form of coke).

Most of the cast-iron pigs produced in this manner were converted into bars of wrought iron at separate establishments called forges. As the name implies, these were oversized blacksmith shops in which the hand hammer was replaced by a large and heavy water-driven forge hammer. Here cast-iron pigs were heated in a reverberatory furnace and then hammered on the forge to remove the carbon (which combined with the oxygen in the air to make either carbon dioxide or carbon monoxide) that had made the cast iron so brittle. The wrought iron was then normally sold in small lots to local blacksmiths, who later heated and hammered the metal into various shapes, such as horseshoes or spades, for local use. Some wrought iron bars, however,

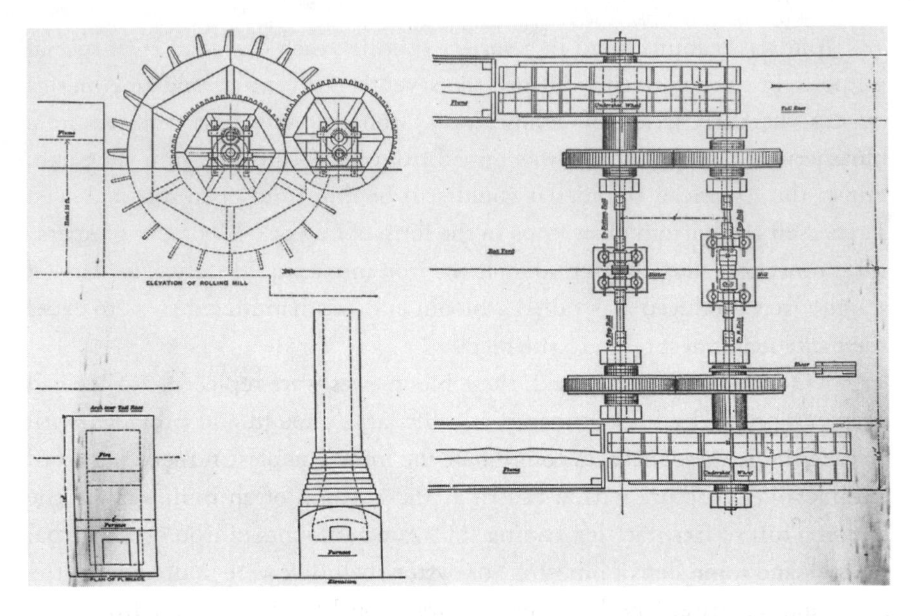

Rolling- and slitting-mills (to roll wrought iron into plates that could then be slit into nail rods and other products) were common in the colonies, though frowned upon by British policy. This plan, drawn nearly a century and a half later, was said to be that of "a rolling-mill built at Middleboro, Mass., in 1751." *Popular Science Monthly,* 38 (January 1891), 320.

were next sent to rolling and slitting mills. Here they were again heated and passed between water-driven rollers (smooth ones were used to produce plates and grooved ones to produce rods and strips). The resulting product was used to make tires for wagon wheels, barrel hoops, nails, and other items. This type of mill was fairly rare in the colonies because its products could, if necessary, be produced by hand by a blacksmith. In addition, the British commercial acts of 1750 expressly forbade the construction of more slitting mills, on the grounds that their production reduced the export of finished goods from England. Cast-iron pigs and wrought bars, on the other hand, were considered to be partly processed raw materials, which, if sent to England, would enhance the manufacturing interests of that nation since England already had to import much of its iron, particularly from Sweden.

In England, however, profound changes were already taking place in the iron industry. The earliest breakthrough came with the introduction of mineral coal, either in the raw form taken from the ground or, if too contaminated with impurities, in the form of coke, which was merely coal baked to drive off impurities. Coal had for many years been used in increasing amounts in England when iron-working and other industrial and domestic uses for wood progressively denuded the countryside. The new fuel was used primarily for those industrial processes in which the flame did not make direct contact with the product, as in glass-making, brewing, and pottery-baking. In these cases foreign material in the coal, such as sulfur, did not enter into the product itself. In the blast furnace, the fuel and the ore were deliberately mixed and impurities in the fuel easily contaminated the iron. In 1708 Abraham Darby I began using coke to make iron at Coalbrookdale. He was fortunate in having available a type of coal that was relatively free of harmful contaminants, and the experiment was a success.

Coke had three advantages: first, being made from mineral coal, it was more abundant and cheaper (at least in England) than charcoal; second, since it was less easily crushed, larger charges could be made in larger blast furnaces without compacting the charge and stifling the draught; and third, coke could take a stronger blast without quickly turning to ash, so that higher temperatures were achieved, the iron was more liquid, and smaller and more complicated shapes could be cast. The disadvantage of coke was that it often still contained some sulfur, phosphorus, and other contaminants. Not being as lucky in the quality of coke as was Darby, many ironmakers simply could not use the fuel until coking techniques were improved and a process for reheating the coke-made iron was developed. By 1790, however, 81 out of 106 blast furnaces in Britain were coke-fed.

Blast furnaces using coke required a stronger blast to make them operate well. An improvement in blast machinery came about 1762 when John Smeaton (the first person to style himself a "civil" engineer) introduced an improved blowing cylinder. The next step came in 1776 when the first Watt steam engine to be used for anything other than pumping water was set up at John Wilkinson's ironworks to operate his blast furnace.

Steam power was gradually substituted for water in other processes as well. Wilkinson was also the first to replace waterpower with steam for driving the large forge hammer that pounded out wrought iron. From 1782 on the hammer thus driven weighed about 800 pounds and was lifted 2–3 feet into the air by cams on a revolving shaft, then let fall onto the anvil.

Unfortunately for the British iron industry, cast iron made with coke proved too brittle to be easily forged into wrought bars, the iron flying apart when struck. Instead of hammering the metal under a forge, Henry Cort in

1783 used patented grooved rollers to knead and compact the metal. In this way, about 15 tons of iron could be made in twelve hours, in comparison with 1 ton by the old method. Then in 1784 Cort received a patent for his famous puddling process. His furnace was much like a common reverberatory furnace, in which the burning coke was kept in a separate compartment. The shape of the compartment directed the flame over an obstacle into the next compartment that contained the heated iron. This "puddle" of iron was stirred with a rod until the decarbonizing effect of the air converted it into malleable wrought iron, whereupon it was fed between the rollers.

The effect of coke, steam, puddling, and rolling on the manufacture of iron was no less revolutionary than those changes then going on in the textile industry. Between 1788 and 1796 the output of the British iron industry doubled, then doubled again from 1796 to 1804. Although most of this increased supply was still wrought iron, a growing percentage was cast. The spread of the steam engine (the first great machine to be necessarily made from iron) and the large number of cannon being demanded by the Anglo-French imperial wars of the eighteenth century, the Revolutionary War in the American colonies, and the French Revolution and Napoleonic conflicts—all greatly stimulated the use of iron, particularly cast iron. Furthermore, as early as 1754 John Smeaton had specified that a cast-iron rather than wooden main shaft be used in a windmill of his design. In 1760 the Carron Ironworks of Scotland, home of the famous carronade cannon, began to cast gear wheels of iron. In 1779 Abraham Darby III built a completely cast-iron bridge, which is still standing. By 1800 many mills in England were being built with iron beams and columns. In short, the increasing availability of cheap iron, both cast and wrought, made it possible to move from the wood technology of time immemorial to modern iron technology.

This was not a simple progression from one development to another. Immediate and critical feedback reinforced the change and made it irreversible through a ratchet effect. The use of coke had, for example, made iron cheaper and available in larger quantities. As a result, it became economically feasible to use steam engines in many more industries. When John Wilkinson, who cast the iron cylinders for Boulton and Watt's great engines, installed one to power his blast furnace, the increased blast further improved the quality, quantity, and cheapness of iron that he then used in improved engines. The steam engines were widely used to drain coal mines, and this application made coal cheaper and more readily accessible. This in turn

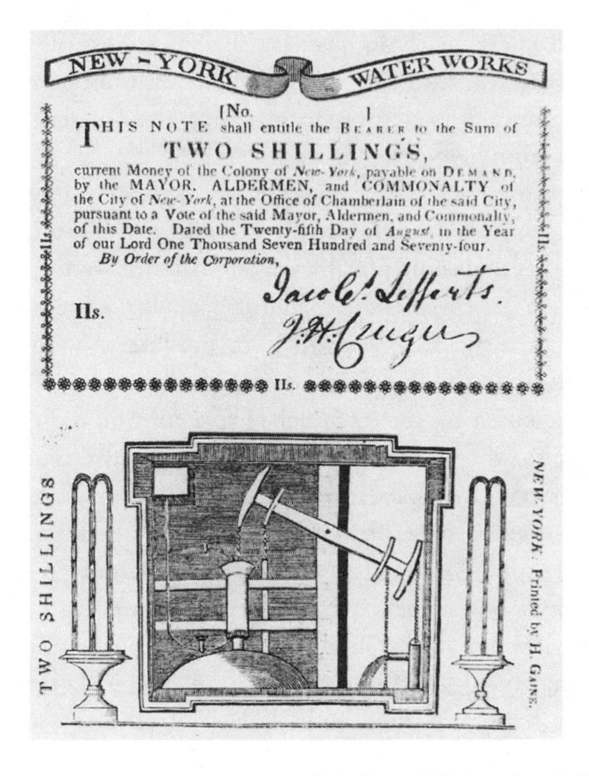

NEW-YORK WATER WORKS

[No.]

THIS NOTE shall entitle the BEARER to the Sum of

TWO SHILLINGS,

current Money of the Colony of *New-York*, payable on DEMAND, by the MAYOR, ALDERMEN, and COMMONALTY of the City of *New-York*, at the Office of Chamberlain of the said City, pursuant to a Vote of the said Mayor, Aldermen, and Commonalty, of this Date. Dated the Twenty-fifth Day of *August*, in the Year of our Lord One Thousand Seven Hundred and Seventy-four.

By Order of the Corporation,

IIs.

Jacob Lefferts.

J.H. Cruger

IIs.

TWO SHILLINGS

NEW-YORK: Printed by H. GAINE.

In July 1774, the Irish-American "Engineer and Architect" Christopher Colles proposed to build and erect a steam engine in New York to supply the city with water. The engine was manufactured the next year and placed in operation in 1776. Meanwhile, the Corporation of New York issued paper notes, called Waterworks Money, to finance the scheme. Courtesy National Museum of American History, Smithsonian Institution.

encouraged the greater use of steam engines that drew on coal for fuel. And so it went.

The rolling mill developed by Cort offers another case in point. The effective rolling of large quantities and dimensions of iron plate depended on, among other things, the use of steam power to operate large rollers. Once puddled iron could be rolled into quality sheets, these quickly replaced the old forged plate of uncertain dimensions and strength that had previously been used for steam boilers. With improved boilers, steam could be used at higher pressures (a pressure of 100–200 pounds per square inch was not uncommon by 1815), thus giving more power to engines more efficiently. The ability to roll iron efficiently in quantity also made possible the use of iron rails for railroad lines, thus spurring on an innovation that was to be the largest single consumer of iron products during the nineteenth century and included steam engines for locomotives.

Another example can be taken from the extremely important field of machine tools. To be most useful, iron had to be worked into useful shapes. The only machines that could possibly accomplish this were themselves made of iron. Thus each improvement in metallurgy made it easier to cut and work iron, and this in turn made it possible to produce more and better iron products. Here, too, Wilkinson was a pioneer innovator. Until his time, metal had to be bored by cutters attached to a turning rod extending from the shaft of a waterwheel to the workpiece. On large jobs, the head tended to droop and thus bore a crooked cylinder. When James Watt developed his steam engine, several people, including Smeaton, advised him that although workable in theory (and in small models with cylinders of tin hammered into shape around a wooden core), it could never be machined with sufficient accuracy to work well in full scale and under industrial conditions.

In 1775, however, Wilkinson built his celebrated boring mill, which allowed him to bore all of Boulton and Watt's cylinders with considerable accuracy. To gain this new accuracy, he fixed the hollow-cast cylinder more firmly while it was being bored, and more important, he ran the boring rod through the entire cylinder so that the cutting blades could be placed in the middle of the rod and both ends could be supported, so as to eliminate the droop. In 1776 Matthew Boulton wrote with delight that "Wilkinson hath bored us several cylinders almost without Error; that of 50 inches diamr for Bentley & Co. doth not err the thickness of an old shilling in no part."

In 1800 the city of Philadelphia, committed to building a modern waterworks, contracted a New Jersey steam engine builder to construct two engines for the project. The importance of the boring process to the general adoption of steam power, and especially the importance of Wilkinson's improvements twenty-five years before, were clearly brought out in a report left by Frederick Graff, a member of the city's Water Commission:

> Took passage [from Philadelphia] in the stage for Soho Works, near Newark, New Jersey, on the morning of 3d of July, 1800; and arrived there about noon of the next day.
>
> Soho is named after the works of Boulton and Watt, in England, and is situated about three-quarters of a mile northwest of the Passaic. . . .
>
> The large cylinder for the engine to be used on the banks of the

Benjamin Henry Latrobe's project to water the city of Philadelphia with steam engines was the boldest use of that technology yet in the United States. His neoclassic engine house, set in the Centre Square of that city (1799), suggested the compatibility of the new industrial technology with both Republican Virtue and rustic repose. Courtesy National Museum of American History, Smithsonian Institution.

Schuylkill at the water works was cast in two pieces, and united by copper, the joint being secured externally by a strong band of cast-iron, eighteen inches broad, weighing 1,200 pounds. Seven thousand five hundred weight of metal was used for the cylinder; it is six and one-half feet long, and about thirty-eight and one-quarter inches in the bore; about ¾-inch throughout was at first to be cut away; one-half inch has been accomplished, two men are required; one almost lives in the cylinder, with a hammer in hand to keep things in order, and attend to the steelings [cutters]; the other attends to frame on which the cylinder rests, which is moved by suitable machinery; these hands are relieved, and the work goes on day and night; one man is also employed to grind the steelings; the work is stopped at dinner-time, but this is thought no disadvantage, as to bore constantly the cylinder would become too much heated; the work also stands whilst the steelings are being changed, which required about ten minutes' time, and in ten minutes' more work they were dull again; I examined some of them and found them worn an eighth of an inch in that time. Three of these steeling (or cutters), about three and one-half inches on the edge, are fixed in the head piece at one time. The head piece is a

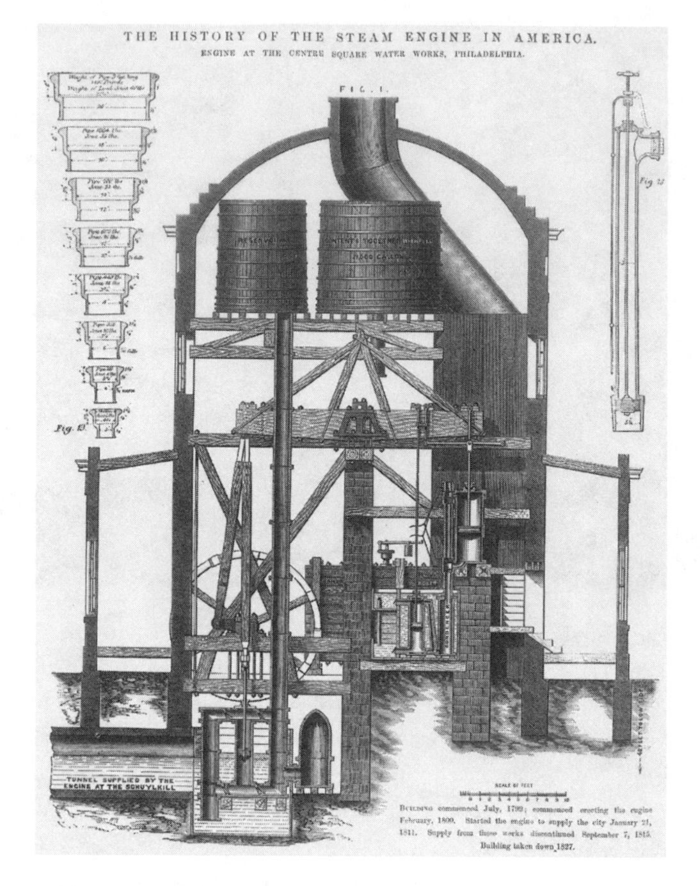

Latrobe's great Newcomen-style engine, housed at Centre Square, was built at the Soho works in New Jersey in 1799 and was the largest yet attempted in America. It proved to be less cost-effective than waterpower but established steam as a viable option for engineers in the country who needed large amounts of power. Courtesy National Museum of American History, Smithsonian Institution.

little less than the diameter of the cylinder, and six inches thick, secured upon a rod of iron eight inches in diameter, which forms the shaft of the water wheel.

The workmen state that the boring was commenced on the ninth of April, and had been going on ever since, three months, and about six weeks more will be required to finish it.

The wrought iron for the flue of the boiler over the fire will be imported from England, and is in sheets 38 by 32 inches. That yet made in

> this country is clumsy stuff, of different sizes, the largest being 36 by 18 inches, with rough edges which have to be cut smooth by the purchaser.[16]

Although the invention of the steam engine may have been the key element in improving industrial power, it had to be manufactured before it could be used, and that proved to be difficult in the extreme.

One further anecdote in connection with the Philadelphia waterworks is of some interest. The prospect of wood-burning steam engines in the city alarmed one particular citizen, who warned that it "must be taken into consideration, that a fire engine to raise water sufficient for the city will consume an immense quantity of wood or coals; and fuel seems already sufficiently high, without increasing it by such a scheme as this."[17] It was in fact many years before mineral coal was successfully and widely used for steam boilers, but the search for some method of utilizing it was no doubt greatly stimulated by the increased fuel needs of such engines as these.

It was only gradually that the great improvements in iron making that took place in England between 1750 and 1800 came to be accepted in America. With respect to mineral coal, for example, Tench Coxe wrote in 1810 that what might be interpreted as American backwardness in this area was in fact a virtue: since there were vast forests and therefore plenty of charcoal in America, there was no need to resort to the use of coal, which many believed produced an inferior product. Three years later, the expatriate English chemist Thomas Cooper wrote that whereas nine-tenths of English iron was made from "coke of pitcoal," charcoal of wood was still prefered in the United States. The main reasons for this, as Coxe had indicated, were that charcoal was cheap and widely available in America, because of extensive forest reserves, and that iron made in that fashion was still of a higher quality than that made from coal. As late as 1835 a medal offered by the Franklin Institute of Philadelphia to anyone who could smelt 20 tons of ore with anthracite coal went unclaimed. Gradually, however, the disappearance of the forests in the iron region led to the adoption of coal: by 1841 there were a dozen anthracite furnaces in eastern Pennsylvania and New Jersey. Coke was first successfully used there in 1837, though it did not enter into general use until after the middle of the century.

Americans were equally slow to replace forge hammers with rolling mills in the manufacture of wrought iron. An anonymous writer in the

Coal Mine at Mauch Chunk, Pennsylvania, 1831. Miners working a surface outcropping of soft anthracite coal, which was only then coming into use for fueling steam engines, blast furnaces, and not coincidentally, the economy of eastern Pennsylvania. *American Journal of Science,* 19 (January 1831), facing 1.

magazine *Agricultural Museum* in 1811, some twenty-eight years after Cort had patented his rolling mill, argued that "the manufacture of *bar iron* (or refined iron) by means of the steam powers and rollers, ought to receive the early and effectual attention of the people of the United States."[18] Two years later, however, Thomas Cooper noted that still "in this country, tilt hammers are universally used for drawing the iron out of the loops into bars. In England they are almost exploded. The work is done by means of cast iron rollers."[19] In 1817 the story was the same, an observer claiming that "nowhere that I know among our iron works, have iron cylinder rollers been substituted for the tilt hammer, although in Great Britain, from motives of economy in point of power, the tilt hammer has been greatly superseded by rollers."[20]

It was also during 1817 that the first rolling mill using puddled iron began operating in western Pennsylvania. Puddling and rolling did not always go together in these early mills. Many of the first generation of rolling mills were simply modifications of the old colonial slitting mills, now used to produce plate for iron boats, steam boilers, roofing, and an increasing number of other uses. Such mills simply rolled out into sheets bars of wrought iron already produced at forges. Puddling was quickly adopted in

England because it allowed the production of wrought iron with mineral coal. In this country, the need was not as great because charcoal was still dominant. Only gradually, as coke replaced charcoal, did puddling join rolling to produce a large number of modern rolling mills.

The fact that Americans kept close track of developments in the British iron industry and that friends of technological improvement periodically called for such innovations in this country is clear evidence of the excitement that surrounded the Industrial Revolution. In the area of iron production, the great increase in the quantity of production and the shift from wrought to cast-iron use, were made possible only by a series of technological changes: the shift from charcoal to mineral coal and coke; the introduction of steam engines for greater power; the use of puddling and rolling; numerous improvements in machine tools and the ability to work iron; and a host of other improvements such as hot blast, which made use of the fire in blast furnaces to preheat the air blown into the hearth. Between 1820 and 1840 the American iron industry, in the face of disappearing forest resources and increased demand for product, adopted this whole range of British innovations and thereby underwent sweeping changes.

The founding generation of the Republic had committed itself to the creation of a new American empire through economic growth. From Alexander Hamilton's *Report on Manufactures* to Thomas Jefferson's grudging realization that industry would have to join agriculture and commerce to safeguard the nation's liberty and prosperity, technology was seen as the key to growth. Many Americans enthusiastically joined in the general effort to invent a new technology for the nation, but it was universally acknowledged that the hardware and know-how of the British Industrial Revolution would also have to be imported, by fair means or foul. In this they were successful.

The chorus of enthusiasm was not quite unanimous, however. As would be the case two centuries later, some in the new nation questioned whether the ills of the Industrial Revolution, already visible in England, could be kept from these shores and, if not, whether they were a price worth paying for the enrichment of a few. Some of the founding generation thought rural environments destroyed by "dark Satanic mills," a work force alienated from their own tools and tasks, a new wealthy class of capitalist masters, and a sweeping away of the "republican" virtues of frugality, simplicity, and sturdy independence were too great a price to pay for technological "progress."

3

IMPROVING TRANSPORTATION

IT HAS BEEN SAID THAT "among the new industries to which the eighteenth century gave rise perhaps the most important was engineering."[1] No branch of that field was more in demand than that associated with transportation. In 1750, on the eve of the Industrial Revolution, neither Great Britain nor its American colonies had a transportation system capable of sustaining the great burgeoning of manufactures that was about to take place, first on one side of the Atlantic, then on the other. A hundred years later, both sides had an excellent network of interconnected canals, hard-surfaced roads, and railroads. Critical to this transformation was the rise of a new profession, civil engineering.

Like the rest of the Industrial Revolution, this great transformation in transportation started in Great Britain. The techniques and some of the people then migrated across the Atlantic to America when the need became sufficient. In the British Isles, the once excellent system of Roman roads had long since been allowed to deteriorate, and overland transport, except on short and heavily used routes, was virtually impossible. Pack trains of animals, people on horseback, walkers, and in some places stagecoachs were used, but whenever possible heavy agricultural and manufactured goods moved by water: down the rivers to the sea, along the coast, and back up other rivers. The same situation pertained in America.

In 1852 the new clipper ship *R.B. Forbes* was declared "a noble vessel." Designed for the Sandwich (Hawaiian) Island trade, it was "built in Mr. Hall's best style": 156 feet long, 32 feet wide, registering 750 tons, it was "rigged in the usual style, and looks finely aloft." *Boston Almanac for the Year 1852* (Boston, 1851), 42.

Then, in 1759, Francis Egerton, the second Duke of Bridgewater, took up again a project that his father had abandoned, that of digging a canal from the family coal mines at Worsley to the nearby city of Manchester. The actual design and supervision of the work was undertaken by the unlettered millwright James Brindley (1716–72). The canal was opened in 1761 and was a great success, cutting the cost of transporting coal by half. In 1767 the duke opened an extension of the canal that carried it all the way to the mouth of the River Mersey, thus providing a good and cheap transportation route between the growing textile center around Manchester and the great port of Liverpool. The success of this important canal set off a canal craze that lasted until the 1830s.

During this same period many private companies applied to Parliament for permission to build turnpikes, which were improved roads over specified routes for the use of which travelers and shippers had to pay a toll. In the 1750s and again in the 1790s, investors rushed in to build such roads. The two engineers who commanded most of the building were Thomas Telford (1757–1834), first president of the (British) Society of Civil Engineers, and John

Loudon Macadam (1756–1836), who gave his name to a kind of road surface made of broken stone or flint, compacted and given the cross section of an arch. Such surfaces, impregnated with asphaltum, are now called asphalt in America but still tarmac in Great Britain.

Then George and Robert Stephenson, father and son, ushered in the railway age in the 1820s. On September 27, 1825, the Stockton & Darlington Railroad was opened, with a Stephenson steam locomotive, the *Locomotion,* providing at least part of the motive power. It was not quite the first railroad in the world, but its innovations and great success made it, as one visitor claimed, "the great theatre of practical operations on railways." With the spectacular success of Robert Stephenson's *Rocket* at the Rainhill Trials in 1829 and the opening of the Liverpool and Manchester Railway the following year, the Railroad Age was fairly begun. Its birth, tragically, was not without incident. On the opening day of the line, William Huskisson, a former cabinet minister, was struck down and killed by a train—the first railroad casualty.

In the improvement of transportation, as in other aspects of the Industrial Revolution, the United States followed the British example, albeit with some delay and modification of circumstances. The political freedom of the American colonies unleashed a storm of economic activity, fueled by patriotism and imperial dreams and redirected away from transatlantic commerce in favor of what came to be called "internal improvements." "The Americans," wrote the British civil engineer David Stephenson in 1838, "have not rested satisfied with the natural inland navigation afforded by their rivers and lakes, nor made the bounty of Nature a plea for idleness or want of energy; but, on the contrary, they have been zealously engaged in the work of internal improvement."[2]

In America, as in Great Britain, canals and turnpikes were both turned to before the end of the eighteenth century, though here the most hectic aspects of the canal craze were delayed until the successful completion of the Erie Canal in 1825. Along the East Coast most rivers ran eastward to the ocean, and north-south travel was handicapped both by being necessarily overland (except for coastal sailing) and by numerous mountain chains that ran more or less east-west as well. Roads, such as they were, could hardly support the economical passage of loaded wagons, though it was estimated that horses could only carry on their backs one-tenth of the load that they could pull.

Despite the fact that Northampton, Massachusetts, for example, was only 100 miles overland from Boston, it was cheaper to send grain 36 miles overland to Windsor, on the Connecticut River, and thence 250 miles by water down that river, through Long Island Sound, and around Cape Cod to Boston. Even so, the land part of the journey cost thirteen times more than the much longer water leg of the trip. When the British blockaded the East Coast during the War of 1812, the cost of sending flour from New York to Boston by land rather than water went up from 75 cents to 5 dollars a barrel.

The early roads of the American colonies had been constructed along traditional lines, with local labor and materials. The first road law, that of Virginia in 1632, stated simply that "highways shall be layd out in such convenient places as are requisite accordings as the Governor and Counsell or the Commissioners for the monthly corts shall appoynt, or accordings as the parishioners of every parish shall agree." In 1639 Massachusetts passed a law specifying that roads in that colony, "in common grounds, or where the soil is wet or miry, they shall lay out the ways the wider, as six, or eight, or ten rods, or more in common grounds." The 1664 law in New York was even more specific: "the highways to be cleared as followeth, *viz.,* the way to be made clear of standing and lying trees, at least ten feet broad; all stumps and shrubs to be cut close by the ground. The trees marked yearly on both sides—sufficient bridges to be made and kept over all marshy, swampy, and difficult dirty places, and whatever else shall be thought more necessary about the highways aforesaid."[3]

During the colonial period the British government had made occasional attempts to build roads to answer the needs of empire. The most famous of these was the ill-fated 1755 effort of General Edward Braddock to carve the Winderness Road through western Pennsylvania in his pursuit of Native Americans. A young George Washington was a surveyor on that expedition. More often, as the colonial laws indicated, road building was a local matter only loosely controlled by minimal standards of design. As in England, it was privately chartered turnpike companies that first attempted to improve overland transport on routes selected for their likely profit.

The Philadelphia and Lancaster Turnpike Company was the first such firm. It operated along a key stretch of road, opened in 1794, carrying passengers to the west and south. In 1826 Josiah Quincy took the route, leaving "Philadelphia by the Lancaster stage, otherwise a vast, illimitable wagon, with seats without backs, capable of holding some sixteen passengers with decent comfort to themselves, and actually encumbered with some dozen more. . . . The roads seemed actually lined with Conestoga wagons each drawn by six stalwart horses and laden with farm produce."[4] Most vehicles were not so large and had not such a good surface to travel over. As late as 1853 it was noticed that "many of the carriages, especially those technically called 'waggons,' are made of an exceedingly light construction, and are intended generally to carry two or sometimes four persons. . . . It would seem as if the elasticity of these carriages peculiarly fitted them for the very bad roads on which they in general have to run, and it is evidently a principle with the Americans to use their light carriages and save their horses."[5] It was not the last time that American technology was to be characterized as "light and flexible."

The Philadelphia and Lancaster turnpike had been an immediate financial success, and by 1800 there were seventy-two such corporations, at least on paper. A decade later there were several thousand. By 1820 all the cities along the Atlantic corridor in the northern and middle-Atlantic states were connected by improved roads. As early as 1807 Secretary of the Treasury Albert Gallatin advocated that the federal government accept responsibility for land transportation between the states and planned a rational, coordinated network of internal improvements to meet the commercial needs of the growing national empire. The proposal ran afoul of the constitutional scruples of narrow constructionists in the Congress, however.

One exception had already been made. Because of the importunings of Westerners, Congress in 1806 authorized the building of what was at first called the Cumberland, and later the National Road. Construction began in Maryland in 1808 and reached the Ohio River nine years later. It was 1820 before surveying began on the route through Ohio, Indiana, and Illinois and not until 1825 were appropriations sufficient to begin heavy work in those states. The road was built 80 feet wide and covered with stones a foot deep, at least as far as Indiana. By the time it had reached Columbus, Ohio, in 1833, it was obvious that canals were in some ways a better method of transportation for the time, and the rest of the road was left with a dirt surface as far as Vandalia, Illinois, which terminus it reached in 1852. During the intervening years, the right and duty of the federal government to plan and finance a national transportation system were hotly debated and never totally accepted.

Indeed, roads and highways remained primarily a local responsibility until World War I. Only then did the National Road become a part of a newly designated U.S. Highway 40.

In the face of federal reluctance to undertake a coordinated transportation program, states were forced to fend for themselves. By 1816, with the end of the War of 1812, each major seaboard city—Boston, New York, Philadelphia, Baltimore, and Richmond—wanted desperately to tap the western hinterlands and make that trade its own. Thus both Boston and New York looked to Lake Erie as a link with the old trans–Appalachian West, while Philadelphia, Baltimore, and Richmond all sought to connect with the Ohio River. The prospect of emigrants and manufactured goods traveling westward, and farm produce eastward, was a powerful imperative.

Virginia handled its transportation problem in a most rational manner. In early 1816 the General Assembly passed an act creating a fund for internal improvement and a board of public works to look after it. A year later it passed the General Turnpike Law to give the board some guidance. Under this law, the board could give its approval to certain turnpike companies and allow them to draw on the fund for up to 40 percent of their capital. The lynchpin of the entire enterprise was the chief engineer. Col. Claudius Crozet, an artillery officer in Napoleon's army, had come to the United States in 1816, after the Battle of Waterloo, and taken a teaching position at the American military academy at West Point. It was at the time America's only engineering school and a veritable haven for French engineers who had brought with them a knowledge of the best and most scientific of European civil, as well as military, engineering practice. Crozet went to Virginia to be chief engineer in 1823.

His job was not an easy one. Whereas in Europe civil engineering was rapidly becoming an applied science, in America most work was still being done by local individuals using local materials and local knowledge. In fact, some builders and contractors lacked even traditional skills, having never before tried to build canals, bridges, turnpikes, or other such structures. One contemporary observer, commenting on a bad habit shared by the Virginia road builders, noted that "the route [of turnpikes in England] usually winds around the bases of high hills, in place of adhering to the absurd and unpicturesque mathematical straight line, so much admired in the United States, under the erroneous idea of saving distance, and consequently labour."[6]

There were 180 turnpike companies in Virginia, and for each, someone had to survey the proper route (a controversial matter indeed), estimate the costs involved, and draw up and let proper contracts embodying these facts. Crozet had no way of forcing companies to follow his advice, even when it was asked. Most errors pertained to location, drainage, and grades (roads went up and down hills rather than around them, for example). Bids were advertised only in local papers, and all contractors were from within the state. Crozet commented on the general level of competence in 1842 in discussing wooden bridges of a lattice design: "Probably this plan has been favored on account of its being within the comprehension and capacity of any common carpenter, which I would myself rather consider an objection, as it not unfrequently becomes a guarantee of inferior workmanship."[7]

The technique of constructing roads developed as slowly as the political will to build them. The common want of stone suitable for macadamized surfaces, and the expense of breaking it up, led to the use of local materials for surfacing in a few cases: oyster or clam shells near the ocean, or perhaps furnace slag in iron-producing districts. In most places, no surfacing at all was provided. The great abundance of wood in many parts of the country combined with a lack of capital and a feeling of urgency led to the creation of the plank road, on which rough sawn planks were laid out on timber stringers. The first of these, perhaps, was built in Toronto, Canada, in 1835–36, and the type was introduced into the United States within the decade, one being built near Syracuse, New York. One had been chartered in Virginia as early as 1833, but only ten were ever built, all between the years 1850 and 1853. These plank roads were fast and cheap to build but deteriorated quickly owing to the weather and wear and tear of traffic. Nonetheless, such a road was used as late as the 1920s to allow automobiles to navigate across the sands of the Imperial Valley east of Los Angeles.

Corduroy roads were even cheaper and faster to lay down. These consisted of trees felled, trimmed, and laid side by side directly on the ground. They were especially desirable for marshy ground but rotted quickly and were extremely rough unless provided with parallel tracks of planks laid end to end along their surface. Macadamizing became more practical in 1858 when Eli Whitney Blake, the nephew of the cotton gin inventor, designed an effective rock-crushing machine. The following year the steam roller was introduced, adding to the ease with which good, hard-surfaced roads could be constructed.

A stone-crusher, designed by Eli Whitney Blake (nephew of the cotton gin inventor), had been in "extensive use both in this country and abroad" for several years when it was pictured in 1876. Manufactured by the Blake Crusher Company of New Haven, Conn., it lowered the cost of constructing macadamized road surfaces. *Scientific American,* 34 (April 29, 1876), 275.

As the history of the National Road hinted, turnpikes, as in England, were soon rivaled by canals. Native Americans, and then the colonists after them, had depended on rivers for quick, cheap transportation. These had severe disadvantages, of course, from flooding in spring, freezing in winter, and low water in summer. All up and down the East Coast there was also a fall line marked by rapids and actual waterfalls where the rivers left the Piedmont Plateau to run more tranquilly across the coastal plain to the sea. The fall line running north-south was a precious resource of waterpower for manufacturers but a fatal obstacle to transportation for shippers. In 1772 Benjamin Franklin, then in England, wrote to the mayor of Philadelphia concerning the new British solution to the problems of water transport. Referring to some "Canal Papers" previously sent, Franklin revealed that "here they look on the *constant Practicability* of the Navigation, allowing Boats to pass and repass at all Times and Seasons, without Hindrance, to be a Point of the greatest Importance, and therefore they seldom or ever use a River where it can be avoided. . . . Rivers are ungovernable Things, especially in Hilly Countries: Canals are quiet and always manageable."[8]

Canal projects were often dreamed of before the Revolution. In several places coastal shipping was put to great pains and waste, traveling far around such land masses as Cape Cod, New Jersey, and the Delmarva peninsula separating the Delaware and Chesapeake bays. Capital, know-how, and a clear political responsibility were all missing until after the war, however. Then in 1785 the government of Virginia authorized the building of a 7-mile canal to avoid the James River falls at Richmond. This first American canal was soon followed by a handful of other, relatively short, undertakings: the Dismal Swamp Canal in Virginia and South Carolina (1794), one around Patopwick Falls in Massachusetts (1797), the Santee Canal in South Carolina (1802), the Middlesex Canal in Massachusetts (30 miles long, with 20 locks, finished in 1808), the Bow Canal in New Hampshire (1812), and the very ambitious Schuylkill River navigation in Pennsylvania, begun in 1815 and finished in 1826, 108 miles long with 129 locks. Although the technology was new, many of the routes around river obstructions had long been used by native peoples for the same purpose.

Important as these canals were, the Erie dwarfed them all and truly ushered in a canal-building craze. Private parties had tried and failed to build a canal in the Mohawk Valley of upper New York State when the legislature stepped in in 1817 and authorized a bold undertaking that eventually encompassed a canal 40 feet wide, 4 feet deep, and 363 miles long. Eighty-four locks raised the water 62 feet up from Lake Erie to the summit and then 630 feet down to the Hudson River. The first official boat, called the *Chief Engineer* after Benjamin Wright who held that position with the company, finally made passage from Rome to Utica in October 1819, to the accompaniment of "the ringing of bells, the roaring of cannon, and the loud acclamations of thousands of exhilarated spectators, male and female, who lined the banks of the new created river."[9] This celebration was eclipsed in 1825 when the canal was opened for its entire length.

Parades, fireworks, and speeches declared it New York's, and perhaps America's, finest hour. Cadwallader Colden, the official historian of the Erie, asked in 1825: "Who that has American blood in his veins can hear [the sound of cannon announcing the opening] . . . without emotion? Who that has the privilege to do it, can refrain from exclaiming, I too, am an American citizen; and feel as much pride in being able to make the declaration, as ever an inhabitant of the eternal city felt, in proclaiming that he was a Roman."[10]

The Erie Canal (opened for its full length in 1825) not only set off a canal craze but established records for excavation and construction in the antebellum period. Perhaps the most dramatic site was the great excavation at Lockport. Cadwallader D. Colden, *Memoir . . . at the Celebration of the Completion of the New York Canal* (New York, 1825), facing 298.

The building of the canal marked a new stage of sophistication for American engineering. The entire route had been divided into sections, each of which was under control of an engineer, the actual work being undertaken largely by local surveyors and contractors with masses of labor provided by both immigrant and native workers, for the most part using hand tools such as shovels, picks, and wheelbarrows. British engineering expertise was consulted when the opportunity arose, for the two chief engineers, Benjamin Wright and James Geddes, had had experience before only as lawyer, judge, and, more usefully, surveyor. Working with and under them, however, a whole generation of American engineers found their experience on the canal to be a kind of engineering college in the field. For many years, a cohort of engineers traced their education and introduction to the profession back to the Erie Canal.

New Yorkers did well to celebrate the opening of the canal in 1825. Goods could now move by water from the Midwest directly to New York City, and that metropolis soon forged ahead to become the undisputed commercial center of the nation. The center of grain-growing and milling moved decisively from the Delaware River and Chesapeake Bay to upper New York State, Rochester becoming the "flour city" and holding that position until wheat farming in the trans-Mississippi West forced a shift to Minneapolis and St. Paul. The cheaper, more certain passage by water now freshened the flow of emigrants to the West, in a chain of settlements from Buffalo to Cleveland to Chicago and beyond. And in New York State also, all along the path of the canal, a rebirth of economic activity, population growth, and new enthusiasms led to its being called the "burned-over district" for the many religious and social reform movements that began there: the evangelism of the Second Great Awakening, Mormonism, abolitionism, women's rights, and others.

Mules and horses to pull the boats remained for many years the principal motive power on American canals, but on the nation's rivers the steamboat quickly came to the fore. The notion of steam propulsion for boats was nearly as old as the steam engine itself, but the proper proportions of engine to boat and the best way to apply the power—whether by water forced out the back in a jet, screw propellers, paddle wheel (on the stern or on the sides), or by banks of oars or poles—was not obvious. Engines had to be made powerful enough, boilers strong enough, and all connections adequate. Perhaps even more difficult, money, mechanical talent, and political support had to be marshaled to conduct the trials that would establish the success of the experiment.

Before the eventual triumph of Robert Fulton on the Hudson River in 1807, more than a dozen Americans had experimented with such boats. In every case an inventor—James Rumsey, John Fitch, John Stevens, and others—were underfunded and had virtually to reinvent the steam engine before they could even begin to work. Even so, the progress was striking. Fitch, in the 1790s, had a steamboat running on schedule as a common carrier from Philadelphia across the Delaware River to New Jersey and back. Fulton's celebrated success was not in inventing the steamboat but in overcoming the handicaps that had hamstrung previous inventors. First he bought a Bolton and Watt atmospheric engine from England, finessing the whole problem of building such a machine in American shops with imperfect knowledge of its design. Next he lined up powerful financial and political support, particularly in the person of Robert R. Livingston, chancellor of the state of New York. The successful voyage of the *Clermont* up the Hudson in 1807 marked

Artist's conception of John Fitch's steamboat of 1788–90, which operated on the Delaware River. The "oars" aft turned out not to be the best way to propel such a vessel. Thompson Westcott, the *Life of John Fitch, The Inventor of the Steamboat* (Philadelphia, 1857), 285.

the beginning of regular commercial steamboat use in this country, and in the world.

The Fulton interests, which had a monopoly on routes in New York, continued to use the low-pressure Bolton and Watt type of engine for their boats, and their first boat on the Ohio and Mississippi rivers, the *New Orleans* of 1811–12, was of this same type. Soon, however, beginning with Daniel French's *Washington* in 1816, high-pressure steam was used on the western waters. Low-pressure engines had used steam at about 16 pounds per square inch (psi), whereas high-pressure steam was already being used at over 100 psi. With the latter engines, therefore, greater power could be achieved without necessarily having to make the engines larger. The *Washington* had a 100-horsepower engine designed by Henry M. Shreve, a man with two years' experience on western steamboats who was later to receive great acclaim for designing a boat that could pull snags from the shallow waters of the Mississippi and its tributaries. Shreveport, Louisiana, was named for him in gratitude for these efforts. In 1816–18 Oliver Evans built a high-pressure engine for a boat to be used on the eastern waters, the *Aetna.*

Soon his engines, and others operating on high pressure and without condensing their steam, displaced the older type. By 1830 the standard western steamboat engine had been established, and it changed little over the next sixty years. Unlike the sedate, floating palaces of the eastern waters used mainly for passengers, the western boat was designed for freight and was derisively described by one contemporary as "an engine on a raft with $11,000 worth of jig-saw work." David Stevenson wrote in 1838, "On minutely examining the most approved American steamers, I found it impossible to trace any *general* principles which seem to have served as guides for their construction. Every American steam-boat builder," he asserted, "holds opinions of his own, which are generally founded, not on theoretical principles, but on deductions drawn from a close examination of the practical effects of the different arrangements and proportions adopted in the construction of different steam-boats, and these opinions never fail to influence, in a greater or lesser degree, the build of his vessel, and the proportions which her several parts are made to bear to each other."[11]

The steamboat of Mark Twain fame, however ungainly looking, was a ubiquitous and powerful influence in the western United States. By the period 1848–50, an average of 3,629 boats called at the city of Cincinnati each year—nearly 10 per day. By 1838 some 348 boats had been built on the western rivers to ply their trade on the Mississippi, Ohio, Missouri, and other rivers. In 1860 alone, a total of 52 new boats were built in the three cities of Cincinnati, Louisville, and St. Louis. That same year, 168 boats were owned and operated out of St. Louis. The demand for and construction of the engines for these boats had the spillover effect of spreading and strengthening the ability to make engines for stationary use by mills and factories as well. Cincinnati, for example, became a center not only of steamboat building and repair, but also of general machine work and of machine-tool manufacture.

Tragically, by 1850 a total of 520 steamboats had been destroyed on the rivers—because their boilers had exploded, they had hit snags, or suffered some other accident—and with them a large number of human lives were lost. The federal government, in an attempt to find the cause of boiler explosions, gave its first grant for scientific research to the Franklin Institute of Philadelphia in 1830. That pioneering effort was inspired, at least in part, by the destruction, earlier that year, of the *Helen McGregor,* which exploded on its run between New Orleans and Louisville with a loss of fifty lives and at least as many injuries.

Turnpikes, canals, and steamboats all made significant contributions to the improved transportation facilities of the country, but it was the railroad that most captured the imagination and resources of the nation. In 1830, just five years after the opening of the Erie Canal and the same year the Liverpool and Manchester Railway was opened, the United States already had 23 miles of railroad in operation. During the next ten years this mileage grew to 2,818; by the eve of the Civil War it stood at 30,626, and by 1890 it had reached 166,703 miles. It represented an enormous commitment of national treasure and technical resources over two generations.

American railroads began with British technology, and the borrowing was both direct and prompt. Even George Stephenson once thought of moving to America. His sister had emigrated in 1807–8. As he explained in later years: "Well do I remember the beginning of my career as an engineer, and the great perseverance that was required for me to get on. Not having served an apprenticeship, I had made up my mind to go to America, considering that no one in England would trust me to act as engineer."[12] He stayed, of course, and Americans had to be content to import his innovation. In 1824 (a year before the Erie Canal had been opened) a group of Philadelphians, mostly connected with the Franklin Institute, formed the Pennsylvania Society for the Promotion of Internal Improvement as a part of that city's effort to find an effective commercial route to the West.

The next year they sent the architect and engineer William Strickland to England to discover what he could about the new railroads that were being discussed. Their instructions to him were explicit: "It is not a knowledge of abstract principles, nor an indefinite and general account of their application to the great works of Europe [that is wanted]. . . . These we possess in books. . . . What we earnestly wish to obtain, is the means of executing all those works in the best manner, and with the greatest economy and certainty. . . . We desire to obtain working plans . . . so that those works may be executed in Pennsylvania, without the superintendence of a civil engineer of superior skill and science."[13] It was not that the society was against such experts, it was simply that they were so scarce they could not be depended on.

Strickland returned with his information. Better yet, a locomotive, the *Stourbridge Lion,* was brought over from England and set up at the West Point Foundry in New York. In 1830 David Wright and the same work people who

Peter Cooper's "Tom Thumb" racing, and beating, a horse-drawn car on the Baltimore and Ohio railroad on August 28, 1830. William H. Brown, *The History of the First Locomotives in America* (rev. ed., New York, 1874), facing 119.

set up the *Lion* also built the first locomotive to be made in America, the *Best Friend of Charleston.* Eight years later David Stevenson noted that "the first locomotive engines used in America were of British manufacture, but several very large workshops have lately been established in the country for the construction of these machines, which are now manufactured in great numbers."[14] He mentioned the Lowell Engine-works, at the great textile mills, and four shops in Philadelphia.

The work in one of these, that of Matthew Baldwin, he described in these terms: "I found no less than twelve locomotive carriages in different states of progress, and all of substantial and good workmanship. Those parts of the engine, such as the cylinder, piston, valves, journals, and slides, in which good fitting and fine workmanship are indispensable to the efficient action of the machine, were very highly finished, but the external parts, such as the connecting rods, cranks, framing, and wheels, were left in a much coarser state than in engines of British manufacture."[15]

Baldwin and a partner had set up a manufacturing machine shop in Philadelphia about 1825. Two years later, at what appears to be the initiative of Baldwin, they added stationary steam engines to their line of products, and his partner soon left the firm. By 1834 he had built only six engines, but that experience tempted him to take the next step toward manufacturing steam locomotives, and by September 1839 he had already completed 140 of them, which represented no less than 45 percent of the American-built engines at work in the United States.

Nor were all American engines kept at home. By 1840, just ten years after the first engine was built in this country, they were already being shipped to Russia, Austria, some German states, and even to Great Britain. Of the first 145 engines made by William Norris of Philadelphia, 41 had been sold in Europe, including 10 to the Birmingham and Gloucester Railway. The railway line built by I. K. Brunel between London and Bristol was so level that it was called "Mr. Brunel's billiard table," but as an American testified before a Parliamentary committee of inquiry: "[in the United States] we aim to economise capital, and we therefore are not so particular about reducing the gradients as you are, therefore our roads are not on so dead a level as yours are; our roads are made with considerable ascents, greater than in Europe, and particularly in England, and the engines that are made for that purpose are made to overcome these steep ascents."[16]

There had been a general fear that locomotives, with their smooth wheels running on equally smooth tracks, could not climb a grade of more than 4 percent. In a series of three trials in 1836, a Norris locomotive handled the challenge easily. Two years later the Birmingham and Gloucester promised to buy ten engines if they came up to expectations. The first was shipped in 1839 but failed to live up to specifications: it is still not clear why, since it was of good construction. Rather than have the engines rejected, Norris agreed to lower their price. Eventually the railway company bought seven more locomotives from Norris and built an additional nine themselves on Norris's design. The entire transaction was a matter of some satisfaction to Americans, and surprise and disgruntlement to many British.

A larger and more significant sale was made to Russia, where Americans undertook to build the entire railway along the important route between Moscow and St. Petersburg. In all, 162 American locomotives and 2,700 freight and passenger cars were built for the line. American, rather than British, engineers and equipment makers appear to have been chosen for this enterprise at least partly because they were better adapted to the Russian climate and terrain, which was much more like that in America than in England. Among the many American engineers working on the project was the artist William McNeill Whistler's father, an engineer trained at West Point.

Roadbeds as well as rolling stock differed in America. Stevenson noted that American railroads were cheaper than English ones because "*first,* they are exempted from the heavy expenses often incurred in the construction of English railways, by the purchase of land and compensation for damages; *second,* the works are not constructed in so substantial and costly a style; and, *third,* wood, which is the principal material used in their construction, is got at a very small cost." Part of the cheapness came in not fencing the right-of-way. This in turn led to the invention of the cow-catcher: again Stevenson noted that "from the unprotected state of most of the railways, which are seldom fenced, cattle often stray upon the line, and are run down by the engines, which are in some cases thrown off the rails by the concussion, producing very serious consequences."[17] The cow-catcher neatly scooped the animal up and threw it to one side.

Minus Ward, a Philadelphia inventor and observer of mechanical progress, spoke of "the bolder geological features of our own country" and "the severity of our winters," both of which required Americans to depart from British practice. Since rugged country made the straight lines of English railroads impractical, a freely turning front bogie was designed to help American engines "move safely and freely on a serpentine rail-way."[18] Since the quarried-rock foundation stones used to support British rails were thrown badly out of line by the freezing and thawing of American winters, another technique had to be found. As Stevenson had often noticed with other forms of technology, "the Americans have put numerous plans to the test of actual experiment, in their endeavours to form a structure for supporting the rails, adapted to the climate and circumstances of the country. There are hardly two railways in the United States which are made exactly in the same way, and few of them are constructed throughout their whole extent on the same principles; but although great improvements have undoubtedly been effected, it is doubtful whether a structure perfectly proof against the detrimental effects of frost has yet been produced."[19] Eventually, the gravel roadbed with wooden ties to support steel rails became standard American practice.

Nothing did more than the railroad to connect the cities and farmlands of the East and open up the West for conquest and settlement. Great cities like Chicago grew large and prosperous because of its presence, whereas others, now unknown, died because the gift of rails was withheld. During the Civil War the industrial superiority of the Union was nowhere better shown than in the ability of General Ulysses S. Grant to move whole armies virtually overnight, and keep them supplied, by railroad. After the war, and after the betrayed reforms of Reconstruction, a sectional reconciliation was accomplished when Southern railroads, in one single week between May 26 and June 1, 1886, moved the rails of 9,000 miles of line 3 inches closer together so that trains from the North and South could use the same roadbeds.

Bridges of many sorts, like this Chicago lift bridge, were built by the thousands and facilitated both highway and railroad connections within as well as between cities. Postcard from the author's collection.

Less dramatic but no less important, the giant size and scale of operations of railroads by midcentury led them to become the first modern corporations, in which scores of the details of modern life were worked out to suit the needs of bureaucratic organization. Not the least of these was the adoption, in 1883, of standard time zones for the nation. Not until the railroad, with its far-flung empire of rails, huge bureaucracy, unprecedented speeds, and complex scheduling, was it necessary to subordinate local to national time. It was an omen of new standardizations and uniformities to come.

Canals, and more so railroads, quickly captured the attention of the nation's artists and writers. The railroad was a signal example of what Leo Marx has called "the Machine in the Garden." Nathaniel Hawthorne found excitement, even erotic fantasy, on a canal boat. But Henry David Thoreau's rude awakening by a railroad whistle in his cabin at Walden Pond reinforced an earlier fear about the coming of this representative of industrial technology and capitalism. As early as 1844 Ralph Waldo Emerson declared that the railroad was a fit subject for poetry, and although Walt Whitman could celebrate this part of America's "vast seething mass of *materials*," the sense of unease was not so easily dismissed. In 1855 George Innes painted his celebrated canvas "The Lakawanna Valley," showing a railroad in peaceful harmony with an only partly altered rural setting. Ominously, the painting had been commissioned by the Lakawanna Railroad Company itself. Eventually, Hart Crane and a host of other modernist poets and artists would celebrate the Brooklyn Bridge for its symbolic and mythic qualities, and even the railroad became a popular, sublime, and romantic symbol of the triumph of people over nature, especially of technology's ability to annihilate space and time.

As America's transportation infrastructure rose layer by layer, railroad upon turnpike and canal upon river, during the nineteenth century, its central purpose was development. In England, such transportation technologies linked one place with another, manufacturing center with deepwater port. In the new United States, however, the lines ran from somewhere to nowhere. Towns and cities bid against each other for the chance to become a terminus, or at least a way station, along newly planned routes. Land speculators tried to get the railroad or turnpike to pass through their holdings, knowing that land prices would rise sharply as people and goods began to flow along the route. Today's Interstate highway exits have the same effect. That was the economic armature around which was wrapped a romantic tradition of Americans "on the road."

II

THE DOMESTICATION OF THE INDUSTRIAL REVOLUTION

4
THE EXPANSION OF AMERICAN MANUFACTURES

THE INDUSTRIAL REVOLUTION was born in Great Britain, but once transplanted to the United States, it flourished and took a significant new shape. So distinctive was the way in which Americans began to make goods that within a generation it was being called "the American system of manufactures." Under the older system, manufactured items such as muskets or locks were made one at a time by skilled craftspeople who, perhaps with the help of an apprentice, saw the entire process through to its end.

The essential idea of the American system of manufactures, or armory practice as it was also called, was to achieve uniformity of product by the transfer of skills from workers to machines. Using jigs, fixtures, and gauges, specialized machines could produce a large number of similar parts, the accurate dimensions of which were determined by the design and setting of the machine tool, rather than the skill and experience of the worker. In this way, semiskilled labor could produce parts that only experienced machinists could produce with general-purpose tools. The copying lathe, invented by Thomas Blanchard, was a good example. The ancestor of the key-grinding machines now found in every hardware store, this lathe allowed an operator to copy a pattern and thereby produce an exact imitation.

Thomas Blanchard's gun-stock lathe for making identical copies of a single pattern was one of the key inventions of the American system of manufactures. Artist's conception from Benjamin Butterworth, *The Growth of Industrial Arts* (Washington, 1892), 200.

The idea is said to have originated with Eli Whitney, the inventor of the cotton gin. In 1798 war with France seemed imminent, and the federal government, in a panic over its lack of arms, was prepared to grasp at any straw for new supplies. At the same time, Whitney had just lost an important court case in his continuing effort to secure his patent on the cotton gin and desperately needed the credit and prestige that a government contract would give him. The government and Whitney were brought together by their shared need, and only the cool head of Tench Coxe, purchaser of supplies for the government, coaxed order out of chaos. He signed gunsmiths up for weapons contracts, among them Whitney, who promised to deliver 10,000 muskets: 4,000 on September 30, 1799, and 6,000 more on September 30, 1800. For this he was to receive $5,000 in advance and a total payment of $134,000 at the completion of the contract, a certain and dramatic end to his poor credit rating.

The desperation of both the government and Whitney can be measured by the fact that the latter knew nothing of making guns (he had already failed in his attempt to manufacture his relatively simple gins), had no work force,

The Springfield Arsenal was, more than any other single site, the birthplace of what was called "arsenal practice." The almost domestic (and religious) style of architecture helped reconcile Americans to the birth of the industrial age. *Boston Almanac for the Year 1852* (Boston, 1851), 47.

and owned no factory. Despite his vague references to "machinery moved by water," Whitney apparently had no idea of how he was going to proceed and delivered nothing on his contract until early 1809, by which time he had gotten advances of all but $2,450 of his total payment. It was a record of delayed delivery and cost overruns that was to become familiar among defense contractors in more recent times.

Apparently Whitney had somewhere picked up the idea of making the muskets from "interchangeable" parts, although there is no evidence that he himself succeeded in doing so. The idea went back at least to Christopher Polhem, a Swedish technologist who made clocks by this method in the eighteenth century. By 1788 the Frenchman Honoré Blanc sought "the greatest economy and the most precision" in making small arms from interchangeable parts for the French government. Thomas Jefferson certainly knew of these efforts, and he may have been the one who told Whitney about the successful experiment. In England, blocks for the Royal Navy were uniformly made in large numbers and to standard size by special-purpose machinery at the Portsmouth shipyard. Although it was not widely practiced, this was hardly a new idea.

In the United States, the idea of making arms with interchangeable parts took root in the military, where the dream of being able to repair weapons on the field of battle was enormously attractive and expense was a secondary consideration. Not for the last time, the army proved willing to press for and finance a level of technological performance that seemed excessive from the viewpoint of civilian needs. By 1799 the federal arsenal at Springfield, Massachusetts, was using machines that reportedly cut the number of person-days to make a musket from twenty-one down to nine. Although no one knows what these machines were, it is clear that Whitney was planning to lure workers away from Springfield for his own factory.

Sometime during the first decade of the nineteenth century, John Hall planned out the production of guns from interchangeable parts, made by machines, and he took out a patent for his system in 1811. By 1817 he was installing his machines—with the enthusiastic support of army ordnance officers but the resistance of many of the skilled workers—in the federal arsenal at Harpers Ferry, Virginia. When an official commission visited this operation in 1827, its members declared the system was unique and further advanced than any other they had seen. The master mechanic, E. K. Root, installed a similar system in the private arms factory of Sanuel Colt in Connecticut, where the famous Colt revolver was made.

A British delegation visiting the federal arsenal at Springfield at midcentury put the new system to the test. Ten muskets, made between 1844 and 1853, were selected and torn down, and their parts thoroughly mixed. When reassembled at random by regular workers, they fit together and worked perfectly. It was a triumphant proof of the technical excellence of the system but of course gave no hint at the cost involved. Although the military had spent large sums of money over the years to achieve this result, Springfield was able to produce only a small number of such arms, and those only at higher-than-prevailing costs. Interchangeability proved, in short, to be an expensive luxury for most manufacturers. Yet over the years the basic notion of uniformity spread to other industries, particularly to the manufacture of sewing machines, bicycles, and eventually automobiles.

As always, the spread owed much to the migration of skilled workers from one industry to another. Such changing of jobs was typical of American workers in general but proved particularly important when Henry M. Leland, for example, quit his job at a federal armory to work for a machine-tool manufacturing firm that also produced sewing machines and finally wound up manufacturing automobiles. Leland was the founder of both the Cadillac and Lincoln automobile companies. The machine-tool companies contributed a great deal to this process because they made the specialized machines that solved production problems in one industry and then applied those solutions to others.

Interchangeability was never a precise term, however. It made a great deal of difference whether tolerances of one ten-thousandth of an inch were required, or whether a close approximation was good enough. Standardization and uniformity encompassed interchangeability but also applied to a greater range of activities. The growing shift from wrought iron to cast iron no doubt promoted standardization, since the use and reuse of patterns to make molds inevitably led to that result, even if it was not specifically sought.

Sewing machines, typewriters, bicycles, and other such consumer goods developed in the mid- to late nineteenth century were assembled in large numbers from a myriad of small metal parts. The Singer Sewing Machine Company was producing 43,000 machines a year by 1867 and a half million by 1880. By the latter date the McCormick reaper plants were turning out 21,600 machines a year and 5,000 binder attachments. Both of these successful firms priced their products at the top of the market and depended primarily on advertising and other marketing devices to keep up sales and profits. Singer, for example, took pride in the high quality of its machines, clinging to the "European method" of having each unit "finished" by hand. Only gradually, and not until the 1880s, did Singer attempt (with only limited success) to move toward the armory system of production. It was also during the 1880s that the McCormick reaper works moved in the same direction.

More than any other product, it was the bicycle that forged the link between early nineteenth-century armory practice and early-twentieth-century mass production, as pioneered by Henry Ford. At the peak of the bicycle boom in the 1890s, Albert A. Pope was producing 60,000 of his Columbia models a year. Like Singer and McCormick, these were high-quality, high-price machines. Although Pope disdained the innovations, other manufacturers were taking advantage of two important new manufacturing techniques: sheet metal stamping and electric resistance welding. Both of these were carried over into automobile manufacture by Ford and contributed significantly to his mass production system. Indeed, by 1912 all the technologies for mass production except the moving assembly line were

The making of clocks and watches was one of the earliest and most important industries to which the machines and methods of the "arsenal system" were transferred. The Seth Thomas Watch and Marine Shop, at Thomaston, Conn., was located along a source of waterpower, but obviously used steam power as well at the turn of the century. Postcard from the author's collection.

widely spread among American manufacturers, and even that had its prototype in the flow-process techniques of flour mills and the so-called disassembly lines of giant slaughterhouses in Cincinnati and Chicago.

Henry Ford made the next great leap forward in production technology. Early in its history, the Ford Company mastered the armory idea of interchangeable parts and carried over from the bicycle industry the use of punch and press work on sheet steel. In 1913 the Ford Company added a powerful new way to assemble parts. The moving lines of subassemblies and final chassis units raised productivity from 50 to 1,000 percent with a relatively small additional monetary investment in machinery. Ford did produce large numbers of Model Ts at an ever diminishing cost to the consumer. The miracle of Fordism (*Fordismo* in Spanish, with similar variations in other languages) spread rapidly around the world as a modern and specifically American ideal of technological behavior.

Although the manufacture of consumer goods like sewing machines, clocks, and bicycles was dramatic in its adoption and extension of the

American system of manufacturing, such goods did not account for the great bulk of the manufacturing industry in the country. Wherever large runs of similar items were not possible, as, for example, in the furniture or jewelry industries, products continued to be made in batches rather than flow patterns. Even in some traditionally flow industries, the methods used could hardly be described as mass production.

The innovations in automatic flour milling introduced by Oliver Evans dominated the technology of that industry for nearly a century. With the westward movement of population and the improvement of transportation for bulky raw and processed goods, milling establishments followed the center of grain growing from east to west. As settlers moved into the trans-Mississippi West, they planted hard Spring wheat from eastern Europe in the northern states of the Middle Border region. Soon, Minneapolis replaced Rochester as the center of milling activity, accounting for 5.1 million barrels of flour in 1884. Ten years before, the roller process of milling had been introduced, also from eastern Europe, where it had been in use for half a century. The ancient millstones were replaced by steel rollers ("rolled oats" is a still familiar reminder of that shift) which improved the appearance and conserved the nutritive value of the wheat. It also proved to be a more effective means of crushing the harder kernels of Spring wheat. For the industry as a whole, the wheat flour milled increased from 39.8 million barrels on the eve of the Civil War to 105.8 million barrels in 1900.

In 1860, the nation's per capita consumption of iron was nearly 120 pounds a year, five times what it had been half a century before. A major innovation in the industry, just after midcentury, was the Bessemer process for making steel, discovered in England. This metal had been produced for millennia in small and expensive batches. Then, in 1855 Henry (later Sir Henry) Bessemer developed a method of burning carbon out of molten iron by blowing cold air through it under high pressure. His attempt to gain an American patent was successfully challenged by the Kentucky ironmaster William ("Pig Iron") Kelly, who had anticipated him and in 1857 received the American patent for the process.

A syndicate purchased the rights to Kelly's patent in 1861 and then proceeded to obtain the American rights to Bessemer's process also, believing it to be superior. William F. Durfee, a cousin of one of the principals, was hired to build a chemical laboratory at the site of the new plant in Wyandotte, Michigan. Having "very little knowledge of an exact character as to

Two-high trains of rolls in this rolling mill shaped red-hot iron for a variety of purposes. Small hand trollies hardly mitigate the heavy (and skilled) lifting and positioning that was required to present the ingots to the rolls. *Popular Science Monthly,* 38 (March 1891), 593.

what had been done by others," he later wrote, "but a very clear idea of the rationale of the new process, supplemented by an absolute faith in the great future before it, I proceeded to evolve from my own internal sense of the fitness of things, apparatus and methods suited to the general idea and environment of the proposed experimental works."[1] Partly because of the Civil War, success was not achieved until September 1864, when the first American batch of Bessemer steel was produced at the plant. Durfee's laboratory was then abandoned, a useful but aborted example of the harnessing of science, or at least something like the scientific method, to improve technology.

Hard upon this success, the open hearth method of making steel was introduced, also from abroad. This process was developed in 1856 by William Siemens. It used the hot waste gases leaving the furnace in new ways to heat incoming air for the blast. Unlike the Bessemer furnace, it could take scrap steel mixed in with the iron pigs. Since it was more expensive than the Bessemer process, and for some important purposes, such as making steel rails, not markedly superior, the older process remained widely used.

For a wide range of manufacturing and other industries, one of the most significant changes continued to be increasing mechanization. A government

Within the Philadelphia exposition, the great twin-cylinder Corliss steam engine was the most celebrated exhibit. Here President Ulysses S. Grant and the emperor Dom Pedro of Brazil greet the crowds. Ironically, this single, powerful engine animated and controlled all the other machines in what was billed as a celebration of "independence." Frank Leslie's *Illustrated Historical Register of the Centennial Exposition, 1876* (New York, 1877), 79.

survey of the mid-1890s comparing the data on "hand and machine labor" discovered that "scarcely an article now in use is the exact counterpart of the one serving the same purpose forty or fifty years ago." It was easy enough to discover current machine practices, but some industries had changed so completely that little could be learned about the previous hand process. "In certain cases," the investigators wrote, "it was ascertained that the article, though produced quite largely by machinery in the more densely populated sections, was still made by the primitive method in some rural community. More often it became necessary to hunt for employers or workmen, long since retired from active life, who had been engaged in the making of the article in question by the old-fashioned hand methods, and draw from them the needed facts."[2]

Among the 672 firms visited, some startling examples of improvement were discovered. In the manufacture of "Men's cheap grade, kip, pegged boots, half-double soles," for example, the hand-made operation in 1859 had involved 83 different operations performed by only 2 workers. In 1895 the machine process involved 122 operations carried out by 113 workers, a startling increase in the division of labor. The work involved in the hand-making of 100 pairs of these boots, however, had taken 1,436 hours and 40 minutes, at a cost of $408.50, while that for the machine operation involved 154 hours and 10 minutes, at a cost of $35.40. "Ordinarily," reported the investigators, "the greatest efficiency is obtained in the production of the cheaper grade of shoes."[3] The division of labor in this example was said to have reached its peak about 1880. After that new efficiencies were sought in the separate production of the different parts of the shoes and boots by specialty firms. Only in the "upper-cutting department" did hand methods continue to be important.

Dramatic improvements were also found in processes performed "by common laborers, and in which muscle and endurance are the chief requisites." One example was in the transferring of 100 tons of "coal from canal boat to bins 400 feet distant," where "the methods by which this work was accomplished have undergone a complete change. Under the primitive method the coal was shoveled into baskets and carried to the dock, where it was dumped into wheelbarrows and conveyed to the coal bins, while under the modern method, with the aid of a coal elevator, which consists of an endless chain rigging, with iron scoop cups attached, and constructed so as to be lowered into the hold of the boat, the coal was fed into the cups and elevated to a receiving bin, where automatic cars were stationed to receive and convey it to the storage bins."[4] The time needed for the unloading by the modern method was 10 hours, compared with 120 under the old, and only ten instead of twelve workers were required. A century after Oliver Evans installed elevators in his gristmill, this simple mechanism was still finding new and useful applications.

And so it went, in industry after industry. Sometimes astonishing increases in productivity resulted from doing more with less—or more accurately, with fewer workers. Only an economy undergoing massive growth itself could have displaced so many workers, found new work for them, and at the same time accommodated an increasing army of immigrants. The nation would not always be so fortunate. Even in periods of deep recession, it was widely accepted that continued increments of "efficiency" were somehow a good thing.

One signal accompaniment to the mechanization of industry was the introduction of large numbers of women into the factory work force. At least since the days of Giles's card-making factory in Boston, large numbers of women (and not a few children) had found employment in the nation's workshops. Sometimes, as in the case of hand cards, new machines first created work for women, then later machines destroyed their jobs. In some cases, as with cigar making, women were brought in with new machines to replace well-organized and politically aware male hand-workers. In other instances, as in papermaking in the Berkshire region, women's work was the last to be mechanized, at least in part because of the skill and discrimination of the workers, which could not easily be built into mechanical devices. The only common denominator seemed to be that the workplace remained heavily gendered, with men's and women's tasks strictly defined and enforced, and with a patriarchal hierarchy that often found men cooperating across class lines to maintain their masculine advantage in the face of technological change.

A presumed encouragement to all this technological change was the tradition of patenting, carried over from the colonial period. Each colony had allowed its own patents, and the Articles of Confederation, adopted in 1781, left this situation as it was, in keeping with the general tendency to delegate only limited powers to the central government. As a result, inventors like Eli Whitney and Oliver Evans, with their cotton gins and steamboats, were forced to secure patents from each of the states individually, hoping to cover all bases and secure equal protection in all areas of the country. The confusion and expense attendant upon such a policy was obvious to everyone. The only real progress made during these years came when South Carolina, in 1784, passed a general patent law to protect machines. Intended to place the issuance of patents on a regular bureaucratic basis, it was the first such law in the nation.

When the new Federal Constitution was drawn up in 1787, the powers of Congress were enumerated under Article I, section 8. Among these was the power to "promote the Progress of Science and useful Arts, by securing for limited Times to Authors and Inventors the exclusive Right to their respective Writings and Discoveries." Sentiment was strongly against monopolies at the constitutional convention, but such members as James Madison argued fervently for this exception. In his first annual message to the Congress, President George Washington urged that body to implement this section of the Constitution: "I cannot forebear intimating to you the expediency of giving effectual encouragement, as well to the introduction of new and useful inventions from abroad as to the exertion of skill and genius at home."[5]

At Washington's urging, the Congress passed the first U.S. patent law in 1790. A special examining board was set up, composed of the secretary of state, the secretary of war, and the attorney general. Thomas Jefferson, as secretary of state, headed this board and looked personally at every one of the patent applications filed during the next three years. If he and his colleagues decided that the invention was in fact both novel and useful, a patent could be had for a $4.00 fee. During the three years this law was in effect, fifty-seven patents were granted.

This law proved burdensome both to Jefferson and to the many disappointed inventors who had their applications turned down by the indefatigable secretary of state. The members of the board were said to be philosophically opposed to the industrial classes and therefore overly parsimonious with patents. One enthusiastic supporter of patents, who had been associated with James Rumsey through his steamboat fight, wrote a pamphlet in 1792 entitled a *Treatise on the Justice, Policy, and Utility of Establishing an Effectual System for Promoting the Progress of Useful Arts, by Assuring Property in the Products of Genius.* In it he stated:

> Free governments have been always found most disposed to encourage the rise and progress of science, and the arts; for this obvious reason, that in republican governments *merit* alone is, or *ought* to be the *standard* for character; which necessarily *excites* that *laudable* spirit of emulation, so essential in society; and which never fails to produce celebrated philosophers, statesmen, husbandmen and artists; the necessary effect of which, is, not only, the promotion of science and useful arts, but, the *happiness* of men—the primary, the grand *object* of their existence.[6]

In 1793 these criticisms were answered by a second patent law, which wiped out the examination system and set up a system of registration. Now an inventor needed only to allege that a device was original and pay the prescribed fee—the government made no attempt to determine whether the invention was, in fact, either new or useful. These questions were left to the courts, which soon found themselves burdened with a large number of suits for infringement. It was not uncommon for someone to patent an already known device, then go to the frontier and sell it as novel. By the time the real inventor discovered the operation and brought suit in whatever district court was involved, the cost of litigation was prohibitive and the damage already done.

One further feature of the new law revealed the problem of an underdeveloped nation trying to establish itself in the shadow of an already industrialized world power. The act of 1793 absolutely forbade the issuance of patents to foreigners. In 1800 this was amended so that aliens resident in the country for two years who had filed an intention of becoming citizens could apply for patents. This policy was softened somewhat by the act of 1836, which allowed alien patents but levied a discriminatory fee of $500 for British citizens and $300 for other aliens, as opposed to the $30 required of American citizens. Finally, in 1861 all discrimination against foreign patents was removed.

The confusions caused by the law of 1793 finally led, in 1836, to the establishment of the modern patent system. During the previous forty-six years, 9,957 patents had been issued. Now that the pace of applications was picking up, the off-handed system of registration, watched over by two or three clerks, was proving inadequate. The act of 1836 was without precedent in history and has since served as a model for other patent systems around the world. A patent commissioner was appointed, and a bureau set up as a part of the Department of State. They were given quasi-judicial powers and the responsibility of once again examining all patent applications for the virtues of originality and usefulness. From this date on, a patent was considered prima facie evidence of its validity.

In addition, a separate Patent Office building was authorized. Even before the building was completed, the old headquarters of the bureau was destroyed by fire on the night of December 14, 1836, along with all of its patent specifications, drawings, and models. Those who cared to were allowed to resubmit documents, but many never did.

The English author Francis Trollope provides a view of the old patent office in 1830, before the fire:

> The patent office is a curious record of the fertility of the mind of man when left to its own resources; but it gives ample proof also that it is not

> under such circumstances that it is most usefully employed. This patent office contains models of all the mechanical inventions that have been produced in the Union, and the number is enormous. I asked the man who shewed these, what proportion of them had been brought into use, he said about one in a thousand; he told me also, that they chiefly proceeded from mechanics and agriculturalists settled in remote parts of the country, who had began [*sic*] by endeavoring to hit upon some contrivance to enable them to *get along* without sending some thousand and odd miles for the thing they wanted. If the contrivance succeeded, they generally became so fond of this offspring of their ingenuity, that they brought it to Washington for a patent.[7]

One particular result of the nation's patent system was the rise of patent agencies, patent lawyers, and patent brokers. Perhaps the most famous of these operators was Orson Munn, who edited the journal *Scientific American* (founded by Rufus Porter in 1845) from 1846 to 1907 and was the principal in the patent agency of Munn & Co. A less well-known firm was the American Patent Agency of Cincinnati, Ohio, founded in 1878. This company published a magazine, the *American Inventor,* to advertise its clients, maintained thirty-five branch offices from Boston to Virginia City, and employed nine traveling sales agents to cover the territory between. The agency found clients because it was increasingly difficult for the individual inventor to catch the attention and financial backing of those who might be interested in their new devices and processes.

For one thing, inventors were no longer the sole source of new technologies. Over the course of the nineteenth century, engineering developed into a new profession, to join the ancient ones of law, medicine, and the cloth, and increasingly engineers were seen as those with the education and experience to combine the best of traditional practice with the most promising of scientific possibilities.

Until the late eighteenth century, most engineers were with the military, constructing battlements and attending to artillery. Early in the Industrial Revolution some engineers working on civilian problems of manufacture and transportation began to style themselves "civil" engineers. As their numbers increased and they applied their practice to a wider range of problems, specialists in mining, mechanical, electrical, chemical, and other types of engineering split off to pursue their own fields of practice. Until the Revolution, Americans had little need for professional engineers, either military or civil.

The experiences of the Revolutionary War, with its need for engineers, met only in part by French officers, led finally in 1802 to the establishment of a military academy at West Point in New York. Until the opening of the Erie Canal in 1825, the growing need for civil engineers to build the infrastructure in cities, create a transportation network, and provide for the nation's expanding industrial plant was answered by a combination of transplanted professionals from France or England, by West Point graduates detached from military service, or self-taught Americans. Indeed, the Erie Canal proved to be a great school for engineers, training through practice a whole generation of American practitioners.

Although on-the-job training and in New England a kind of apprenticeship system that flourished in the antebellum period provided the bulk of practicing engineers for most of the nineteenth century, one by one other schools joined West Point in providing professional instruction. Rensselaer Polytechnic Institute became the world's first private engineering school in 1829, Harvard College added an engineering school in 1842 and Yale in 1847, and with the setting up of a system of federally funded land-grant colleges in 1862, every state and territory established an institution to teach the sons and daughters of "farmers and mechanics" the science and practice of agriculture and mechanics. Special curricula were set up by the Columbia School of Mines in 1867; by Yale and the Massachusetts Institute of Technology, which graduated their first mechanical engineers in 1868; and by the Masachusetts Institute of Technology and Cornell University, which began their electrical engineering programs in 1882. Although these and a host of institutes of technology (Case School of Applied Science was opened in Cleveland in 1880 and Lehigh University in Bethlehem, Pennsylvania, in 1865) were available, 150 American students were still attending German engineering schools as late as 1889, because many still considered them the world's best. By the end of the century, a growing cadre of young engineers, increasingly trained in both science and business, were forming a social stratum between the mechanic and the entrepreneur. It was not always a comfortable social role, but an increasingly common one as the nation developed a trained and bureaucratic middle class.

As was so common in America, the new engineers sought to organize for their own advantage. Learned societies went back in the American experience at least to Benjamin Franklin's American Philosophical Society, founded in Philadelphia in 1743. A generation later, in 1780, John Adams was instrumental in helping found the American Academy of Arts and Science in Boston. Both of these still exist, but now, as then, they broadly reflect a wide range of interests that includes but is not limited to engineering. Over the next century specialized societies arose that centered on either particular branches of engineering (civil, mechanical, and so forth) or particular regional centers of engineering activity (Chicago, San Francisco, Cleveland). These societies served parochial interests well but eventually gave rise to overarching unity organizations hoping to represent all engineers.

In 1852 engineers living in the vicinity of New York City organized the American Society of Civil Engineers and Architects, thus joining two fields now thought distinct but then often linked. Indeed, the society thoroughly mixed interests by welcoming into one grade of membership all "Civil, Geological, Mining and Mechanical Engineers, Architects and other persons who, by profession, are interested in the advancement of science." Even with its net cast so wide, however, the society languished, and from 1855 to 1867 no meetings were held.

The society was reorganized in 1867, with stricter membership requirements. The highest category of "Member" was limited to those who had been in the profession for five years and at some time had been "in charge of some work in the capacity or rank of Superintending Engineer." Although more professional and elite, the society tried to broaden the geographical base of its membership and in 1872 began to hold some annual meetings in cities other than New York. Membership, which stood at only 55 in 1853, rose to 160 in 1869 and 2,018 in 1896.

With the growth and specialization of American technology, other societies sprang up whose members more nearly shared the same concerns and enthusiasms. The American Institute of Mining Engineers was established in 1871, in response to the growth of coal and iron mining in the East and hard-rock mining of precious metals in the West. In 1880 the American Society of Mechanical Engineers was founded, marking the new muscle of manufacturing industries. The American Institute of Electrical Engineers, founded in 1884, grew out of an International Electrical Exhibition held that year in Philadelphia at the Franklin Institute. In 1908 the American Institute of Chemical Engineers was established, rounding out the group sometimes referred to as the Founder Societies.

These engineering societies, like the scores of other, more specialized groups that grew up around them, shared a number of important goals. For one thing, they sought to set standards for practice in their fields, tried to enforce those standards, tried to limit entry into the field, exchanged information, provided a venue for socializing and "networking," and addressed issues of public policy. Some of these goals were controversial, and not all were forcefully pursued at all times, but nevertheless they represented the ideal.

As seen in the example of the American Society of Civil Engineers, these groups tended to evolve into organizations of, by, and for the elites in each field. In part this was a straightforward attempt to put experienced and successful people in leadership positions so that standards would be not only high but also credible. In part it grew out of the fact that New York City members and those independently wealthy or supported by large corporations had the resources in time and money to devote to society business. One result was that the larger, New York–based societies tended to take quite a conservative approach to political and economic issues affecting both their membership and the nation as a whole.

Another result was that young people, and those far removed from New York City, often formed local engineering societies to meet their own needs. In San Francisco, the Technical Society of the Pacific Coast flourished from 1884 until 1915, serving as a focal point for the professional needs and contacts of engineers from as far away as Hawaii. The Cleveland Engineering Society, self-consciously democratic and attuned to the needs of younger members, served during the Progressive era as a kind of brain trust for the city's reforming mayors Tom Johnson and Newton D. Baker.

The exclusion of women from membership in these societies was only one of many ways in which female engineers, or those women who aspired to reach that status, were kept from being effective members of the profession. Julia Morgan, born in Oakland, California, wanted to be an architect but since the nearby University of California had no such curriculum, she entered the engineering college and took her degree in civil engineering in 1894, the first woman to do so at that school. Later she was able to study architecture at the prestigious Beaux Arts school in Paris, but only after running the gauntlet of male prejudice and hindrance. Back in California, she found herself much in demand after the devastating San Francisco earthquake and fire of 1906. It did not escape notice that buildings constructed of the new reinforced concrete material withstood the disaster better than others, and she was one of the few architects to have a strong professional engineering and design background. Her best-known commission was undoubtedly the William Randolph Hearst "Castle" at San Simeon on the California coast.

Another pioneer engineer, Kate Gleason (1865–1933), entered the mechanical engineering department at Cornell University in 1884 but withdrew later that year to help her father in his machine shop. She learned the trade in the traditional manner of an apprentice and worked closely with her father in designing gear-cutting machinery. In 1914 she was elected the first female member of the American Society of Mechanical Engineers and was also the first woman to gain membership in the Verein deutscher Ingenieure.

Nora Stanton Blatch was born in 1883 and in 1905 took a degree in civil engineering from Cornell University. She began her career in the drafting room of the American Bridge Company, then moved to become an assistant engineer with the New York Board of Water Supply. While there she met and married Lee de Forest, the radio inventor, and quit her work to assist him in his. The marriage ended in divorce in 1912, and Blatch, a granddaughter of Elizabeth Cady Stanton, began to work full time for the cause of female suffrage. She was allowed to become a junior member of the American Society of Civil Engineers (ASCE), but when she applied for associate membership in 1915, she was denied the right. She sued the ASCE in court, but her petition was also denied. She later became an architect and building contractor.

Of the small number of women engineers who struggled against the sexism of their profession, Ellen Swallow probably made the greatest institutional contribution. Born in 1842, she attended Vassar College and then applied to the Massachusetts Institute of Technology. She would have preferred to take the engineering curriculum but decided to reduce the odds against her by electing a chemistry course. Fearful of admitting a woman as a regular student, the institute granted her special status, and she took her degree in 1873, the first woman to matriculate there.

While a student and after, she worked on the problem of industrial wastes in public water supplies. After marrying Robert Richards, an MIT mining and metallurgical engineer, in 1875, she helped with his work and was elected the first female member of the American Institute of Mining Engineers. In 1876 she opened the first women's laboratory at MIT. By 1882 women were being admitted as regular students at the school (four had already graduated), and the following year the separate women's laboratory was abandoned. In 1884 she was made assistant at the new sanitation labora-

Although women were in most cases systematically blocked from full participation in the growing culture of scientific technology, the chemical laboratory of the Massachusetts Institute of Technology in 1880 seems to show as many female as male students. Frank Leslie's *Illustrated Newspaper* (1880), reproduced in Karl T. Compton, *Massachusetts Institute of Technology* (New York, 1948), 16–17.

tory at MIT, and in 1890, when the school inaugurated the nation's first program in sanitary engineering, she taught courses there.

In 1890 Richards was one of the women to establish the New England Kitchen, copied from the public kitchens of Europe and designed to teach (presumably ignorant) working-class immigrant women how to cook cheaply and nutritiously. She also took the lead, in 1899, in calling the Lake Placid meeting of what was to become the core of the home economics movement. When the American Home Economics Association was founded in 1908, she was elected its first president.

Ellen Swallow Richards's career highlighted both the increasing opportunities for and the continuing roadblocks to female participation in the nation's engineering profession. In both education and professional membership, she opened doors, which, however narrow, were never quite closed again. At the same time, her career was deflected from engineering, her first choice, into science, and within science, toward its application to domestic concerns. Although domestic science provided an opportunity for many

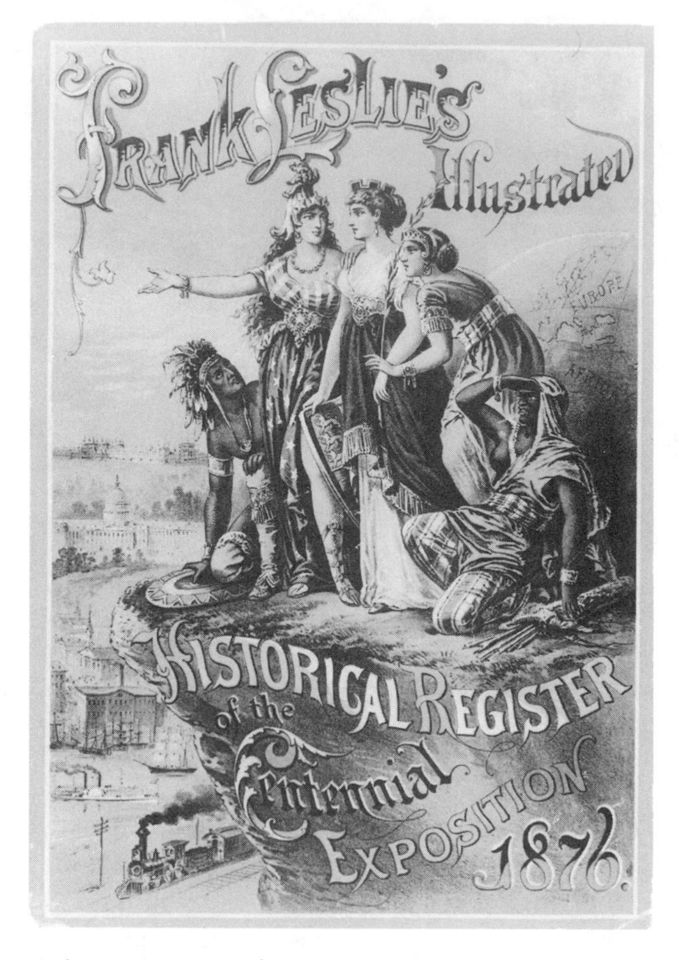

In this allegorical scene, America exhibits a "century of progress" to Africa, Asia, Europe, and, surprisingly enough, Native America. If the foreground suggests importance, the nation was most proud of railroads, telegraph, steamboats, sailing ships, factories, religion, and government, in that order. Frank Leslie's *Illustrated Historical Register of the Centennial Exposition, 1876* (New York, 1877), frontispiece.

women to take up scientific careers in the twentieth century, it also proved to be a carefully constrained field, ghettoizing its practioners in a dead-end and less-respected profession. Engineering as such remained for another century a largely male, and masculine, activity. No less than their brothers on the shop floor, engineers enjoyed and protected the advantages of gendered privileges in the technological trades.

At the dawn of the nineteenth century, Americans were only beginning to use some of the machines and processes that were already creating an industrial revolution in Great Britain. By the end of the century, not only had the machines been begged, borrowed, or stolen, but the revolution itself had been firmly transplanted to this side of the Atlantic. Under the spur of mercantile and then industrial capitalism, and blessed with a plenitude of natural resources and an expanding labor force, the manufacturing sector of the American economy expanded. As it did so, it was served by a set of social inventions, from the modern patent system to the new profession of engineer. By the end of the century, England was no longer the undisputed workshop of the world.

5
THE MECHANIZATION OF FARMING

AGRICULTURE IN AMERICA in 1800 was the occupation of 80 percent of the American people. It was largely a hand operation, lightened only by the use of horses or oxen in certain limited, though important, tasks such as hauling and plowing. By 1900 a large percentage of the farmer's task was mechanized, and tools were often powered—in most cases still by horses and oxen, but steam, electricity, and internal combustion engines were already available for those who chose them. The primary characteristic of agricultural progress then, at least as far as technology was concerned, was the coming of machines. The second most important, and closely associated change, was the increasing use of horses to power those machines.

Farming was a hard and grinding vocation. Before 1850 farms were often carved from the virgin forests of the East Coast and Appalachian regions. A family that concentrated on growing a season's crops might be able to clear 5 acres of additional land a year. Some farmers, however, counted on the clearing itself, rather than the growing of crops, to make a living: the market demand for land rather than for foodstuffs would provide the reward. Even for such a person, however, a lifetime of toil would scarcely add up to 200 acres of cleared land. For the country as a whole, land clearing was the major task of farm improvement and, until 1900, it absorbed much more work than building, fencing, and ditching combined.

Inventiveness did not keep pace with the importance of this activity. By 1850 land clearing had changed but little since the earliest days of settlement. In both the North and the South trees were killed, dried, and burned. In the North, however, they were chopped down then piled up to dry, to be burned later, a practice probably introduced by the Swedes into colonial Delaware. In the South, the trees were girdled, then the field was used for crops. The trunks and limbs were piled up and burned as they fell. This left only the stumps, which were a great problem in the North and South alike. One had only the axe, the lever, and teams of oxen to help with the job of removing stumps, which took as long as the rest of the process of clearing together; in 1841 in Massachusetts it took an estimated thirteen days for one person to clear 1 acre of stumps.

By midcentury, more than 100 million acres had been made ready for the plow—an incredible investment in human effort. Between 1850 and 1910, however, half again as much virgin forest was cleared as had been in the previous two hundred years. Simple lever or screw devices, powered by horse traction and designed to help pull stumps, began to appear on the market in 1850, but as late as 1889 it was declared that no truly effective device had as yet been developed. The only real aid was blasting powder, which reportedly cut the expense of ridding a field of stumps by one-half.

After 1850 forests were not the only problem. The prairie had already been encountered in Indiana and Illinois. With the traditional plow, it took three to five yoke of oxen and two people an entire day to turn just one-half acre of prairie sod. The common plow, as it had come down through the eighteenth century, was made entirely of wood, with perhaps some iron added at the share or with iron strips covering the moldboard. This was not referred to as a wooden plow (although of course it was) until the introduction and spread of wrought iron plows in the early nineteenth century. In part this change was due to the increasing amount of iron available in the early years of the Industrial Revolution, in part it was a result of the same spirit of improvement that led Thomas Jefferson to try to work out mathematically an improved design for the moldboard.

The manufacture of plows and something of their character in the mid-nineteenth century were described by two British mechanics who visited the United States in the early 1850s:

> Labour-saving machines are most successfully employed in the manufacture of agricultural implements. In a plough manufactory at Baltimore, eight machines are employed on the various parts of the woodwork. With these machines seven men are able to make the wooden parts of thirty ploughs per day.
>
> The handle-pieces are shaped by a circular cutter, having four blades, similar to those of smoothing-planes, fixed on a horizontal axis, with about 2 inches radius, and making nearly 4,000 revolutions per minute. The work to be shaped is fastened to a pattern, which is pressed against a loose roller on the axis of the cutter as the workman passes it along, and it is thus cut of exactly the same shape as the pattern [this was, of course, the Blanchard copying-lathe, borrowed from small-arms manufactories].
>
> All the ploughs of a given size are made to the same model, and their parts, undergoing similar operations, are made all alike. Some of the sharp edges of the wood are taken off or chamfered by a cutter revolving between two cones; these guide and support the work as it is pressed down edgewise on the cutters, and passed along by the workmen.
>
> The other machines in use consist of a circular and vertical saw, and machines for jointing, tenoning, drilling, and for making round stave rods, and giving them conical ends, the whole being of a simple and inexpensive character.
>
> The curved handle-pieces of the ploughs, which require to be steamed and bent, are obtained already shaped from the forests where they are cut, and are advantageously supplied to the large manufacturers. The prices of the ploughs vary from $2 1/2 to $7.[1]

It is particularly significant that the machinery described was all for dealing with the wooden parts rather than the iron, that much of it had arisen from the practice of making small arms, and that the plows sold for a small price in a day when most Americans needed one. It seems likely that the system of small, democratic, and self-sustaining farms in the United States owed at least as much to the low cost of the technology available as to the availability of free land, or the economic and political systems under which the farmers worked.

Probably the most important breakthrough in plow design, at least for the western states, was the introduction of what came to be called the "steel plow." Prairie plowing involved two separate problems: the breaking of the mat of roots (sod) and the tendency of moist soil to stick to the moldboard. It seems probable that during the 1830s blacksmiths in the prairie states began, here and there, to nail strips of saw-steel to the face of wooden moldboards to prevent the soil from sticking. Then in 1837, at Grand Detour, Illinois, John Deere began to manufacture the first of his famous "steel" plows.

The name was somewhat misleading. Apparently what Deere did during these early years was to attach a share of steel to a wrought iron moldboard and to polish the iron so that soil would be less likely to stick. He made ten of these plows in 1839 and 400 in 1843. Then in 1847, he moved to Moline, Illinois, a site that had better transportation to the growing West. By 1857 he was turning out 10,000 plows a year. It was probably not until about this time—that is, sometime late in the 1850s—that Deere actually began to make plows with a steel moldboard. The fact that he referred to his plows as steel years earlier merely shows his own flare for advertising and the importance attached to having a share that could cut sod.

Gathering the grass for hay was another of the great bottlenecks of agricultural technology in the early nineteenth century. To take the hay rake as an example, a whole stream of devices had made their appearance by the time of the Civil War. By hand, it took one day to cut an acre of hay and another half day to pile it into windrows where it could dry without molding; turning the hay to cure it and then picking it up required still more labor. Under favorable conditions, one person in one day could harvest, with hand tools only, 1 ton of hay; under more usual conditions, a third of a ton was more likely.

Devices designed to dispense with the pitchfork in gathering hay began to appear in the late eighteenth century, and it has been reported that George Washington used a drag rake made of wood with wooden teeth. This type was not efficient and was little used. About 1830 drag rakes with wire teeth appeared on the market. Both types were in use until about 1850, but neither played a significant role in agricultural technology. A more useful device, the revolving horse rake, was apparently of American design and began to appear about 1820. Within thirty years it was standard among progressive farmers who grew large quantities of hay. Costing only $6 to $12, it was an important implement in the progressive application of horse power to agriculture.

The revolving rake eventually evolved into the sulky rake, which a farmer could ride. One of these was noted as early as 1837, but it did not receive wide attention until an improved version was patented by Calvin Delano of Maine in 1849. This machine was essentially perfected by 1870 and

This spring-toothed, self-dumping hay rake dated from 1856 and was typical of the horse-drawn and labor-saving devices available to farmers with enough acreage in grass and grain to warrant the purchase. Benjamin Butterworth, *The Growth of Industrial Art* (Washington, 1892).

was a great labor-saving device: with it one person and a horse could gather 20 to 30 acres of hay a day. Its comparatively high cost, however, made the older revolver still an attractive implement, especially if money was in short supply or the hay acreage insufficient to justify the sulky rake.

Harvesting was a great challenge even for mechanical ingenuity. Reporting the claim of a Cincinnati inventor, who in 1833 supposedly developed a machine for cutting wheat or other small grains by horse power, *Niles' Weekly Register* added with some resignation that "this, if it proves fully successful, is an important invention; but so many ingenious novelties have been brought forward of late, and so few have answered the expectations at first held out of their utility, that we are disposed to be somewhat cautious and incredulous."[2]

Niles had a right to be skeptical, but perhaps the most famous agricultural implement to come out of America in the nineteenth century was the reaper, especially as perfected by Cyrus Hall McCormick. He patented his horse-drawn reaper in 1854, a year after Obed Hussey had patented what is generally accepted as the first successful machine of that type. The attempts to solve the problem, however, had been going on for some time: Cyrus's father, Robert McCormick, had spent years working on a reaper that he never

An artist's conception of McCormick's original reaper of 1831, arguably the single most important agricultural machine of the nineteenth century. Benjamin Butterworth, *The Growth of Industrial Art* (Washington, 1892).

brought to perfection. In 1847 Cyrus moved from Cincinnati to Chicago, where he made an enormous fortune and came to dominate the industry. His great horse-drawn machine, with its revolving blades slicing off the wheat, provided the primary mechanical icon of nineteenth-century American agriculture, and his manufacturing firm, protected by its monopoly on the device, was an early giant among American corporations.

The land given over to farming in the United States more than doubled between 1850 and 1890 and trebled by 1910. The farm population of the nation rose by 50 percent between 1880 and 1910, from 21.9 million to 32.4 million souls, a high figure that was never significantly exceeded. The percentage of farmers among the total population was 43.8 in 1880, and it declined steadily after that until it stood at only 4.8 percent in 1970. During the last half of the nineteenth century, this great community of farmers filled up an area in which the environment made much of traditional farming techniques useless, or at least less than optimal. A band of land running from the western part of Minnesota, through Iowa, and taking in most of Nebraska, Kansas, Oklahoma, and Texas, was subhumid. To the west, another band running from Montana and the Dakotas down to New Mexico and western Texas was semiarid, and the rest of the country westward to the great Central Valley of California was very arid. To settlers from the East and Midwest the land represented a great paucity of trees and water, a land that the geographies of the day sometimes called "The Great American Desert."

Official government policy had been to fill up these lands with family farmers, and it was to this end that the homestead laws were passed. Although slightly more land was actually sold than given away to farmers, it was relatively cheap. So, too, it turned out, was the technology to work it. Had this not been so, the government policy would certainly have been thwarted. In the 1870s, a wagon cost only $75 to $125, and the team to pull it another $150. Plows could be had for less than $30, and a harrow was even cheaper. These would suffice for getting in the first crop, but the less well-off farmer could actually begin with less. In one township in Minnesota in 1870, farmers reported that their machinery, per farm, had a total value of $10 to $200, and that most had invested less than $100 in equipment. If necessary, one could have the plowing done for $2.50 to $4.00 an hour, then plant and harvest by hand.

The great herds of buffalo that roamed this grassland indicated that it would be suitable for a range-cattle industry, and livestock ranching gave the West some of its most glamorous experiences. Wheat was the other crop that suggested itself because of its suitability for dry farming. Once new types were introduced to survive the harsh winters on the plains, the Hungarian process of rolling wheat into flour became popular, replacing the mill stones of ancient vintage.

In areas such as the Red River Valley and the central valley of California, the simple pioneer with perhaps $100 worth of machinery gave way to "bonanza farming," so named for its great scale and rich rewards. Many of these large landholdings were obtained directly from railroads, which, in turn, had obtained them free from the government as an inducement to build. One farm syndicate in the Dakotas in the mid-1870s was 6 miles long and 4 miles wide. In 1879 it had 10,000 acres planted to wheat. In 1877 this farm was equipped with 26 special plows to break the sod, 40 plows for turning over the earth once broken, 21 seeders, 60 harrows, 30 self-binding harvesters, 5 steam-powered threshers, and 30 wagons. Some idea of the expense of such an operation is suggested by the costs incurred by a similar farm during 1878–80: it had taken $60,100 of investment to prepare for

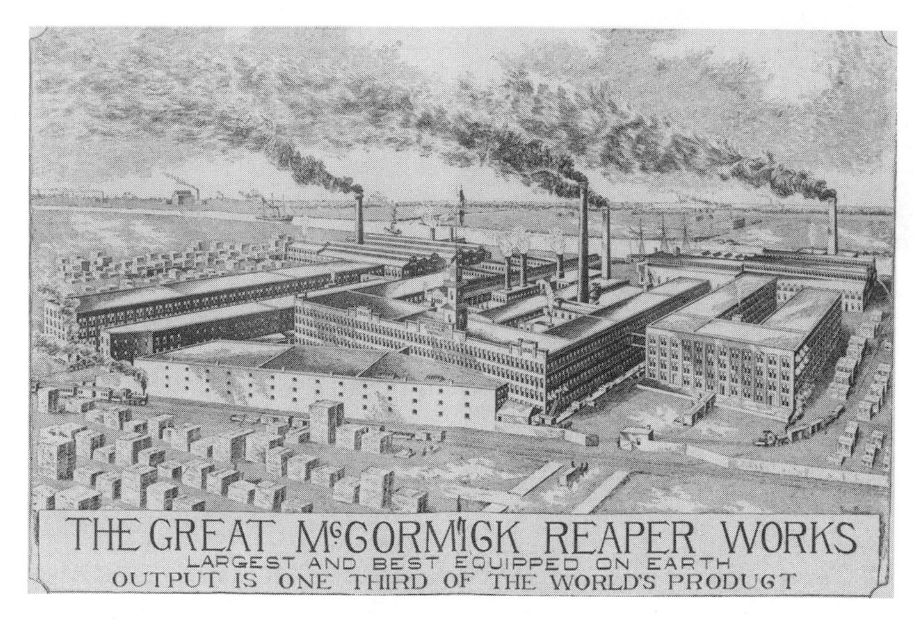

The extravagant success of McCormick's reaper worldwide made his company one of the largest and wealthiest of the century. The industrialization of agriculture and manufacturing reinforced each other. The Great McCormick Reaper Works, illustrated from the company's 1894 catalog. Courtesy American Museum of American History, Smithsonian Institution.

production: $17,000 was for breaking the sod, $9,000 for machinery, $6,500 for buildings, and the rest for miscellaneous costs.

Such farms had appeared in California much earlier. One farm of 8,000 acres near Sacramento in 1856 had 1,000 acres under fence. A nearby farm had 1,600 fenced acres in addition to large unfenced tracts. A thousand acres of wheat and barley were under cultivation, and farm equipment included 20 wagons, 50 plows, 25 harrows, 2 threshing machines, 7 reapers and mowers, and 4 hay presses. Farms of 11,000 acres were not unknown. McCormick's reaper was a necessity on such farms, of course, as was the steam-powered thresher to remove the grain from the dry stalks. Late in the century, repeated efforts were made to join these two together in what came to be called a combine. *Scientific American* described one of these monsters at work on a California bonanza farm in 1898:

> The harvester cuts a swath 28 feet wide and thrashes, cleans, and sacks the grain as it moves along. . . . While the combined harvester is not a new

> feature in the handling of crops on the coast, only recently, and not until the [steam] traction engine became a success in the field, did they even attempt a cut wider than 18 feet; 16 feet being the standard machine, requiring from 30 to 40 head of stock to handle them. . . . While the horse power machines are still in the majority, yet for extensive harvester work, where large acreage is to be dealt with, the steam rig will undoubtedly become the favorite. . . . When the writer visited the ranch to obtain data regarding this monster of California farming, they were cutting, thrashing, cleaning, and putting in sacks at the rate of three sacks per minute of barley, each sack weighing one hundred and fifteen pounds, requiring two expert sack sewers to take the grain away from the spout, sew the sacks, and dump them on the ground. Seven men constitute the whole crew, including engineer and fireman.[3]

It is not possible to say exactly when the first steam engine was used on a farm, although by 1838 several hundred were at work grinding sugarcane, sawing wood, and being used for other tasks of this kind. Such stationary engines did not provide adequate power for American farmers, however, who needed something more flexible. The first portable engines were built by two firms in 1849. These could be moved about from barnyard to field and were available to do various kinds of belt work. About a dozen firms were making them during the 1850s.

Since plowing accounted for about 60 percent of the labor involved in raising grain, agricultural societies, journals, and inventors concentrated on

In the West, the "combine," which combined the machines for reaping and thrashing wheat, became an indispensable tool for exploiting the vast bonanza farms of the area. This machine, used in North Dakota in 1892, was pulled by seventeen mules, traveled 23 miles in ten hours, and cut, thrashed, and bagged 61.87 acres of wheat in a day. *Scientific American,* 67 (September 24, 1892), 198.

Men cutting the grain with cradle-scythes were accompanied by the rest of their families: women tied the cut stalks in sheathes and children carried water to the other workers. The scene took place in Virginia. *Harper's New Monthly Magazine,* 88 (December 1893), 13.

developing a self-propelled traction engine. Another incentive was that threshing machines of increased capacity (the acreage of wheat doubled between 1866 and 1878) were beyond the efficient capacity of horses to operate. Portable engines built to provide this power cost about $1,000.

Before 1873 scores of companies and individual inventors had tried to make engines self-propelling, but to no avail. It was not until the builders of threshing machines began to make their portable engines self-propelling that real progress was made. These earliest traction engines still depended on horses to steer them, and it was not until 1882 that an Ohio firm produced the first self-steering, self-propelled traction engine. This success greatly stimulated the use of steam on the farm: in 1880 it produced 1.2 million horsepower; in 1890, 2 million; and in 1910 it reached its highest level at 3.6 million. The largest number of steam traction engines in use at any one time was not reached until 1913, when 10,000 were counted. By then, of course, the popularity of the gasoline tractor was eating into this number. In 1920 only 1,700 steam engines were built, and by 1925 their manufacture was largely abandoned. The remaining examples of this machine are still impressive: some of the largest weighed 20 tons (developing 50–100 horsepower) and consumed 3 tons of coal and 3,000 gallons of water a day. They cost anywhere from two to six thousand dollars.

The productivity of farm workers rose dramatically during the nineteenth century, and the time spent on different tasks changed as well. In 1800 it took an estimated fifty-six person-hours per acre to produce a crop of wheat, and forty of these were taken up by harvesting. By 1880 the total number of hours had dropped to twenty, and of that only twelve were used in harvesting. In 1900 the total required per acre of wheat was fifteen person-hours, of which only eight were for harvesting. Not only did the number of hours fall dramatically, from fifty-six to fifteen, but the proportion needed for harvesting dropped from nearly four-fifths to only slightly more than half. It was a signal example of the oft-boasted fact that farm workers were becoming more productive, which was part of the reason that the food one farm worker could raise increased from an amount for only 4.1 persons in 1820 to 7.0 persons by 1900.

The appearance of labor-saving devices on American farms worked to push down "hired help" from a class of apprentice farmers, often the children of other farmers who were considered almost one of the family, to a permanent class of migrant and part-time workers, socially stigmatized in many districts as "tramps." Some improved machines, like the self-binding harvester, significantly cut labor needs. This device, which first became available in 1878, displaced an estimated 75 percent of the workers needed in harvesting grain.

Not only that, the machine significantly widened the gap between farmer and hired hand. Now the former tended to do the lighter work, and the latter the heavier. Total, or at least seasonal, unemployment meant that hands had to travel to find work, thus weakening local ties and creating a floating body of workers who were strangers wherever they went. One worker testified: "Of one thing we are convinced, that while the improved machinery is gathering our large crops, making our boots and shoes, doing the work of our carpenters, stone sawyers, and builders, thousands of able, willing men are going from place to place seeking employment, and finding none. The question naturally arises, is improved machinery a blessing or a curse?"[4]

In some places strikes were held in an attempt to keep up wages. During the 1878 harvest season, some farmers in Ohio, Indiana, Michigan, and in parts of other states had their machines destroyed. Farmers, usually described as wealthy or well-to-do, received threatening letters warning, as one put it, "if you use one of these G—d—d machines we will burn your wheat stacks."[5] Sometimes the machines themselves were burned, and in other cases farmers who had already purchased self-binding reapers were too fearful to actually put them into the fields, preferring to go back to cradle scythes for that season. A few farmers voluntarily clung to hand methods, acting on an ethic other than that of maximizing profits no matter whom it hurt. "A large class of farmers," one newspaper noted, "has steadily refused to purchase binders for the reason that it deprives the laboring class of work." The other extreme was represented by a Minnesota farmer who shot and killed two men setting fire to his harvester. He surrendered to local authorities but was released without charges, and neighbors formed a rifle club and put up signs declaring "Tramps wanted as a top-dressing [i.e., fertilizer] for the growing crops." Both the rising capital requirements for entering into farming and the disappearance of year-round work opened a widening gulf between those who could successfully farm and those who could never hope for better than inadequate seasonal wage-labor.[6]

Not only the farm laborer but the farm owner, too, began to find technological improvement a two-edged sword. Transportation had always been critical to successful commercial farming, and before the canal era even farms in the East and Midwest went through a pioneer period in which crops that could not be easily transported were grown only for family consumption and cash was provided by the butter and egg business of the farm women. In the West, large crops and vast distances made farmers entirely dependent on railroads for getting produce to market. Refrigerated cars helped cattle-raisers carry beef, but the building of giant storage silos provided yet another opportunity for monopoly practices that cut farmers off from customers and channeled the profit to those who controlled the mediating technology. The power of railroads over farmers was dramatically conveyed in Frank Norris's novel *The Octopus* (1901), which fictionalized an actual violent confrontation between California wheat farmers and the Southern Pacific Railroad. An increasing dependence on technological systems and markets, often international (and linked by telegraph and shipping lines), over which they had no control, provided the stimulus for some of the most effective political reform efforts of the late nineteenth century.

Although horse power was the most significant and widely useful form of energy for the farm, new forms other than steam were also appearing by the

Giant grain elevators were built along the Mississippi and other rivers of the West, as harvesters, thrashers, and combines turned the Middle Border into the nation's latest bread basket. *Lippincott's Magazine,* 17 (February 1876), 145.

end of the century. Electricity was not easily tamed for farm work, and it was not at all clear just what it could and could not be expected to accomplish. During the century innovators sought to apply it to belt work and plowing, to use it for illumination, and use it as a substitute for fertilizer to stimulate plant growth. By the 1870s and 1880s electric plowing was finally accomplished in England and on the European continent; by the end of World War II there were an estimated 1,600 electric plows in Germany alone. In 1890 an electric traction engine was displayed at the California State Fair, but nothing came of it. Electric plows never became common in the United States.

Electricity had greater success in belt work, general stationary power supply, and illumination than it had in field work. In 1892 it was first used for belt work at the Agricultural Experiment Station at Alabama Polytechnic Institute, where a 10-horsepower motor was used to operate a thresher, feed cutter, cotton press, cotton gin, and cotton-seed crusher. In that same year,

In the southern states, the cotton gin was as necessary to cotton culture as the combine was to wheat growing in the West. *Lippincott's Magazine,* 17 (February 1876), 144.

the Crystal Hill Dairy in Pennsylvania began to use electricity for dairy equipment.

California led the nation in the use of electricity for farm purposes, consuming 200,000 horsepower in 1918—more than all the other farms of the United States combined. Nevertheless, the acceptance of electricity was quite slow, in part because of rural conservatism, in part because the equipment in the early days was not flexible, and in part because of the high cost. In 1892 a 25-horsepower motor, the steam-generating plant to power it, and the wiring and attachments cost about $7,800, the price of an already improved farm. Indeed, well into the twentieth century American agriculture generally lagged behind that in other countries on this score. As late as 1935 only 10 percent of American farms had access to electric power lines, as compared with 90 percent in Japan, 95 percent in France, and almost 100 percent in the Netherlands.

Such technological advances as steam plowing and electric milkers were relatively exotic in American agriculture before the twentieth century. Rather, the technological aid received by the average farmer came in the form of better hand tools and a wider range of horse-powered equipment, both manufactured more cheaply and in greater abundance by the adoption of the American system. Touring the United States at midcentury, British observers noticed that a manufactory at Buffalo made mowing machines in large numbers, and that one such machine, drawn by two horses, could mow an average of 6 acres of grass per day. At Worcester, they found 250 hands employed principally in making ploughs, hay-cutters, and churns: "Templates and labour-saving tools are used in the manufacture of these implements, which are sold in very large numbers."[7]

In 1849 a Scottish chemist attended an agricultural fair at Syracuse, New York, where he observed that

> The general character of the implements, was economy in construction and in price, and the exhibition was large and interesting. . . .
>
> Ploughs, hay-rakes, forks, scythes, and cooking-stoves, were very abundant, and many of them well and beautifully made. American ploughs are now exported in considerable numbers. . . . The potato grips and forks, of various kinds, cut out of sheet-steel, were very elastic, light, strong, and cheap. They seemed to leave nothing in these articles to be desired. The cradle-scythes were also excellent: an active man was said to be able to cut four to six acres of wheat a-day with them. . . .
>
> Of reaping machines there were several varieties on the ground, and several . . . actually in use in the Western States. Hussey's, which I saw on the ground, was said to cut twenty-five acres of wheat a-day. My friend, Mr. Stevens, who went round the yard with me, assured me he had seen one of them cut sixteen acres. McCormick's machine, I suppose, must be a good one, from the information here given me that as many as fifteen hundred of them have been made at Chicago . . . this year, and sold for cutting wheat on the prairies of the North-Western States.[8]

Technical improvement in agriculture, as in other areas of American life, covered a broad range of devices and processes. As a predominantly agricultural nation in which farming was spreading rapidly, market conditions generally tempted inventive activity in that field. As late as 1837 only twenty patents were taken out for agricultural implements that year. The tide changed in about 1851, and in 1860 alone, 507 patents were issued. Again, it is important to note that major innovation came in only a few areas. Because the labor that went into harvesting and threshing was exhausting and seasonal in nature, it tended to be concentrated in these areas, but even here progress was selective: these machines became common in the Middle West after midcentury but were not found in large number in the East until the 1870s, or in the South until the twentieth century. Furthermore, improvement tended in many cases to be entrepreneurial and managerial rather than technical. In the wasteful bonanza areas of the West there was a tendency to increase the size of machines, and the areas they could cover in a day, rather than the efficiency with which they did their jobs. It was thought better to have a big machine that would thresh more acres in a day than an improved model that would save a larger percentage of the grain from a smaller acreage. This was another example of the general American belief that labor was the resource most worth saving.

As technology progressed largely through cut-and-try methods, some observers began to suggest that science would prove a useful guide to mechanical improvement both in agriculture and in other branches of practical activity. There was little evidence of such a relationship. Jefferson might have tried to apply mathematical modeling to the design of plows, but it is now difficult to imagine how that would have proved effective. At the end of the nineteenth century Benjamin Butterworth, a commissioner of patents, remarked sadly, "It is strange, in view of the antiquity and importance of the plow, that its construction should have received so little attention from scientists, and that the principles of its construction should have been so little observed by those who used it."[9] The plow, of course, was not alone. Only in a few isolated areas, such as electricity, did science provide leadership for technical improvement during the nineteenth century.

Instead, scientists interested in agriculture concentrated on chemistry, first of soils and then of plant and animal nutrition and health. American farm journals saw in soil chemistry an important antidote to the growing problem of worn-out farmlands. When an American edition of the German chemist Justus Liebig's classic book *Organic Chemistry in Its Applications to Agriculture and Physiology* appeared in 1841, it was widely heralded as a panacea for soil exhaustion. Three years later, Liebig's *Familiar Letters on Chemistry* fell on fertile ground and a soil chemistry craze followed.

Inspired by Liebig's work to improve farming, a number of future American scientists went to Europe to study with him and other chemists in Great Britain and on the continent. Several of these became associated with the beginning of engineering and scientific education at Yale University, and in 1875 Samuel W. Johnson established an agricultural experiment station in Connecticut, the first in the nation. The enthusiasm for Liebig's work had initially led to inflated hopes that could not be fulfilled, and as practical farmers demonstrated that dirt farming was more complicated than book learning allowed, the notion that science could be useful took a beating. Johnson, however, was able to use chemical analysis to alert Connecticut farmers to the many fraudulent commercial fertilizers then on the market. His success in institution-building grew directly from the friends he won with this practical bit of science.

Since most Americans were farmers, it seemed obvious to many that governments at all levels should aid agriculture through the employment of scientists. At midcentury a Scottish chemist noted that "the Legislature of Maryland has not been unmoved by the recent contributions of science to the progress of agriculture, and has been among the first to recognize the especial usefulness of chemistry by creating the office of 'State Agricultural Chemist.' . . . Part of the duties of this officer," he continued, "is to visit the different counties of the State, to give private advice and public lectures to the farmers, to collect soils, marls, and other substances, which it may be desirable to analyse, and to *analyse them on the spot.*"[10] The visiting chemist sniffed that "a peripatetic laboratory is inconsistent with correct analytical research," but as historian Margaret Rossiter has pointed out, American scientists discovered that in the United States institutions for bringing science to bear upon technology had to be democratically acceptable, which meant both practical and available. "Correct analytical research" meant little if it was not funded and readily available.

Calls for federal government support of agriculture, through the encouragement of science, resulted, before the Civil War, in only a few small programs: the search abroad for plants that might be usefully cultivated at home, the distribution of free seeds by members of Congress, and the keeping of agricultural statistics by the Patent Office. With the removal of Southern members from the Congress after the outbreak of the Civil War, the way was cleared for the central government to make a more forthright stand in favor of supporting scientific research on a permanent basis. In 1862 the nation's farmers received institutional support for a scientific attack on their problems through the establishment of a department of agriculture and the passage of the Morrill Act for the establishment of land-grant colleges in each of the states. These land-grant colleges, set up to educate the sons and daughters of "farmers and mechanics," quickly evolved into centers of education in science and engineering, which provoked more "practical"-minded advocates.

The proper institutional framework for the U.S. Department of Agriculture (USDA) did not evolve overnight, nor did activities translate themselves immediately into useful results. The political tradition of party patronage and spoils worked against trained competence and the multiyear funding that good science often required. Appropriations for the new department stood at $199,770 in 1864, and were at the same level fifteen years later. Worse yet, the department organized itself rather like a university with separate units defined by discipline: chemistry, horticulture, entomology, and statistics. One even dealt simply with microscopy, even though microscopes were needed as research tools by all departments. Since problems in the real world did not come neatly packaged by academic discipline, bureaucratic lines of division worked against an effective attack on such problems as Rocky Mountain Spotted Fever, which was soon to strike a devastating blow to the range cattle industry of the plains states.

Gradually, however, the department pioneered that powerful modern tool for solving problems, the scientific bureau, with an interdisciplinary staff organized around the "mission" of solving a class of practical problems. In 1881 the old division of entomology was reorganized into the department's first modern bureau of this type. By 1899 not all existing divisions had been so converted, but there were already the Weather Bureau, the Division of Biological Survey, the Office of Road Inquiries, and the Division of Forestry, among others.

While the department was working out its own proper internal arrangements, the passage of the Hatch Act in 1887 gave it the opportunity to create a nationwide network of agricultural experiment stations, coordinated from Washington. Connecticut had had the first such establishment, but many states had departments of agriculture that the federal department saw as problematic. In California, for example, the state department of agriculture was busy searching for insect species that could be introduced to control agricultural pests. Such importations as the ladybug smacked of the very rule-of-thumb agricultural improvement that the USDA was committed to upgrading. In part because chemistry was clearly and undisputably scientific, whereas biological control seemed to owe more to popular lore and traditional prejudice, the scientists in Washington tried to use their newfound central power to impose their own view of improvement on local practice.

In 1906 the Adams Act increased federal support for the state experiment stations and at the same time restricted some of the funds to the execution of original research. These growing relationships between the department and scattered land-grant colleges and experiment stations not only strengthened the resources of the department but also served to complicate federal-state relations generally and offered the most compelling precedent before World War II for federal support of nongovernmental research and involvement in higher education.

Although the department had always had an interest in farm machinery, its main activities have never been so concerned. By and large the development of new devices and improvement of old ones had been left to the vagaries of private enterprise. It was not until 1921 that any attempt was made to centralize agricultural engineering activities, and it was not until 1931 that the separate Bureau of Agricultural Engineering was established.

In all regions of the country, but especially in the Northeast and South, there was a lingering tendency to rely on hand tools down to the twentieth century. Whole areas of farming—such as dairying, poultry raising, horticulture, local transport, and housework—received little mechanical aid. The greatest single factor in this retardation of mechanization was the lack of any cheap and flexible power. What progress was made in the nineteenth century was largely the result of harnessing the horse to perform what had previously been done by hand. It was not until the appearance of the small rubber-tired, gasoline-powered tractor and the Rural Electrification Agency of the 1930s that adequate power was finally available to the nation's farmers.

The most important results were that farming remained a hand operation to a large extent, and the relatively simple technology remained fairly inexpensive to acquire. Patents for farm devices had been among the most numerous during the nineteenth century, and a wide range of small hand-operated machines were available. Small horsepowers, which used an animal on a treadmill to produce mechanical power, were widely available. In 1862 it cost only $968 to equip a medium-size general farm. The increasing application of something like mass production to simple hand and horse-drawn equipment brought prices down, and by 1907 one could equip that same farm for only $785.

The breaking of the power bottleneck on the vast majority of farms came only in the early twentieth century, with the wide availability of electricity and tractors: these technologies, and the kind of capital-intensive agriculture they represented, brought to a close more than 300 years of small family-owned farms in America. The days when the yeoman farmer was the backbone of the nation had only been made possible by a solicitous political system, an abundance of cheap land, and the relatively slow development of farm technology.

III

THE IMPRINT OF AMERICAN TECHNOLOGY

6

CREATING AN URBAN ENVIRONMENT

ALTHOUGH AMERICANS continued to think of themselves as an agricultural people well into the twentieth century, by 1920 the census showed that slightly more than half the nation's population lived in urban areas. Robert Ridgway took the opportunity of his 1925 presidential address before the American Society of Civil Engineers to declare that "perhaps the most notable effect of the application of those laws of Nature which have been brought to light by the patient investigations of the scientist during the past century and a half is evidenced in the wonderful growth of cities everywhere."[1]

It was a conceit flattering to his auditors but perhaps would have been a surprise to a less interested audience. If the modern city was not so obviously the child of science, it was evident to all that it existed only through the midwifery of engineering. First, every city had an extensive (and expensive) engineering infrastructure of streets, sewers, water supply, and electrical lines that formed the skeleton and nerves of the urban body. Second, no part of this infrastructure was more important than those transportation facilities that were constructed over the years to serve first the commercial and then also the manufacturing activities of the city. Third, this manufacturing itself was only able to exist in an industrial form through the use of transport to bring in raw materials and take out finished goods, and through the application of steam and other forms of mechanical power. Fourth, many of the amenities that made city life exciting and rewarding—for example, newspapers, and parks and sports venues—were made possible on a large scale by modern invention and engineering. And fifth, in contrast to the older, inherited organic city, a newer planned city appeared, in the form of the company towns that often sprang up around new industries or in the form of suburbs, often laid out to take advantage of cheaper land and newly installed rapid transit lines. If the American city by 1920 was much more than a machine for making money, it was also unthinkable without that machine.

In mid-eighteenth-century Philadelphia, one observer was moved to remark of the city that "additions, alterations, decorations are endless. 'Tis one eternal scene of pulling down and putting up."[2] American cities grew because the number of people wanting to live there increased dramatically. In 1790 only 201,655 Americans lived in urban areas (defined as having more than 2,500 people). By 1810 that number had risen to 525,459; on the eve of the Civil War it was 6.2 million; twenty years later, 14.1 million; in 1900, 30.1 million; and in 1920, 54.1 million. Simply to accommodate these rising numbers, the old structures (of buildings, city plan, and even government) were pulled down and replaced with new. Along with the building of the railroads, the building of cities was probably the biggest single economic activity in the country.

No part of the physical structure of cities was more important than its streets. Whether they followed the natural paths of Native American trails or animal migration, as in parts of some eastern cities, or conformed to the commercial grid as was typical in western towns, they were usually unpaved in the early years, places not only of transportation but also of communication, commerce, and waste disposal. It was reported that by 1838, to keep down dust and slow the destructive effects of frequent passage, "most of the streets in New York, and indeed in all American towns, are paved with stones." It was also confessed, however, that owing to the "small size and round form" of these stones, "they easily yield to the pressure of carriages passing over them, and produce the large ruts and holes for which American thoroughfares are famed."[3]

There was little science needed to pave streets: it was done by traditional methods and with whatever material seemed handiest and most abundant. Cost was the chief factor in deciding the matter. Late in the century some cities began using bricks, and during the 1870s asphalt, from California tar pits on the West Coast and those of Trinidad for the East, was used in increasing amounts. Concrete was not introduced until the twentieth century, after experiments with it in the small community of Bellfountain, Ohio, in 1892.

Lighting the streets was another concern. Partly as an anticrime measure, street lighting was considered a desirable thing but was quite expensive and unsatisfactory until Baltimore, in 1816, became the first American city to begin using illuminating gas made from coal. So important did this process become that the Gas House Gang, originally a corrupt Philadelphia ring in control of this utility, became virtually synonymous with both a modern technology and its manipulation for private and corrupt purposes. Sixty years later the manufacture, distribution, and use of coal-gas became the paradigm that Thomas A. Edison copied in inventing his own system of electrical, incandescent illumination.

Before incandescent lighting, however, carbon-arc lighting had been pioneered for city use by Charles F. Brush of Cleveland, Ohio. In 1865, while still a student at a Cleveland high school, Brush managed to produce an arc light (that is, a bright spark bridging the gap between two carbon poles) in his home workshop. Many others had produced the same effect, but since the electrical current depended on the use of a battery, the process was prohibitively expensive for practical use. Brush took a degree in mining engineering from the University of Michigan in 1869 but seven years later was back in his Cleveland workshop, designing an improved electric dynamo for use with an arc-light system. The dynamo won a prize at the Franklin Institute in Philadelphia in 1878, and that same year he demonstrated his dynamo and lighting system at the American Institute in New York and the Mechanics Fair in Boston. Apparently as a direct result of these demonstrations, he sold systems to department stores in Brooklyn, Boston, and Philadelphia promising to provide twice as much light as gas, for a cheaper price (one dollar per hour).

The very intensity of arc lights made them unsuitable for domestic use. Large interior spaces, like department stores, were more appropriate, but street lighting seemed even more suitable an undertaking. Back in Cleveland, he had already tested such lighting in the Public Square. Brush wrote:

> It was shown from a second story window on the south side of the Public Square. That was in the autumn of 1876, just after the Centennial. The

> light was a very small one, of course, but it was concentrated by a parabolic reflector. The occasion was one of a parade of horsemen and foot soldiers, and all that sort of thing, and the light was thrown in their faces as they came up the street. I remember how the eyes of the horses looked like green balls of light. I do not know how well the horses liked it. They did not seem to care for it very much. After a while a big policeman came up and said, "Put out that damn light!" and we put it out.[4]

On the night of April 29, 1879, he put his system into full service at the same place: that night, someone remarked, was "like a day to remember." His success appeared to be entirely due to his own efforts. Brush later recalled that "in the very early years of electric lighting mine was strictly a 'one-man' laboratory. I had no assistant; indeed, no assistant was available. I made all the working drawings for the dynamos, lamps and special shop appliances needed. Wrote all the patent specifications. Tested and adjusted all dynamos and lamps."[5]

Some aspects of Brush's system stand out as particularly ingenious. His carbon rods, for example, burned off at the rate of 2 inches an hour but were automatically advanced to maintain a uniform distance apart by a special mechanism of his design. Nevertheless, it was the whole rather than the parts that ensured its preeminence. As one historian has written: "In analyzing the success of the Brush enterprise, it becomes evident that its outstanding feature was the high degree of integration of the system. Dynamo, arc lights, insulation, and carbon rods were conceived as integral parts of a unified whole by one man. Furthermore, all of the shop machinery and tools for manufacturing each individual piece of equipment were originated by Brush himself."

The mercantile cities of early America had been what can be called walking cities. That is, they were small enough to cover on foot: even by 1840 they were no more than 10 square miles in area. Baltimore and Philadelphia were that size, Boston's population lay within a 2-mile radius of City Hall, and only one-fourth of Manhattan Island accommodated New York's 312,000 souls. As a result, most people in American cities either lived where they worked or were able to walk there.

Transportation was available, however, for those who needed and could afford it. By 1827 a twelve-passenger vehicle ran on a regular schedule up and down Broadway in New York. A larger omnibus was added in 1831, and the next year the horse-drawn New York and Harlem Railroad extended service northward. By 1837 some 25,000 New Yorkers were using omnibuses every day, although the 25 cents required for a round trip kept nearly all but well-to-do merchants from commuting by this method. Ferry traffic back and forth across the East River also increased as people flowed to and from Brooklyn.

Two new housing patterns developed in cities along with the rapid transit. Many of those who could afford it moved into the new suburbs, and businessmen commuted into the city everyday while their wives strove to create the ideal Victorian "separate sphere" at home. In 1848, 20 percent of Boston's businessmen commuted on steam railways into the city, and 118 of the 208 passenger trains entering the city each day were on suburban runs of no more than 15 miles.

Those who could not or would not flee the city were being crowded into ever denser buildings in ever denser neighborhoods. Already at midcentury warehouses in New York were being converted into tenements with up to five or six families jammed into a single 12-by-14-foot room. For people not so desperately poor, New York's first modern apartment house was put up in 1869. In 1880 eight apartment buildings with elevators were built, forerunners of innumerable others. By 1890, some 30,000 tenement buildings in that city contained more than a million people. Those who had work could no longer walk to their jobs. In the 1870s the Rapid Transit Commission recommended that New York provide elevated rail lines on two of its north-south avenues and these, with the electrification of lines, were the major improvements until the building of the first subway.

The first successful electric street railway was built in 1879 for the Berlin Exhibition, but it was introduced in the United States by Frank J. Sprague. After attending the Naval Academy at Annapolis, Sprague visited the Crystal Palace Electrical Exhibition at Sydenham, England, in 1882. Inspired, he resigned his commission in the navy and went to work briefly for Edison. In 1884 he quit and formed the Sprague Electric Railway & Motor Company. Three years later he realized that if electric street traction were to be widely used, some standardized and successful practice would have to be instituted. At the time there were nine lines in Europe and ten in the United States, but they worked on radically different plans and had an aggregate trackage of less than 60 miles. Then in 1887 he received the contract to install a system for the Union Passenger Railway of Richmond, Virginia. He lost $75,000 on the

From the late nineteenth until the mid-twentieth century, the streetcar was the symbol of urban mobility, as well as an instrument of suburban development. *A Popular Treatise on the Electric Railway* (Boston, 1890), 48.

contract, but the success of the line more than compensated him. Later in life he relished the headline for a pro–street transit mass meeting in New Orleans: "Lincoln Set the Negroes Free! Sprague Has Set the Mule Free!"[6]

The mules might have been free, but the trolley cars were not. Speculators soon moved in and began to buy up lines, until at one time they owned the systems of New York, Chicago, Philadelphia, Pittsburgh, and one hundred other American cities. They had more than a billion dollars tied up in these lines, and another three hundred million in lighting plants that helped build electrical demand during the hours when few trams were running. Since there were practical limits to how many tram lines could be franchised by the city, these agreements were lucrative sources of profit to corporations and of bribes to city and state officials. For neither the first nor the last time, corruption followed closely on the heels of a new technology.

For many observers, rapid transit held out high hopes of social reform. San Francisco's *Scientific Press,* in an editorial of 1872, "Cheap City Transit," maintained that "there is at this time no more important question demanding a solution from our inventors and mechanics, than that of cheap transit for the clerk, artisan and merchant from the shop or place of business to the dwelling, which point, in a large city, must of necessity be quite widely separated."[7] Under the heading of "better housing for the working classes," the journal hoped that "the humanizing influence of gardens may be accessible to some of the families of working men, as distance vanishes by mechanical facilities." The opening of New York's first subway in 1904 led the social commentator Herbert Croly to observe that "the people of New York do well to celebrate with trumpets and drums. . . . The event begins the emancipation of the larger part of the city's population from an excessively cramped and uncomfortable manner of living." He also noted, however, that the service on the elevated lines "broke down fifteen years ago" and predicted that "in a few years the subway will doubtless be as crowded as the elevated roads are now."[8]

No more dramatic improvement in the mobility of New Yorkers was provided during these years than the Brooklyn Bridge, which connected that borough with Manhattan. John A. Roebling first drew up plans for the bridge in 1856, and a company to construct it was formed in 1865 and chartered two years later. Roebling was made the chief engineer and after his sudden death, his son took over the project. Everything about the bridge was new or at least pushed beyond the existing structures. Its suspension system, realized through woven wire rope, was a harbinger of the great bridges of the twentieth century, although the massive masonry piers and towers harkened back to a more classical design and construction. Opened in 1883, it did and still does serve to link the cities of New York and Brooklyn, which in that year were merged in theory as well as in fact. Even in its early years the bridge served as a powerful symbol of America's use of technology to embody the ideal, to bridge the physical gap between the possible and the real.

Another obvious urban need was for potable water. The nineteenth century was some decades old before many American cities built proper waterworks, although the religious community of Bethlehem, Pennsylvania, had such a facility as early as 1762. In 1801 Philadelphia became the first major city to invest in a waterworks, largely because of a series of devastating yellow fever epidemics beginning in 1793. The first of these killed one of every twelve people in the city, and it was obvious to many that the disease was somehow connected with "filth," especially in domestic wells contami-

When it was completed in 1883, the Brooklyn Bridge not only united that borough with Manhattan Island, but it fixed itself as a permanent technological icon in the nation's culture, the subject of paintings, poems, publicity stunts, plays, and movies. Courtesy Museum of American History, Smithsonian Institution.

nated by nearby privies. In 1799 Benjamin Henry Latrobe, the English-born architect and engineer, persuaded the city to undertake the building of a municipal waterworks, the most dramatic feature of which was to be two large low-pressure steam engines. One of these was to pump water up from the Schuylkill River to a reservoir and the other up to a second reservoir near the present site of the city hall. From there it was to be distributed through wooden pipes to those homeowners who cared and could afford to pay for the service. For the poor who crowded the alleys and back streets of the city, some free hydrants were to be placed at curbside.

The civic initiative of the project was matched by the technological daring: steam engines on that scale had never been used, let alone built, in America before. The scheme did not bring in sufficient income, however, and a combination of high fuel consumption, leaking hydrants, and rotting pipes kept costs high. In 1811 only 2,127 of 54,000 Philadelphians bought water from the system; the rest used the free hydrants or continued to rely on private wells. After sixteen years, the pioneer Latrobe engines were replaced with two of Oliver Evans's high-pressure engines, but fuel costs remained stubbornly high, and replacing the wooden with cast-iron pipes was expensive.

In 1819, the authorities proposed to place a dam across the Schuylkill that would power a number of waterwheels. These would pump water up to a reservoir on Fairmount Hill from which it would flow by gravity to the city. It took four years and $400,000, but the new plant worked splendidly. The city's population had doubled between 1810 and 1820, and the new works kept pace with the demand. By 1836, six waterwheels were supplying the needs of 196,000 people, and the per capita daily use of water stood at 20 gallons, twice what it had been thirteen years before. In addition, the surrounding Fairmount Park had been expanded into the first of the large urban parks in the United States. The project was a nice combination of social service, technological success (albeit a rather conservative one), and urban amenity.

The Schuylkill Waterworks, along the banks of the river in Philadelphia, had been the site and reason for Benjamin Henry Latrobe's original steam-powered operation at the beginning of the nineteenth century. The water-powered system that replaced it drew its water from the dam on the extreme left. On the right a canal boat passes through a lock and a handsome covered bridge crosses the river just downstream. Courtesy American Museum of American History, Smithsonian Institution.

By the time of the Civil War, the sixteen largest American cities had all built waterworks of some sort, but in the face of continued urban growth and the increase of per capita water use, no city could long assume that it had really solved its water problems. Such large-scale engineering works went far toward accounting for the estimated $200 million municipal indebtedness for the country in 1860. Within twenty years, however, this sum had risen to $725 million, and in 1902 stood at a staggering $1.433 billion. In 1905 it was estimated that waterworks, alone, accounted for more than a billion dollars worth of debt.

Boston had to go farther for its water and expanded its facilities several times during the nineteenth century. New York had the same experience, which included the building of the famous Croton aqueduct. In the West, Los Angeles engaged in a violent and possibly corrupt water war set off when it bought up the water rights of the Owens Valley to the north and killed the agricultural base of the area by moving the valley's water south and across the Mojave Desert in an aqueduct completed in 1903. The project was the particular crusade of the Irish-born water engineer William Mullholand, who, although entirely without academic training, rose to become the chief of the city's water system. It, along with a new rapid transit streetcar system, made possible the development of the dry San Fernando Valley. In San Francisco a long-standing dissatisfaction with the private Spring Valley Water Company led the city, under the leadership of its engineer M. M. O'Shannessey, to reach as far as the Sierra Nevada to flood a valley of the Yosemite and bring it to the city by the infamous Hetch Hetchy system. The battle between the engineers and civic leaders on the one hand, and preservationists, headed by John Muir, on the other, gave rise to the National Park Service, established in 1916. In the nineteenth century, scientists had worked to describe and understand what they saw as the unique, precious, integrated natural environment of California. The defeat of Muir and the building of Hetch Hetchy marked the ascendance of engineers who saw resources as discrete and discontinuous, and who dedicated themselves to the physical reconstruction of the state in order to support a more dynamic and industrial economy.

Even though many American cities moved toward building large systems for supplying potable water during the first half of the nineteenth century, not one made, at the same time, suitable provisions for removing the water from the cities once it had been brought in and used. The problem was

The Gothic Old North Point Water Tower, built 1871–73 in Milwaukee, stored water for that city's system. Such towers were a prominent civic feature of cities ranging in size from Chicago to Fresno, California. Photograph by the author.

an enormous one. Typically, when such water systems were put in place, daily per capita use rose from 2–3 gallons per day to 50–100 gallons per day. Such quantities completely overwhelmed the wastewater resources of urban areas.

Just as city dwellers had previously depended principally on wells for their water, so they had traditionally relied on pit privies or cesspools for the disposal of human waste. Sewers, if they existed, were not connected to homes but were designed to carry storm water away from the city streets. Now, with plenty of fresh water available, often piped directly into the home,

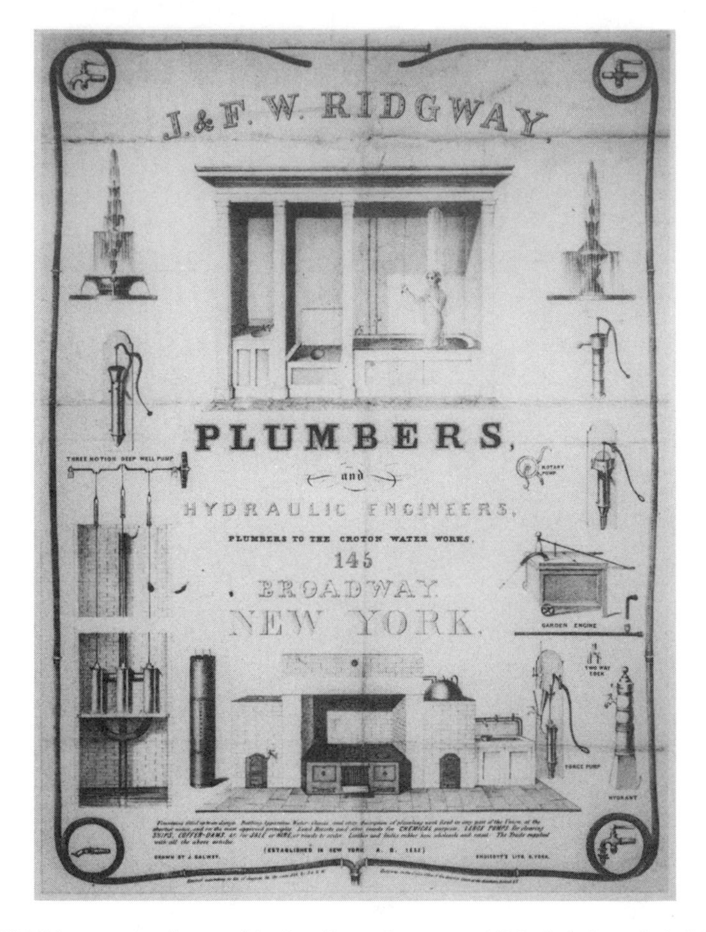

J. & F. W. Ridgway, plumbers and hydraulic engineers, established their trade in New York City in 1835. They advertised their readiness to supply "Bathing Apparatus" and "Water Closets," as well as large hydraulic pumps and leather or India rubber hoses. Courtesy Museum of American History, Smithsonian Institution.

a British invention came to be more readily used in the United States—the water closet. None had been patented in this country before 1833, but once the flush toilet was feasible, many householders chose to introduce it into their homes. The earth closet, which decomposed human waste into fertilizer on the premises, was urged by some officials who worried about the large new amounts of wastewater, but this device never found a significant market. By 1880 about one-third of urban households had adopted the water closet,

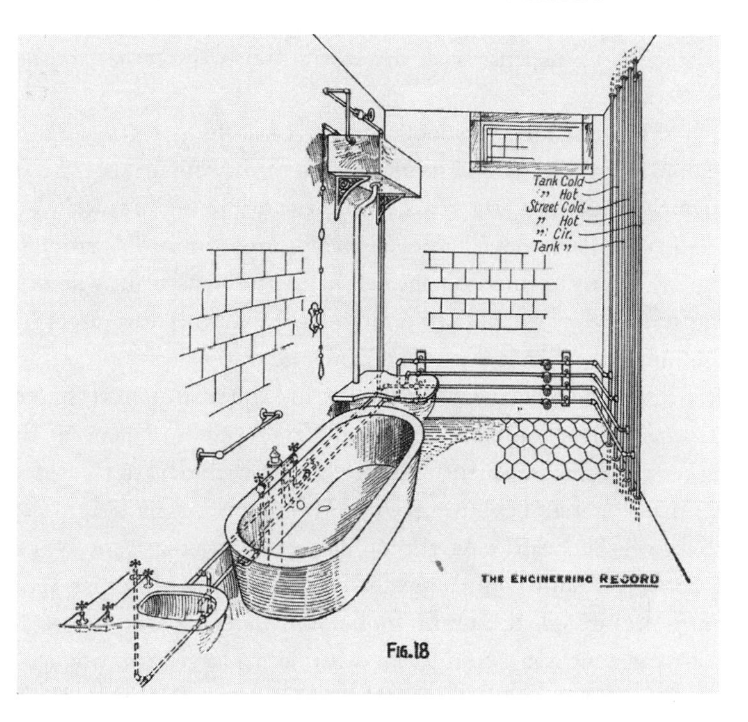

In 1896 a diagram of "water supply in the house of Mr. Cornelius Vanderbilt, New York City," showed a bathroom with elegant leaded window and tiles, as well as a bathtub, toilet, and wash basin. Not surprisingly, new technology was first available to those who were willing and able to pay for it. *American Plumbing Practice* (New York, 1896), 63.

though as usual poorer families could not afford this amenity, and as late as 1900 there were still an estimated ninety thousand backyard privies in Baltimore, and in the 1930s Washington, D.C., still had thousands.

The new problem was that waste from water closets overwhelmed cesspools, which had to be emptied much more often, and saturated the earth around them. It made far worse the problem of pollution that had led to the building of the municipal waterworks in the first place. Gradually it became obvious that cities were going to have to either connect the water closets to existing storm drains or build entirely new sewers to carry off the effluent to the nearest lake, river, or ocean. In 1857 Brooklyn, New York, became the first American city to build a combined storm and wastewater sewer system, and this became the standard method. Cities with a population of less than thirty thousand tended to dispense with the storm drains altogether and deal only with the sewage.

With the overloaded cesspools and privies, the pollution had at least been local. With the water transport of waste, the problem spread to a much larger area. During the early years of the twentieth century, cities debated whether to treat wastewater before it was dumped into the nation's waterways, usually to create enormous health hazards downstream when drinking water was removed by the next city, or to let the downstream users filter and otherwise purify it when it was taken from the waterways.

Because sewage plants were expensive and the treatment technology was difficult, engineering opinion tended to favor the treatment of drinking water rather than the treatment of waste. Faced with devastating epidemics, however, many states, beginning with Massachusetts in 1869, decided to establish boards of health. As the epidemics, particularly of typhoid and cholera, continued and polluted water was clearly identified as the carrier, many states prohibited the dumping of raw sewage into waterways. The first system of treatment was to pump the waste onto large farm acreage and let the soil filter out the impurities. Because this used large areas of land that were not always available, trickle filtering at central treatment plants became a favorite technique. By the turn of the century, the typhoid rate began dropping, as some 28 percent of urban dwellers were being supplied with filtered water.

The case of Chicago was particularly instructive because that city drew its water from the same place it dumped its waste. The city took over its ten-year-old water system during a severe cholera epidemic in 1852. The system, consisting of 2 miles of wooden pipe and a single pump powered by a 25-horsepower steam engine, served only a small fraction of the city's population, the rest of whom depended on private wells already contaminated by privies. In 1854 the city turned, not for the last time, to the waters of Lake Michigan and built an intake 600 feet offshore.

In 1855 Ellis S. Cheesbrough, another of those heroic-statured engineers who helped shape American cities in the nineteenth century, was appointed chief engineer of what turned out to be the first integrated sewage system in the country. His solution, chosen for financial rather than technical reasons, was to collect the city's sewage and dump it into the Chicago River, where it would flow into Lake Michigan. To preserve the city's water quality, the water system's intake was extended to a point 2 miles out in the lake in 1867. A second water tunnel was added in 1874.

This self-defeating game of tag was finally abandoned with the decision to build a canal, as had been proposed by Cheesbrough when he first began working for the city. The canal connected the south branch of the Chicago River with the Des Plaines River, thus reversing the flow of the former, away from Lake Michigan and toward the Mississippi. In 1887 certain parties began to campaign to have the canal made large enough to handle ocean-going ships, and this, too, was eventually done. In 1900 the Chicago Sanitary and Ship Canal was opened to traffic of all sorts. Twenty-eight miles long, 202 feet wide, and 24 feet deep, it had cost $54 million to construct. Sending their waste to people along the Mississippi and its tributaries greatly improved the water quality of Lake Michigan; the last sewer that flowed into it was shut off in 1907. In addition, the water intake was extended once again in 1894, this time to 4 miles offshore. In 1912 the city began to add chlorine to its drinking water, and in 1947 it began to filter the water. All of this helped but called for the development of new technologies, an improved understanding of the sciences of bacteriology and epidemiology, the willingness to invent new governmental entities such as regional sanitary districts, and a determination, however reluctant, to spend large sums of money on the problem.

Sewerage was not the only form of pollution in urban America in the last half of the nineteenth century. The concentration of so many people and so much manufacturing in so small an area created problems of noise, smoke, and garbage for which little provision was made. Typically, noise was ignored, smoke was seen as a powerful symbol of prosperity, and garbage was indifferently collected by private firms, perhaps to be fed to pigs on the city's edge. The city of Cleveland made no provision whatsoever for the collection of household garbage until the late 1880s. The most direct and obvious effect of this environmental disaster was an appalling record of sickness and death from contagious diseases. Toward the end of the century, a coalition of engineers, physicians, club women, and progressive businesspeople began a long but eventually successful fight to ameliorate the worst abuses of corporate pollution and governmental indifference.

The combination of environmental pollution and lack of public services often fell hardest on the working-class housewife. In Pittsburgh, as S. J.

Klineberg has shown, these women at the turn of the century lived closest to the steel mills and their smoke and grime, and farthest from those upper- and middle-class neighborhoods in which new housing and indoor plumbing made household chores less arduous. Improvements came first to those neighborhoods in which homeowners agreed to pay for the new services and last to those in which absentee landlords saw no reason to tax themselves for someone else's convenience. In addition, what water was available to workers living near the mills was often shut off for periods each day, when the mills needed more for their operations.

Until at least the mid-nineteenth century, American cities were primarily mercantile centers, buying, selling, insuring, financing, and transshipping the agricultural produce of their rural hinterlands and the foreign goods coming mainly from Europe. Manufacturing, such as it was, consisted largely of processing those domestic agricultural products for sale back to the farm population, to the city dwellers themselves, or to foreign customers. In 1840, for example, capital invested in manufacturing establishments in New York amounted to only one-fourth that invested in commission houses and less than that invested in retailing activities. Proportions varied from city to city, but of the major urban areas, Philadelphia was the only one in which manufacturing outstripped commission houses, and there retailing was more than four times larger. At the same time, in all cities, the size of manufactories was comparatively small.

Several factors accounted for this relative neglect of manufacturing. There was, for one thing, a lack of readily available workers. In 1840 there were only 1.8 million Americans living in towns of more than 2,500 people, compared with 15.2 million rural dwellers. Surprisingly, New York City had perhaps only three hundred machinists and machine makers employed in that year. Second, the transportation network radiating out from each city was still inadequate. Water transportation by river, lake, or ocean remained the primary means of transport, especially for heavy shipments of raw materials or manufactured goods.

Another constraint was the lack of adequate motive power. Urban manufacture had been done mainly by hand fabrication, since the most reasonable alternative was waterpower, which existed only in small measure and in certain places. Baltimore, in 1822, pointed with pride to the thirteen cotton mills "in our vicinity" and bragged of having "30 of the best and most improved merchant-mills within the limits and environs of the city, that manufacture about 300,000 barrels of flour annually."[9] Dayton, Ohio, in 1836, was said to contain "within its corporate limits water power sufficient for thirty-five pair of mill-stones, or seventeen thousand five hundred cotton spindles."[10] In 1850 *DeBow's Review,* recognizing the lack of waterpower available in New Orleans, urged the building of cotton mills in the city but admitted that "the motive power for manufacturing purposes . . . must be steam."[11]

DeBow's claimed that if steam power "were not a matter of necessity, experience would show it to be a course based on economy," but it was precisely the matter of economy that seemed so much in doubt. Steam engines had been operating in American cities for half a century already, but the question of whether or not they were economical remained ambiguous and controversial. Until coal could be delivered cheaply to the factories, fuel costs were high. Steady improvements in engine design and economic operation made steam engines more attractive, as did the increasing size of manufacturing establishments. The question of introducing steam engines into urban manufacturing establishments remained one that had to be answered each time for each individual case, not in the aggregate based on large generalizations. It was a question answered more often in the affirmative as the century wore on.

By the early twentieth century, engineers were beginning to look at large metropolitan areas as productive machines to be efficiently designed and managed for maximum output. In 1909 it was pointed out that one-seventh of the entire manufacturing production of the United States was located in the 700 square miles encompassing the city of New York and adjacent parts of New Jersey. It was an area, wrote one engineer, "destined to be a most interesting technological center by reason of the establishment within it of the largest works in the world treating raw materials for the production of staples."[12] To make this machine more efficient, he recommended measures already taking shape as a matter more of necessity than policy: regional government rather than local control, "a universal project for water supply," the reclamation of adjacent wetlands, and the provision of abundant electric power.

As the twin forces of manufacturing and population growth forced each other into an upward spiral, the modern industrial city was shaped by their dynamism: a great inflow of people from both small-town and rural America and from abroad; a differentiation of the city into zones of manufacturing, commerce, residence, and so forth; a spreading out from the walking-city of earlier times, facilitated by rapid transit. Traditional worries about the effect of cities on the political and social life of the nation were heightened now by worries about large corporations, hordes of emigrants, and people of color, all as far from the salubrious effects of rural life as could easily be imagined. With the enforced division or separation of work from residence, of producer from consumer, of owner from worker, and of industrial from private time, new forms of social intercourse were formed in the cities.

Early cities had had centrally located common areas: the Commons and Garden in Boston, the Battery in New York, the plazas of Spanish towns in the Southwest. By the mid-nineteenth century, however, some reformers began to believe that more was needed. The Philadelphia Watering Committee had created, as a by-product of their duties, a great urban park at their Fairmount works in that city. Mt. Auburn Cemetery in Boston, the Lakeview cemetery in Cleveland, Chicago's lakefront, New York's Central Park, and San Francisco's Golden Gate Park were all attempts to mute the increasingly technological face of urban life by providing "natural" areas in which values other than those of industrial capitalism could be appreciated, even if only for a few minutes. Ironically, of course, these great parks were themselves the result of careful engineering work: precisely surveyed and plotted, with roads, bridges, lakes, and buildings. Only nature seemed unnatural there.

At the same time, other parks for public amusement were being erected on urban fringes, places celebrating rather than transcending the technology and alienation of modern industrial, urban life. Served by rapid transit lines and crammed with "rides" even more exciting and dangerous, Coney Island in New York, Playland in San Francisco, and a host of others, provided the urban, industrial working class with cheap amusements as mechanical and meaningless as the work they did on company time. The Ferris Wheel, designed by a prominent engineer to provide Chicago's Columbian Exhibition in 1892 with something as symbolically colossal and technological as the tower Eiffel had built for the Paris Exposition in 1889, was a suitable symbol of these new amusement parks.

Even before amusement parks, new technology had begun to create the kind of large-scale, commercial spectator sports that are so characteristic of urban America. First the telegraph and penny press began to bring quick information on results and scores from widely scattered venues. Steamboats and railroads made it possible for teams and individuals to travel over most

By transmitting business news and accounts of sports events, as well as personal messages, the telegraph helped provide the fast communications that facilitated the rapid growth of American cities. The plant on the table and picture on the wall provide an almost domestic context for the female telegrapher. *Harper's New Monthly Magazine*, 47 (August 1873), 332.

of the country, and the rise of large manufacturing cities, linked by telegraph and railroads, led to intercity rivalries. The organization of baseball's National League in 1876 was made possible and indeed logical by these events. Basketball (1892) and volleyball (1895) were both invented by the Young Men's Christian Association to provide wholesome indoor entertainment on winter nights. They required, obviously, large gymnasia and indoor electrical lighting. The mass production of inexpensive sports equipment became a separate industry.

The bicycle, displayed at the Philadelphia Centennial Exhibition in 1876, flowered into a sports mania in the 1890s, by which time it had spawned an industry said to be doing $100 million worth of business a year. Calling the bicycle a "typical American device," the scientist W J McGee claimed in 1898 that

> typical, too, is the bicycle in its effect on national character. It first aroused invention, next stimulated commerce, and then developed individuality, judgement, and prompt decision on the part of the users more rapidly and completely than any other device; for although association with machines

of any kind (absolutely straightforward and honest as they are all) develops character, the bicycle is the easy leader of other machines in shaping the mind of its rider, and transforming itself and its rider into a single thing. Better than other results is this: that the bicycle has broken the barrier of pernicious differentiation of the sexes and rent the bonds of fashion, and is daily impressing Spartan strength and grace, and more than Spartan intelligence, on the mothers of coming generations.[13]

Department stores, too, played a widening role in the culture of cities. Replacing the older single-purpose outlets of an earlier time, department stores moved into the new high-rise buildings of urban downtowns served by rapid transit lines. Providing thousands of jobs for young female clerks, they not only sold the growing numbers of consumer goods made available by volume production but also were designed to show how these goods were to be used in everyday life. Displays of room settings or fashions created, as well as met, a desire for consumption. The stores themselves were models of technological vanguardism: elevators and escalators, electric lighting and central heating, cash registers and cash conveyors provided a lesson in modernity, just as department stores in the twentieth century gave millions their first taste of air-conditioning and doors controlled by electric eyes.

The very shape of the city was changing as well. The oldest surviving city in the western hemisphere, Santo Domingo, in the Dominican Republic, was laid out on its present site in 1502 and was carefully planned from the beginning. With its grid pattern of streets and plazas, it became the pattern for the Hispanic cities of the West, including the oldest one now within the borders of the United States, Santa Fe, founded in 1609. Indeed, as one historian has claimed, "nearly every Western town and city began as a planned settlement whose physical forms were determined in advance by individuals, corporations, colonization societies, religious groups, or public officials." By 1840 the basic matrix of these towns had been established, and every one that had more than 200,000 people in 1970 had already been established by 1890. Western cities, he continued, "lighted streets, supplied water and gas, regulated the disposal of sewage, collected garbage, constructed and maintained streets and sidewalks, operated markets, fought crime, furnished mass transportation, cared for the sick, buried the dead, extinguished fires, controlled nuisances, educated their youth, and provided recreational facilities," all with the aid of networked technologies.[14]

The extreme of this type of comprehensive planning was probably the company town. Driven into the countryside in search of unused waterpower sites, many early textile mills had found it necessary to build tenements for workers near the mills, thus creating de facto villages owned and controlled by the mill owner. Later in the nineteenth century, southern textile mills were notorious for running company towns in which democracy was rare and unions nonexistent. It seems likely that many, if not all, of these towns were especially the responsibility of the company's engineering staff. Along with the U.S. Army Corps of Engineers, which had played a leading role in laying out and building the nation's capital at Washington, D.C., engineers who managed company towns took their place among the earliest city managers during the Progressive Era.

In the West such towns were most often connected with isolated lumber mills, and coal or copper mines. There were probably several hundred such towns with their characteristically uniformly built and painted houses, grander home for the superintendent, and company store. The corporate hierarchy was plainly built into the physical plant itself. One town owned by the Colorado Fuel and Iron Company had its hierarchy of housing: "The inferior houses," one observer described, "are designed to meet the requirements of the lower class of improvident miners who would be indifferent to anything above the crudest type of house and probably abuse anything better. These houses may be built somewhat along the lines of the next class above, but would be finished with wood ceiling inside, have outdoor toilets, and be dependent on an outside spigot for its water supply."[15] As late as the 1930s, the building of Boulder Dam east of Los Angeles created the town of Boulder City, built by the contractors but kept free of dissident elements, including labor organizers, by the authority of the federal government.

Class distinctions were built into more conventional cities in other ways. As Sam Bass Warner has pointed out, "a zoning fight, a subdivision controversy, an open land campaign, a tax assessment scandal, a bond referendum for building a water, sewer, or school addition. These often narrow engineering and fiscal controversies express the class conflicts of American society."[16] His own classic work on the "streetcar suburbs" of Boston shows how the evolution of the walking, mercantile city of earlier years into the industrial city of the late nineteenth century, with factories and the poor within the city and the middle and upper classes in the suburbs, was closely tied to apparently engineering decisions. Railroad commuter trains, serving outlying

By the early twentieth century, Los Angeles, which later became known as that most automobilized of American cities, was being extended in all directions by streetcar lines designed not to fill a need but to create one. Like canals a century before, they were nakedly developmental. Postcard from the author's collection.

suburban stations, created wealthy villages 6 or more miles from the center of Boston. Street railways created two belts, the inner one being for the lower middle class from 2½ to 3½ miles from the city center, and another for the central middle class reaching beyond it out to a 5-mile arc from city hall. The belts were filled in with speculative housing for those who could afford it, leaving the core of the city to the working class, often recent immigrants, who took what second-hand rental housing they could find. A combination of wealth and transportation needs (the two were not unrelated) produced a segregated community. Ironically, the high cost of commuting by railroad kept the middle classes massed close to the city center, where the price of land, and therefore housing, was higher.

Although most early eastern cities were founded and enlarged more according to the dictates of private profit than public good, their unplanned sprawl was sometimes leveled by great fires during the nineteenth century. The resulting ruins could then be pulled down and vast tracts rebuilt according to more rational plans. A number of engineers came to have enormous influence over the fate of certain cities. Perceived as having an indispensable expertise, a civic vision, and, perhaps most important, a corporate though nonpartisan bias, Cheesbrough of Chicago, Mullholland of Los Angeles, and others exercised an influence that they directed toward rationalizing their cities. Municipal engineers liked to believe that even more than mayors, they were the ones to whom good government looked for the care of "physical matters." They were a brotherhood well known to each other, influential in both local and national engineering societies and widely traveled and experienced about the world. As the nation moved into an urban and managerial age, the engineers were in a splendid position to serve both their own interests and these new national imperatives.

For a nation that thought of itself as one made up of yeoman farmers living arcadian lives in rural environments, the United States during the nineteenth century was remarkably urban. Americans might fear cities and denounce them as basically alien, yet they flocked there by the thousands, to join a rising tide of immigrants from Europe, Asia, and the rest of the Americas. Americans also prided themselves on their modernity, and the city was the defining construct of the modern age. If no American city, even New York, quite measured up to the metropolitan centers of Europe, they all, to some measure, aspired to that role. In a land blessed and made unique by nature, the American city was the signal triumph of modern artifice.

7
WESTWARD THE COURSE OF INDUSTRY

When Frederick Jackson Turner proposed his celebrated frontier thesis in 1893, he concentrated on the American West as the site and cause of the repeated renewal of the country's democratic habits and institutions. But besides any reinventing of democracy that might have gone on, those who pushed westward in the years before and after the Civil War found and exploited a vast reservoir of natural resources that were critical in helping Americans become what one historian has called a "People of Plenty."

America had always prospered through the continual bringing in of new resources: machines and tools; techniques and business organizations from the European centers of the industrial revolution; monies for investment in mine, mill, farm, and canal; working people from Europe, Mexico, Asia, and (in chains) from Africa; and natural resources in the form of imported fertilizers, molasses, and similar raw and semirefined products; or, more significantly, resources unexploited and in situ, buried under or laid out on lands annexed to the Republic through treaty, purchase, or conquest. In 1800, most of this bounty lay to the west of the already existing states of the Union.

But resources are defined by their use, either actual or potential, and that use is in turn defined by the technology available for its exploitation. Trees are a certain kind of resource to those who can cut them down; gold to those who can wrest it from the earth; rivers to those who can ply or dam them; land to those who can plow and water it and to those who are willing to survey and map it, making it a commodity to be bought and sold. And before the land could be plowed and planted, the great herds of free-roaming buffalo had to be slaughtered and the native Indians also slaughtered, or driven onto small islands of exile on what was hoped to be the least resource-endowed of the land. And for all this, roads, steamboats, and railways had to push westward to bring in the settlers and take out the wealth of the land.

The "winning of the West," therefore, was not an alternative to the industrialization of the East as it was so often portrayed, but a direct result and extension of it. Not until the revolver and the steamboat, the railroad and the telegraph, were the peoples of Europe able to push inland from that membrane of settlement they had established earlier along the coasts of Africa, Asia, Australia, and the Americas. In the United States, the surveying and mapping of the land was a critical element in its conquest, rooting out its wildness and allowing it to be bought and sold like any other commodity into which, indeed, it could now be converted.

The pursuit of fur-bearing animals had led many of the first adventurers deep into the unknown territories west of the European settlements. Using guns and traps and the superb Indian bark canoes, trappers crossed the Appalachians, fanned out across the Mississippi watershed, and even made their way to the Pacific Coast along the Missouri, Snake, and Columbia rivers. In 1803 President Thomas Jefferson sent the explorers Lewis and Clark to that same area to discover and map the land, describe the flora and fauna, and bring back word of the native peoples. The trappers, though, lived lightly on the land, using technologies and adopting lifestyles nearly as much Indian as European. Nevertheless, even this romantic activity partook, after a pioneer period, of the industrial world. In 1832 the American Fur Company pushed its first steamboat as far up the Missouri River as the mouth of the Yellowstone.

On the north-central headwaters of the Mississippi, and again on the Pacific Coast, trappers found dense forests, home for fur-bearing animals but also valuable in themselves for timber and fuel. By 1869 the lake states were the source of more lumber than any other section of the nation, but a generation later, on the eve of the First World War, more than twice as much was coming from the West Coast. It should be noted that in 1917 the South was turning out more board feet of lumber than the lake states and the West

Coast combined. As the South began to falter, however, the West Coast became, in the years after World War II, the source of approximately half of the nation's lumber.

The first proper sawmill on the Pacific was apparently built in 1822 near San Gabriel, in what is now called Southern California. Combined with a gristmill, it was of a type little different from that found in the East before the end of the eighteenth century. Five years later the Hudson's Bay Company built a sawmill at Fort Vancouver. Most early lumber production, however, recapitulated the experience of the eastern states. Individuals styling themselves as sawyers used saw pits and whipsaws to produce a small number of boards and planks in what was a labor-intensive activity. Even after the construction of the first steam-powered sawmill on the Pacific Coast, near Bodega Bay just north of San Francisco in 1844, most of the early mills were powered by water and of a familiar sash or muley-saw design. In 1847 John Sutter, the Swiss entrepreneur of California, decided to build such a mill, and it was while deepening the tail race in early 1848 that his carpenter, John Marshall, found the gold that began the California Gold Rush.

The whipsaws of the pit sawyers were not significantly different in principle from the sash saws of the later water-powered mills: both produced a reciprocating motion. The first circular saw, which could run at greater speeds, was introduced on the West Coast as early as 1857 when one mill on Puget Sound had a blade 6½ feet in diameter; this style replaced the sash saw in the following decade. The great limitation of the circular saw was that it could only saw a log with a diameter equal to half that of the saw itself; the size of the saw was in turn limited by metallurgical considerations, resulting in speeds less than they might have been. By the 1870s the ingenious double circular saw, consisting of two blades set one above the other, began to replace the single blades. In a related innovation, Nathan W. Spaulding began to manufacture circular saws in Sacramento, California, and soon patented a way of using replaceable teeth on his blades. During the 1880s these, too, became standard.

The band saw, a thin, endless loop of toothed steel, was first successfully introduced on the Coast in 1885. In 1888 one mill in northern California had a band saw 51 feet in circumference that could cut a plank 80 inches wide. The next year a mill on Puget Sound, equipped with two band saws, could turn out a quarter million board feet of lumber per day. The great expanse of forests made such production possible, but it was the newly met species of

In 1873, along the California coast in Mendocino County, a steam sawmill cut redwood logs into lumber for water shipment to San Francisco and the other rapidly growing cities of the state. *Harper's New Monthly Magazine,* 48 (December 1873), 39.

the West, redwood and Douglas fir in particular, that dictated the large dimensions of the saws themselves. New mechanical blowers had to be devised to blow away the vast amounts of sawdust produced, and in lumber ports like Eureka in northern California the sawdust dumped into the navigable waters of Humboldt Bay required dredging by the U.S. Army Corps of Engineers and eventually helped lead to legislation against the dumping of waste into such waters. This in turn, in the 1970s, became the basis for the authority of the corps to prevent the general pollution of the nation's waters.

From the 1840s to 1880s only axes were used to fell trees in the woods, but the fallen trees were then often cut into handy lengths with crosscut saws. When, in the 1880s, anonymous lumberjacks began to use crosscut saws on the standing trees as well, in conjunction with axes, it was found that felling time could be reduced by four-fifths.

With more and more trees being cut, ways had to be found to get them to the mills. The first trees exploited were those growing along the water's

In the redwood groves of California, the pioneer axe provided the only tool (before the crosscut saw) for bringing down the giant trees. *Harper's New Monthly Magazine,* 48 (December 1873), 40.

edge, and the logs were floated to the mill. As these were decimated, splash dams were constructed, first in 1881, to hold the logs until sufficient water built up so that when the dam was breached, the rush of water would carry the logs down to a larger body of water. Also in the 1880s, flumes were built where water was sufficient, and logs would hurtle down many miles of wooden trough to the mill. Where no water was available, skid roads were built. A path was cleared through the woods, small trees were cut down and laid crossways at close intervals, and these were greased. Teams of draft animals then were able to pull the logs along these trees from the logging site to the mill. In 1881 the West Coast saw its first steam donkey engine, essentially a stationary steam engine with winch that could pull logs by cable and thus replaced oxen or other draft animals. The larger companies also built logging railroads into rich timber areas, and inventors experimented with track-laying, steam traction engines, precursors of the later Caterpillar tractor. During the last three decades of the nineteenth century, American lumber production rose slightly less than three times, whereas that of the states of the West Coast multiplied nearly six times. It was the fast-disappearing forests of the Midwest and Far West that led to the Forest Reserve Act of 1891, one of the earliest indications that the scientists and engineers in the federal government were beginning to put together what would come to be called the conservation crusade.

The impact of miners on the land was also great. Colonial America had seen scattered iron ore pits along the East Coast from New England to Virginia, and an occasional copper mine. In the Southwest, Spanish, then Mexican, miners had long sought and exploited precious metals found in various parts of Mexico, including what later became New Mexico and Arizona. Copper was found in present-day New Mexico late in the eighteenth century, and gold in Arizona in 1828 and 1833. At about this same time coal was becoming a significant resource in both Pennsylvania and Virginia, gold was found in Georgia and North Carolina, and lead in Missouri. None of this mining activity, however, prepared Americans for the great gold rush to California.

Miners and prospectors were the real pioneers of the Far West, analogous to the farmers of the Atlantic coast and cattle-raisers of the Great Plains. Precious metals were found in the mountainous parts of the West, far from eastern cities. It took from early spring well into the summer to go overland to California in 1849, and normally eight months to travel by sea around Cape Horn. A fortunate few, using the clipper ships that were built in Baltimore and New England from 1830 to 1860 for the China trade, could make the passage in much less time. One of these long, narrow-hulled ships with a vast sail area, the *Sea Witch,* made it from New York to San Francisco in ninety-seven days in 1850, and Donald McKay's *Flying Cloud,* built in 1851, set the all-time record with a mere eighty-nine days. A decade later in the great rush to Colorado, it took a week by stagecoach or five to six times that long by wagon to travel to the mines.

Not only were distances great and the land unknown to Americans, but those who joined the rush were almost totally ignorant of the technology of gold and silver mining. Mexicans had been engaged in these pursuits for centuries and their experience was helpful, but Americans had the good luck of finding the earliest California gold, including that found at Sutter's mill early in 1848, in placers—that is, in a form loose and free from impurities. Over the ages the gold in rocky veins had eroded loose and been washed down the many swift streams into the foothills of the Sierra Nevada. Thus the earliest American miners could merely scoop up a wash-pan of gravel from the creek bed, swish it about, dump the water and waste, and pick up the heavy bits of gold left at the bottom of the pan. It was a simple, cheap, and effective technology—so much so that the estimated fourteen thousand persons other than Native Americans to be found in California in 1848 increased to nearly a hundred thousand by the end of 1849 and a quarter of a million by the end of 1852.

From Mexican miners and those who had worked the gold fields of Georgia and North Carolina, the forty-niners learned to upgrade their simple pans with cradles or rockers. These simple wooden devices were boxes, open at one end, with strips nailed crosswise along their bottoms, which could be rocked back and forth when gravel was shoveled in and then washed out with water. The heavy bits of gold would be caught in the strips while the waste washed away. It was a device known and used in many parts of the world. The international experience of some miners taught others to use quicksilver to collect the gold, from which it could then be separated by vaporizing the mercury. In many cases, the cradles were replaced by "long toms" of similar design, except that they were stationary and longer and could use a continuous stream of water to process more material faster. This device in turn gave way to "sluices," which were simply long troughs with cleats to catch the passing gold.

The easy placers of the early years could not last forever, of course, and as they played out three technological routes lay open to the miners: river mining, the working of deep gravels, and the exploitation of quartz veins. The first of these, tried as early as 1849, was merely a larger-scale version of the damming of creeks. A cooperative of miners would agree to dam a river and work the gravels on the bottom. Even when the season was unusually dry, as it was in 1855–56, the river beds could only be worked for a few weeks from the end of the summer until the winter rains began again.

Working the deep gravel was more productive. From former ages gravel had been laid down in such depths that it now formed whole mountains. At first tunnels were dug and blasted back into these, and by the late 1850s tunnels up to 2,000 feet deep were common. This, like all the technologies applied to mining in California up to then, was learned the hard way or from miners coming from other mining regions of the world. To work the gravels, however, an entirely new technique was quickly developed: hydraulic mining.

In this method, water was collected by dams high in the Sierra and brought by ditches, flumes, and iron pipe to the base of the mountain. There, under very high pressure, it was directed through nozzles called "monitors" at the mountainside, striking it with such force that the whole face was washed down and passed through sluices to collect the loose gold particles. First begun in 1852, this technique was really a form of mass production, for it involved all the common features of modern, capital-intensive industry: corporations trading shares on international markets, engineering expertise applied to the construction and installation of large and expensive works and equipment, and the hiring of laborers for wages. The waterworks required to operate this industry were extensive: it took 200 million parts of water to get one part of gold. By 1867 more than three hundred separate water systems had been constructed, extending over 6,000 miles of ditches through the mountains. In the 1880s the Army Corps of Engineers estimated that some $100 million had been spent to build this system. A generation later, municipal waterworks, irrigation districts, and utilities seeking hydroelectric sites all made use once more of many of these mining water systems. The industry itself did not survive its own success. The legislature in 1878 established an office of state engineer in part to wrestle with the environmental damage done to downstream rivers clogged with mining debris. Before a solution to the problem of silted rivers and consequent flooding could be found by the engineers or the legislature, the U.S. Circuit Court, in 1884, in effect, banned the industry as too destructive. It was an early and notable instance of successful environmental concern.

The third solution to the disappearing placers was to follow the gold back to the veins locked in quartz rock deeper in the mountains. In 1856 the *Scientific American* reported that sixty-three factories in different parts of California were using machinery to grind quartz and extract the gold: "Thirty of these are driven by steam engines, the others by water wheels. The gold quartz mining and crushing is rapidly on the increase in California."[1] This required the prospective miner to find the vein, follow the direction of its course, dig shafts to expose it, wrest it free from the surrounding rock, get it to the surface, pulverize the ore, and then separate the gold from the waste.

The machines and techniques required were beyond the experience and knowledge of California miners except for those who had come from hard-rock mining areas in other parts of the world: notably, Mexico, South America, Germany, and Cornwall.

Early on the grinding of the ore was aided by the use of the Mexican *arrastra.* Being made of stone, this device was not unlike the familiar rolling mill that had been used since colonial times on the East Coast to grind and mix ingredients like gunpowder. Rock crushers, such as those used by road builders, were of some help in breaking the rock down to a proper size for the *arrastra,* but complicated and expensive crushers brought from the East were conspicuous failures. A last resort was made to the old and familiar bank of pestles, with iron ends, which pounded ore in mortars to which quicksilver had been added. The resulting dust was run through sluices to separate the heavy gold. A long series of small but significant, and now anonymous, improvements produced the "California stamp mill," which became well known in the far-flung mining areas of the world. Even more than hydraulic mining, the extraction of gold from quartz required engineering expertise, substantial capital, and wage labor.

The continuing influence of the California stamp mill is impossible to measure with precision but may well have been surprisingly great. One mining engineer wrote in 1899 that "there is no more perfect instance of automaticity than a fully-equipped [California] gold stamp-mill; and I cannot help thinking that it was the example of the gold stamp-mill that, more or less, influenced our manufacturers of iron and steel and other metals to introduce the same principle into their works, until to-day, machinery, rather than manual labor, effects, from first to last, all our great metallurgical operations."[2]

The need for larger, heavier, and more complicated machinery than could be made on the site, for large sums of money to be invested long before any return could be expected, and wage-earning miners drawn frequently from abroad, particularly Cornwall, all brought an end to the brief but romantic era of the forty-niners. San Francisco quickly became a cosmopolitan metropolis, supplying not only capital from its stock exchange but, in 1867, more than $2 million worth of castings from its fifteen iron foundries, which employed twelve hundred workers. Chronicling the activity, both mechanical and financial, was the well-respected journal, the *Mining and Scientific Press,* edited from San Francisco after 1860.

All the mining expertise and mechanical and financial resources of California were brought to bear on the next, great, western mining bonanza, the Comstock strike of 1859–60. A half-million-dollar macadamized road was pushed 101 miles over the crest of the Sierra from Placerville to Virginia City, Nevada, and over this poured workers, engineers, and financiers from California. Between 1860 and 1880 they took $300 million in gold and silver from the Comstock.

It soon became evident that the hard-won California experience would not be sufficient for the scale and nature of the Comstock silver mines. For one thing, the veins were wider, in some places 65 feet wide, and of a soft and crumbling nature. Normal timbering of the shafts was not sufficient to prevent cave-ins. It took a young German engineer with California experience, Philip Deidesheimer, to come up with a system of timbering by using beams, mortised and tenoned, and fitted together into cubes, each interlocked with the next. Called "square sets," they were copied all over the world.

Conditions underground were further complicated by the great heat encountered: 170 degrees Fahrenheit at the 3,000-foot level. Large pumps and blowers were installed, and ice water was provided for the workers. Improved drills, the compressed-air drill and diamond-studded rotary drill, both of French origin, were improved by American licensees in the East and introduced into the Comstock. From Sweden came Alfred Nobel's dynamite and nitroglycerine, both brought to California in the late 1860s. A. S. Hallidie, later the inventor of San Francisco's cable-car system, introduced a flat, woven-wire cable to use in lifting loads up shafts. By the 1870s some three thousand workers from all over the world were toiling in 190 miles of shafts.

The California stamp mill, too, proved inadequate without modification. Silver came in chemical combination with other metals and had to be separated by some further process. Again, Mexican experience proved valuable. The *cazo* and *patio* processes, both of which featured the mixing of silver ore with quicksilver and salt in the presence of heat, were experimented with. By 1862 these had been mechanized and speeded up in what came to be called the "Washoe pan processes" or "Washoe pan amalgamation."

By the time the Comstock began to fail in the 1880s, it had been responsible for teaching western miners a great deal about hard-rock mining and the treatment of ores. From there people spread out in all directions, even back to California where a revival of quartz mining took place, but also to Montana, Utah, and other parts of Nevada's Great Basin.

The success of the silver boom in Colorado demonstrated the new and difficult conditions under which western mining was progressing, and the growing importance of scientific knowledge and engineering in meeting these conditions. Placer gold had been discovered in Colorado in 1859, shortly before the Comstock. In the late 1860s Nathaniel P. Hill, a professor of applied chemistry at Brown University, was given financial backing by New England investors to undertake gold mining in Gilpin County. In 1868 he erected a smelter, based on a modification of those found in Swansea, Wales, and to run it he hired Hermann Baeger, a graduate of the great mining school at Freiberg, in Saxony. In the 1870s, silver became important in several areas, particularly in Leadville. There, in 1877, two American engineers educated in European mining schools, August R. Meyer and James B. Grant, both built successful smelters.

The basic problem with Colorado ores was that so many of them were unoxidized and chemically combined with sulfides. Practical miners considered such ores to be "refractory," that is, hard to purify, and neither California nor Nevada experience really spoke to the problem. The answer lay in smelting (reducing the ores to a liquid state in the presence of other chemicals), but the precise nature of the reaction, and the appropriate machinery to promote it, were not known. It was partly a matter of engineering but first required scientific investigation into the needed metallurgy.

The U.S. Geological Survey (USGS) had been established in 1879 in part to tackle just such problems. From 1879 to 1881 Samuel F. Emmons of the USGS studied the Leadville ores and prepared a report that greatly clarified the nature of both the problem and the solution. The mining historian Rodman Paul has declared that "more than any other event, the publication of this scientific study convinced skeptical mining operators that they could learn something of cash value from university men."[3] In Denver in 1882, Emmons organized the Colorado Scientific Society, the leading members of which were government scientists from the Geological Survey; but it also included "metallurgists, assayers, chemists, geologists, mining managers, mining engineers, and one minister." The work of the USGS was a powerful social contribution by the government to the development of private enterprise in the West.

While government scientists did the fundamental investigation, its practical application rapidly fell into the hands of engineers. There was more than a grain of truth to the old definition of an engineer as a person who could do for one dollar what any fool could do for two. As early placer mines played out and remaining ores proved refractory, mining properties tended to be bought up by Easterners, or even foreign capitalists, who wanted engineers first to appraise the likely value of the property and then decide on a process and design machinery (often a smelter) capable of working the ore at a profit. A cosmopolitan group, these mining engineers were drawn from all the regions of the world where mining was practiced on a large scale. One British engineer, Philip Argall, had begun in Ireland, then made his reputation in Australia, New Zealand, and France before going to Colorado in the employ of English capitalists. The German, Philip Deidesheimer, had invented the square-set timbering for the Comstock. Most of the engineers were probably American, though many of these had studied abroad, particularly at Freiberg in Germany.

Although many of the earliest mining engineers had learned their trade on the job (as was true, indeed, of other kinds of engineers as well), the increasing complexity of mining had led to the establishment of curricula in mining engineering in a growing number of American colleges. By 1921 it was estimated that six out of seven American mining engineers were college trained. The first American school to provide a full curriculum was the Columbia School of Mines, which opened its doors in 1864, at a time when some leading German mining schools were drawing half their students from the United States. Columbia graduated its first class of mining engineers in 1867, and by 1892 it had graduated 402, with an average of more than 15 a year. The University of Michigan graduated its first class in 1867 as well, the Massachusetts Institute of Technology the following year, the University of California in 1877, and the Colorado School of Mines in 1882. A survey in 1892 discovered that sixteen mining schools had graduated a total of 871 engineers since 1867. In 1871 there were already enough mining engineers to form the American Institute of Mining Engineers, the first specialized organization to break away from the American Society of Civil Engineers and of the so-called Founder Societies the one with the largest and most influential admixture of nontechnically trained managers and owners among its membership.

In 1892 the American mining industry included "not over 6000 persons who may be said to require technical training as engineers," but at the same time there were reported to be "about four times as many openings as there are mining graduates to fill them."[4] This helps explain why mining continued to attract persons of ability, like the young Herbert Hoover, later to be president of the United States, who graduated from Stanford University in 1895 with a degree in geology. The demand for engineers, however, was clearly limited by the fact that mines in America, unlike those in Germany and Austria, for example, were strictly private enterprises, though drawing public subsidies from various federal land policies, federal scientists, land-grant railroads, and graduate engineers of public colleges and universities. Samuel B. Christy, a professor of mining engineering at the University of California, asserted that the "American system has many distinct advantages, particularly in a new country. Routine and precedent are thrust aside, and energy and originality have full sweep. But it must be admitted," he added, "that there are serious drawbacks to this system." Chief among these, he noted, was the tendency to avoid the use of engineers because although these were primarily interested in efficient operations (as were the state-controlled mines in Europe), the American mine owners were too often interested only in quick and ruthless exploitation of the mineral resources. "If a government is ever justified in controlling the business affairs of its people," wrote Christy, "surely the mineral wealth of a country, and the conditions under which it should be removed, are the ones which most need regulation and control."[5] No doubt his was a minority opinion, and though restrictions of mining practices became a part of the conservation reforms of the early twentieth century, when the federal Bureau of Mines was established in 1910 its enabling legislation was at pains to deny most forms of federal jurisdiction over mining operations.

With or without the help of trained engineers, the technology of hard-rock mining changed rapidly during these years. Bartlett L. Thane, an engineer from the University of California with extensive experience in Alaska, wrote in 1899 of "the wonderful advance that has been made in all branches of mine-engineering" and asserted that "by means of the important inventions and discoveries that have been made in mechanics, chemistry, and electricity, we are now able to work ores which, only a short while ago, would have been regarded as worthless."[6]

Thane noticed that, curiously, "the work of the engineer usually ends with the erection and installation of the machinery that he has designed [for handling ore after it is broken from the face of the mine]. His interest rarely reaches to the details of breaking ore; this is supposed to be the peculiar province of hand-labor." Yet hand-drilling could account for up to 75 percent of the cost of mining operations. When dynamite was introduced into the

The power drill (shown here in an 1880 version) helped industrialize western mining, although it became famous as the builder of railroad tunnels that broke John Henry's gallant heart. Benjamin Butterworth, *The Growth of Industrial Arts* (Washington, 1892), 70.

mines in 1866, to replace black powder, its destructive force was so great that it required smaller charges and therefore smaller holes to be drilled into the rock face. The first labor strike in Grass Valley, California, came over the attempt to replace "double-handed" crews, needed to handle the large drills used on black-powder holes, with "single-handed" crews of one miner using a much smaller drill.

A perhaps inevitable innovation was the introduction of mechanical drilling. In 1870 the eastern inventor Charles Burleigh demonstrated his machine drill in a tunnel at Georgetown, Colorado. So successful was it that San Francisco's *Mining and Scientific Press,* in retrospect a quarter century later, declared it to be "the beginning of a new era" in mining. The Burleigh was not the only drill on the market. Thane reported that in Alaska his partner, fresh from the mines in Coeur d'Alene, Idaho, found a "Baby Ingersoll" drill cast aside and covered with weeds. He was familiar with the machine and decided to give it one more trial. Mounted on a tripod and weighing

170 pounds, the drill worked twice as fast as hand work through their 1- to 3-foot seam of quartz.

Machine drills and dynamite were not the only innovations underground. Electricity was used for hoisting, lighting, and other purposes. It was reported in Cripple Creek, Colorado, in 1900 that a miner "may go up to his work from the town on an electric car, go down in the mine by an electric hoist, operated by electric signals, the shaft being kept dry by an electric pump, do his work by an electric light, talk to the town and thence to the world by an electric telephone, run a drill electrically operated, and fire shots by an electric blaster."[7]

And he might even die, when a stray spark prematurely ignited a charge. Productivity per worker in mining rose nearly 50 percent between 1880 and 1900 and almost doubled by 1902, but the rapid industrialization and mechanization of western mining was also creating an escalating danger for workers. In 1896 in England 1.7 miners were killed per 1,000 employed. In Colorado that figure was 5.96 and in Montana 8.28. In the face of such new technological hazards, and more traditional problems of pay cuts and blacklists, it is little wonder that western hard-rock miners organized into some of the most committed unions of the time: the Miners' Union at the Comstock in 1867, the Western Federation of Miners, started in Colorado in 1893, and the later, more radical, Industrial Workers of the World. A work force drawn from many different nations and organized into protective unions was a direct outgrowth of, and as much a sign of progress as, the mining corporations and engineers in the western mines.

The scientists of the U.S. Geological Survey were only the most obvious aid that federal science was extending to the miners and foresters of the Far West. Behind the basic research into metals and silviculture (already taking place in the Forest Service) lay the exploration and mapping of western lands, without which the land itself could not be converted into a market commodity, to be bought and sold like any other desirable good. Nothing better revealed the social and economic values of America than this laying out of a scientifically defined and technologically determined rectangular grid on the irregular, seamless web of the natural landscape.

In the western lands, ceded to the central government after the Revolution or acquired through purchase or conquest, exploration preceded permanent settlement by Americans or European colonial powers. At first this had been limited to expeditions along coasts or up rivers, such as that of

Hernando de Alarcon up the Colorado River in 1540. After the Revolution, President Jefferson's Louisiana Purchase led him to send the explorers Lewis and Clark to penetrate as far as the Pacific Ocean. After the Mexican War, military exploring parties were sent to the West, and in the years after the Civil War a host of civilian parties were dispatched: F. V. Hayden to survey Nebraska and Wyoming, Clarence King to follow the 40th parallel, G. M. Wheeler to run the 100th meridian, and John Wesley Powell and Clarence Dutton to map Utah, Nevada, and Arizona. The establishment of the U.S. Geological Survey in 1879 was an attempt to bring such disparate surveys under one agency staffed by technical experts. Clarence King was its first director, and J. W. Powell its second.

Colonial surveying had tended to be done in terms of metes and bounds, that is, lines that followed the landscape, or human activities were described by physical markers: rocks, trees, rivers, and so forth. Towns in which property was divided into very small parcels were platted; that is, land was measured and maps drawn to great exactitude. In the new western United States, a rectangular survey was instituted, which divided the (for American purposes) empty and unclaimed land into equal and easily identifiable rectangles. The basic unit was the township, measuring 6 miles on each side. These units were then divided into thirty-six sections, each one measuring a mile on a side, and thus contained 640 acres. These mile-wide bounds were often etched deeper still into the landscape by roadways, creating a distinct checkerboard effect across the country. These sections, in turn, were divided into quarters of 160 acres, the basic "homestead" of pioneer days. The federal government granted some of these blocks of land to railroads, others to the states to provide funds for "land-grant" colleges. All in all, of the original public domain of 1,442 million acres, 300 million were sold for cash, 285 million were taken up in homesteads, 225 were granted to states, 95 million went for military bounties and private claims, 91 million acres were given to the railroad corporations, and 35 million were disposed of under special legislation like the Timber Culture Act (1873) to encourage the planting of trees and the Desert Land Act (1877) to encourage irrigation.

Americans still thought of themselves as a nation of farmers, but west of the 100th meridian, in the absence of sufficient rain, dry farming was effective and appropriate for only a limited number of crops, mainly grains. One possible alternative was irrigation, which would allow a wider range of crops to be cultivated and brought a greater dollar yield per acre. One source of

In 1807 the priests at the Santa Barbara Mission in California forced Indian labor to build a dam across Mission Creek to impound water for their water system. Water was brought to the mission through a stone aqueduct to provide drinking water (after filtering), and water for irrigation and the washing of clothing. Photograph by the author.

water was deep wells, of course, and in California by 1889 about two thousand such wells were irrigating 38,378 acres of farmland. Such sources continue to be important, but were eventually dwarfed by large irrigation projects based on dams and canals. Nevertheless, the lonely windmill, pumping water, perhaps for stock, became and remains a symbol of the farming West.

Native Americans living in the Southwest had irrigated their croplands for hundreds of years, and their systems, such as that around the present city of Phoenix, Arizona, were extensive. The Spanish, too, were familiar with irrigation, inheriting both facilities and customs from the Muslims in their own arid south, not to mention still-standing Roman aqueducts. Both the customs and engineering works, including even the aqueducts, were exported to the New World. The extensive irrigation systems of San Antonio, Texas, for example, had their origins in knowledge taken from Muslim Spain to the Canary Islands in the fifteenth century, then brought to the American Southwest by settlers from those islands. Part of the system, built in 1735, was still in use as late as 1906. The chain of missions built by Spanish and Mexican priests over the Southwest and up the coast of California, each had its irrigation system. That in Santa Barbara, California, consisted of two masonry dams high in the foothills, stone aqueducts to bring the water down to reservoirs near the mission, then others to distribute it to either a large washing pool or to the fields. The system even included a small adobe filtering unit. One of the reservoirs, dating from the first decade of the nineteenth century, is still in use by the city water department.

The earliest large-scale irrigation undertaken by Anglo-American settlers in the West was built around Salt Lake City by Mormon settlers beginning in 1847. Through the power and authority of the church, it was possible for the Mormons to overcome capitalistic notions of individual and private enterprise, which were totally inadequate for the expensive and socially critical construction of irrigation works. Significantly, an insistence on having the actual users control the cooperative enterprises meant that a large degree of democracy prevailed.

Elsewhere, the government had to provide the incentive and context for large-scale irrigation. The earliest irrigators were those who had farms alongside waterways; they merely diverted some of the water into their fields. In California, however, years of mining activity had already greatly modified the traditional riparian water law that gave exclusive use of the water to those who owned the abutting land. In that state, the water belonged to whoever first appropriated it for a useful purpose, thus making it possible for farmlands far from the original river and under different ownership to be served. Between 1867 and 1877 the acreage of irrigated land in the state doubled as 611 irrigation ditches served 202,955 acres. By 1874 one system, the Kings River and San Joaquin Company, had a canal 40 miles long to bring river water to the fields. Because of the large investments in engineering works required, most of these early enterprises were under the control of large companies. Typically, dry-farmed wheat acreage would be bought up, water appropriated, and dams, headgates, and canals constructed, and then small farms sold to individual farmers for planting to vegetables or fruit. The Wright Law, passed by the state legislature in 1887, made it possible for individual farmers to join together in irrigation districts to raise funds and build their own works. Severe droughts in the 1880s brought the message of irrigation to the Great Plains as well.

By the end of the nineteenth century, many hundreds of miles of flumes, ditches, and aqueducts had been built in California to provide water for mining, agriculture, or the transportation of logs and other resources. In 1890 the Los Coches trestle in Southern California, 65 feet high and 1,774 feet long, also provided amusement for the local population. *Scientific American,* 62 (March 15, 1890), 161.

The number of farms in the seventeen western states that were benefiting from irrigation rose from 54,136 in 1890 to 109,298 in 1900, 159,801 in 1910, and 215,152 in 1920. The number of farms topped out at 283,089 in 1940, but the number of acres continued to rise for at least another thirty years, a clear measure of the consolidation of such lands into ever larger units. In part,

As observed in chapter 1, the victorious and the vanquished were often pictured as standing together in awe of the new technologies turning "wasted" nature into useful commodities. "How the West Is Putting Its Water Power to Work" depicted, in 1920, the production of hydroelectric power for an appreciative audience that represented both the Native Americans who had been unable to hold on to their lands and the engineers who put it to work for a thoroughly modern culture. *Scientific American,* 122 (May 8, 1920), cover illustration.

the continued increase came from the efforts of the federal Bureau of Reclamation, established in 1902 to further help the desert "blossom like the rose." Beginning with the Roosevelt Dam on the Salt River (serving Phoenix) in Arizona, and extending through Hoover Dam on the Colorado River and Grand Coulee on the Columbia, projects by the bureau brought irrigation and, later, hydroelectric power, to thousands.

The technical expertise needed to build these projects came from a host of engineers who, as in the mines, learned first on the job, then at colleges and technical institutes. In the 1880s the Technical Society of the Pacific Coast was created in San Francisco to provide a social and professional locus for western engineers, many of whom specialized in hydraulics. Working first with earth, then masonry, and early in the twentieth century with reinforced concrete, these engineers built dams of increasing beauty and daring: thin, arched structures that towered above canyon floors, impounding whole rivers. Some, like John Eastwood, were identified with particular varieties of these structures, in his case the economical and efficient multiple-arch dam. Eventually, the building of single canals came to involve the moving of earth in quantities dwarfing those of even the Panama Canal.

Settlers could go West on foot, horseback, or in covered wagons, those broad-wheeled "prairie schooners" that had developed as freight wagons around Conestoga, Pennsylvania, late in the eighteenth century. Large-scale lumbering, mining, and farming, however, required more adequate transportation for the machinery and other supplies flowing in, and the finished or semifinished products moving out. Lumber, wheat, and such base-metal by-products of gold and silver mining as lead, were of comparatively low value for their weight. As in the East, water was the first and preferred method of transport. On the West Coast, lumber mills flourished first on the coast, where boards could be put on oceangoing ships by lighters or "dog holes" (chutes precariously built along ocean bluffs where no harbors were available).

The major rivers of the West—the Columbia, Sacramento, Colorado, and the Mississippi with its tributaries, particularly the Missouri—were all served by steamboats as far upstream as water depths would allow. The first steamboat to appear on the Mississippi was that launched in 1811 by Robert Fulton and Robert R. Livingston. Within a generation, Mike Fink, the legendary river-brawler and keel boat-poller, was driven from the river. By 1831 there were five steamboats operating on the lower Missouri, and by 1842 the number had risen to twenty-six.

In the years after the Civil War, it was the railroad that colonized the West, or at least that part of the West lying between the Mississippi and the gold fields of California. Of the last dozen states of the contiguous United States to join the Union (that is, from Nebraska in 1867 to Arizona in 1912), all had railroads before they saw large numbers of settlers. Most of these roads were heading west toward the Pacific, leading Walt Whitman to sing:

In 1897, the first "offshore" oil wells in the nation were pumping at Summerland, south of Santa Barbara, California. About the same time, a large solar energy installation was put into operation in the town, but the abundance and versatility of oil helped create a twentieth-century culture heavily dependent on the internal combustion engine. Postcard from the author's collection.

I see over my own continent the Pacific railroad surmounting every
barrier,
I see continual trains of cars winding along the Platte carrying freight and
passengers,
I hear locomotives rushing and roaring, the shrill steam-whistle,
I hear the echoes reverberating through the grandest scenery in the world.[8]

In 1853 an all-rail route from Chicago east was finally completed, and the next year Chicago was linked to St. Louis. In 1856 the first railroad bridge across the Mississippi was completed at Davenport, Iowa. The most significant post–Civil War project was the building of the first transcontinental railroad, with the Union Pacific starting from the Missouri and building west, while the Central Pacific worked eastward from Sacramento, California. The two lines met on May 10, 1869, at Promontory Point, Utah. The Northern Pacific became the second transcontinental line in 1883. Nationwide, the miles of railroad in operation rose from 9,021 in 1850 to 30,626 in 1860, to 52,922 in 1870. In 1872 some 7,439 miles of track were added, a record not surpassed before or since, and by 1890 there were a total of 166,703 miles of railroad in operation. It was a massive engineering and manufacturing effort that, among others things, helped integrate the West and tie it closely to the eastern centers of population and manufacturing.

The railroad, along with such technologies as the six-shooter (in 1836, the first effective firearm for use by riders on horseback) and barbed wire (1874), the rectangular survey and the engineering schools, the powered hard-rock mining drill and the self-binding harvester, with a host of other devices and techniques, not only brought the Industrial Revolution to the American West but within a generation turned it into an extension of that Revolution. It was the filling out of a continent, but also the preparation for further extensions of American technology abroad, down the landmass through what was left of Mexico, into Central and South America, and across the Pacific; all a part of America's "Manifest Destiny." Thus Whitman chanted:

> A worship new I sing,
> You captains, voyagers, explorers, yours,
> You engineers, you architects, machinists, yours,
> You, not for trade or transportation only,
> But in God's name, and for thy sake O soul.[9]

8

EXPORT, EXPLOITATION, AND EMPIRE

SCHOLARS WHO HAVE CHOSEN to deal with the transit of technology across the Atlantic have quite properly concentrated on the movement from east to west. There had developed, at least by the end of the eighteenth century, however, a small but significant flow the other way, which, by the end of the nineteenth century had seriously challenged the technical supremacy of even Great Britain itself. By midcentury, that international trade in mechanical ideas that had characterized western Europe for at least two centuries was extended to include much of what are now called the developing areas of the world—Latin America, Africa, Asia (particularly China and Japan), and those parts of Europe, like Russia, in which the Industrial Revolution had been long delayed. This export of Western technology was a key ingredient in the rise of imperialism toward the end of the century and was fostered as much by Americans as by European peoples.

Throughout much of the nineteenth century the primary initiative for the transfer of American technology to Great Britain appears to have been taken by the patentors of new inventions who hoped to introduce them into the presumably lucrative British market. Britain was, after all, the "workshop of the world," and to gain acceptance there for one's invention was to give life to the old saw about sending coals to Newcastle. This process can be seen in one of the earliest and most interesting American Victorian careers, that of Joseph Cheesborough Dyer.

Dyer was born in Stonington Point, Connecticut, on November 15, 1780, the son of Capt. Nathaniel Dyer of the Rhode Island Navy, and died at the age of ninety at Manchester, England, on May 3, 1871, his life spanning those very years during which American technology was rising to world prominence. Dyer showed an early interest in mechanics, but at sixteen entered the countinghouse of a French émigré. As part of his work he visited England, first in 1802, and finally settled there permanently in 1811, to become the very model of a Victorian gentleman.

A nineteenth-century biographer cataloged his technical contributions:

> He devoted himself to mechanics, and was active in introducing into England several American inventions, which became exceedingly profitable to him and others. One of the first of these was [Jacob] Perkins's plan for steel-engraving (1809); then followed fur-shearing and nail-making machines (1810), and the carding engine (1811). [Robert] Fulton sent him drawings and specifications of his steamboat in 1811, and Dyer experienced many difficulties and discouragements in bringing the system into use in England. In 1825 he took out his first patent for a roving frame used in cotton-spinning, invented by Danforth and subsequently much improved and simplified by himself.[1]

In addition, Dyer somehow found time to help establish and write for the *North American Review* (1815) and the *Manchester Guardian* (1821), chair the Reform League, help found the Royal Institution and the Mechanics' Institution of Manchester, help establish the ill-fated Bank of Manchester and the Liverpool and Manchester Railway, and take an active part in the activities of the Anti-Corn League.

Dyer's success as a promoter of American technology began when he took a dozen or so entirely new machines from America to England in 1810, with the idea of introducing them to the manufacturers of that country. One such machine was for cutting furs from pelts, a useful device in the age of beaver hats. Dyer described his experience thus:

> In the year 1810 a model fur cutting machine was sent to me in London by a company in Boston, to be patented in England. It was stated to be the invention of a Mr. Bellows, who was unknown to me. The machine was adapted for shearing fibres from surfaces by the action of spiral cutters

> revolving and acting against a fixed straight cutter, so as to shear or cut fibres from the surfaces to which they are attached. I had a machine made and put into operation at a hat manufactory in the Borough [Manchester]; but the workpeople opposed its being used, which discouraged further attempts to bring it into use in that trade. The principle of it, however, was soon after patented for chopping straw, roots, &c., for which it was found valuable. Two or three patents were afterwards taken out for shearing the nap from cloth by the same action of spiral cutters revolving against a straight fixed edge, and many others have since appeared on the same principle, among which is that for mowing lawns.[2]

One of Dyer's greatest successes was with the machine for making hand cards, invented by Amos Whittemore of Boston. The latter had patented his machine in the United States in 1797, and two years later set out for England to introduce it there. What exactly happened has not been recorded, but his only biographer charged that Whittemore was somehow cheated of his rights and robbed of his reward. As was customary, British workers were blamed for opposing machines that would deprive them of their livelihoods. It was hinted that the machines were made in secret, so that an angry mob of workers could not destroy them before they were finished. All in all, the trip appears to have been a failure.

This machine was one of those with which Dyer began his patent operations in England in 1810. He obtained a British patent in that same year and subsequently patented improvements in 1814, 1825, and 1830, thus extending the period of protection far beyond the original expiration date of 1824. Through Dyer's exertions, the machine was sent over to the continent of Europe, and it was estimated in 1841 that the device was being manufactured by at least fifty different firms in France and Belgium. From these countries, in turn, machines were sent to Russia to equip the imperial textile factory at Alexandrofski. Indeed, they seem to have been sent to most countries other than the United States. Here, Whittemore sold his patent in 1812 for a reported $150,000 and by 1825 the manufacture and use of the machines was general throughout the country.

Without the kind of service offered by Dyer to American inventors, it was difficult to gain acceptance for American devices in the British market. Whittemore's initial disappointment is an example of this difficulty, as was the experience of Oliver Evans with his high-pressure steam engine. In 1794 Evans met a Mr. Joseph Stacey Sampson, of Boston, who had come to

Philadelphia seeking patronage for his improved method of making candles. Evans paid his passage and other expenses for a trip to England to try to introduce his invention there. In exchange, Sampson was to act as an agent for Evans. As Evans explained later, "I sent drawings, specifications and explanations, to England to be shown to the steam engineers there, to induce them to put the principles into practice and take out a joint patent for the improvement, in their names, which they declined, as they could not understand the project and had no faith in it."[3] Unfortunately, Sampson died in London and the information on the steam engine that he carried with him disappeared.

As the nineteenth century wore on, America steadily developed the ability to compete mechanically with the British. One early and important market turned out to be imperial Russia, a continental empire even more undeveloped that the United States but beginning its modernization only a few years later. There were at least two good reasons for this exportation: first, it was still illegal for British machine-makers to export their products from the British Isles. Although this was an attempt to retard the industrialization of other parts of the world, it gave Americans a clear advantage in some developing areas like Russia. The second reason was that the geographical needs and resources of Russia (vast distances and heavy forests, for example) were more similar to those of the United States than they were to those of Great Britain.

During the summer of 1843 it was announced that Henry Burden, of the Troy Iron and Nail works, had signed a contract with the agent of the Russian government for one of his patent spike-making machines. This was only one small example of what was going on on a much larger scale. A week later American newspapers picked up a letter to the *Liverpool Times* from a merchant then traveling in the United States:

> I have just seen a gentleman, who has travelled much in Russia; he was sent on a special mission by the United States government to the emperor of Russia, for the purpose of opening an intercourse between the two countries, for the supply of Russia with machinery for manufacturing. He came to England to order machinery to the amount of $200,000, but found he could not send it away on account of our laws prohibiting its exportation; in consequence this machinery was manufactured in the U. States, and sent to Russia; orders were then sent to the states for $500,000 worth. This was sent, and now they are making an almost

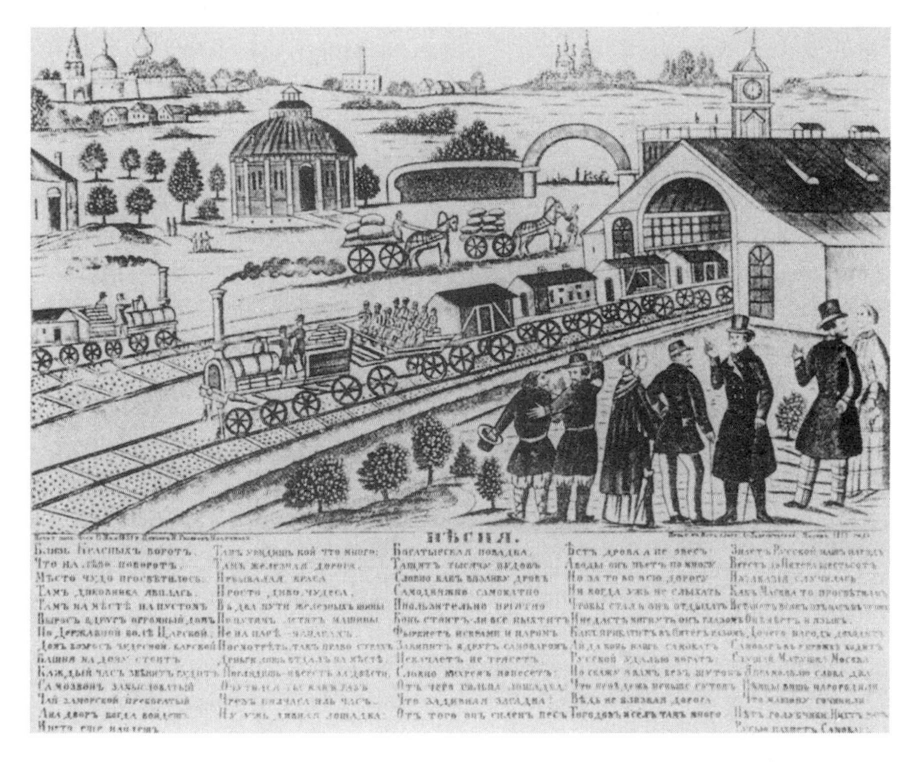

Early in the Railroad Age, American engineers, many trained at West Point, like Whistler's father, were hired to lay out and construct the important rail link between Moscow and St. Petersburg in Russia. American rolling stock was used as well. This contemporary Russian print captures some of the thrill of modernity that accompanied such developments. Albert Parry, *Whistler's Father* (Indianapolis, 1939), frontispiece. Reprinted by permission of Macmillan Publishers.

> unlimited quantity. This is the natural consequence of our absurd restrictive policy; and this is the progress America is making in all those things on which we are in the habit of priding ourselves. The American and Russian governments are on the very best terms; and they are carrying on a trade with each other mutually beneficial.[4]

The note of despair was not misplaced. Not only various machines but whole factories were being sent to Russia. The czar's emissaries decided that American railroad builders, expressing themselves as they did in wood rather than in stone, and using light and flexible rolling stock, were better prepared to conquer the vast reaches of Russia than were British engineers who built in quite different ways. Perhaps because, as Americans often claimed, the British often "overbuilt," the Americans also cost less. A number of prominent American railroad engineers, including George Washington Whistler, the father of the painter, spent years in Russia building lines. On one short but important route, that from Moscow to St. Petersburg, Americans not only built the right-of-way but also supplied 162 locomotives and nearly 2,700 freight and passenger cars.

Even England itself came to patronize American industry in certain selected areas. For one thing, some machines were themselves of American origin. Matthew Curtis, a Manchester machine-maker, told a parliamentary committee in 1841 that

> the greatest portion of new inventions lately introduced to this country have come from abroad; but I would have it to be understood that by that I mean not improvements in machines, but rather entirely new inventions. There are certainly more improvements carried out in this country; but I apprehend that the chief part, or a majority, at all events, of the really new inventions, that is of new ideas altogether, in the carrying out of a certain process by new machinery, or in a new mode, have originated abroad, especially in America.[5]

A further type of American success was represented by the famous American mass-produced clock. Such clocks were originally made of wood and therefore were subject to warping and were not exported overseas. When they came to be made of metal, however, a brisk export trade developed, one firm selling 40,000 clocks abroad before 1843. So cheap were these products of the American system of manufacturing, that they had hardly any competition. The British customs regulations required not only a declaration of worth of incoming products, but contained a clause allowing the Crown to buy the cargo at the declared value if fraud was suspected. The story was circulated that more than one cargo of American clocks was bought by customs officials at the dock, so absurdly low seemed the cost claimed on the manifest.

American technology was also exported to Britain through the purchase of entire factories. A prime example was the famous Enfield small arms factory operated by the British government. Until the mid-nineteenth century, small arms were manufactured in England mainly in small shops by a host of craftspeople working with apprentices. There was, as a result, no significant degree of interchangeability. In 1849 a select committee of the House of Commons was established to investigate the entire matter.

Impetus was added by the showing, at the 1851 Crystal Palace Exhibition in London, of six American rifles assembled from interchangeable parts. When the United States decided to hold its own Crystal Palace Exhibition in New York some time later, Joseph Whitworth and other prominent British technologists were appointed to a royal commission to attend and look particularly at small-arms manufacture in this country. In 1854, the British decided finally to set up a modern, state-run armory at Enfield. A new commission was sent to America to study the American system and to purchase the necessary machines. The initial order was for 131 machines, including 51 of the new American milling machines, valued at $105,000, and was given to Robbins & Lawrence, who made the famous Sharps rifles at Hartford, Connecticut. Before the commission left, it also placed an order for eight more machines from James T. Ames of Chicopee, Massachusetts. Ames was also asked to supply a complete set of jigs and gauges, which "though very expensive, the Committee considered it absolutely necessary to order, as it is only by means of a continual and careful application of these instruments that uniformity of work to secure interchanges can be obtained."[6] The factory was built and proved highly successful.

The Crystal Palace Exhibition that had so dramatically showcased the American rifles made with interchangeable parts was the brainchild of Albert, Prince Consort of the British Queen Victoria. The first true world's fair, the Crystal Palace was opened in London in 1851, housed in a glass and iron greenhouse-like building, which itself reflected both a modern machine aesthetic and, with its modular sections, the logic of the newly emergent American production methods. Some Americans looked forward to this international display of the peaceful arts: the *Springfield Republican* in 1850 predicting that it would be "a great test, full of glorious meaning in truth, and inevitable in the development of facts instructive in the morals, system of religion, modes of government, and intellectual progress of every nation which it may represent. . . . If we mistake not, the English will learn some important lessons from their western children, whom they still associate with savage life and whom many among them regard with dignified superciliousness."[7]

American exhibits, which were without official government support or subsidy, were late in arriving in London and at first made only a poor showing. Before the end of the exhibition, however, some rather startling displays were made. Col. Samuel Colt, in person, exhibited and lectured on his revolver. A. C. Hobbs of New York, who had charge of a lock display by the firm of Day and Newell, managed not only to win a large cash award for successfully picking the lock on the Bank of England but had the pleasure of watching the complete failure of any British lock-pickers to open one of his own locks, manufactured inexpensively by the American system. After the exhibition, British interests bought and set up an entire lock-making factory, complete with machines (what would now be called a "turnkey" project), in England.

One wet day on an English farm, a McCormick reaper took part in a sanctioned contest with other models, including some British ones. Not only did it work best, but it was considered successful by working at all, in view of the inclement conditions. The farm's owner is reported to have jumped up on a platform and said, "Gentlemen, here is a triumph for the American Reaping Machine. It has under all . . . disadvantages, done its work completely. Now let us Englishmen show that we appreciate this contribution for cheapening our agriculture, and let us give the Americans three hearty English cheers!"[8] And finally, on August 28, the American yacht named, appropriately, the *America,* decisively beat the British *Titania* in the first of what came to be called, for the winner, the America's Cup race. In all, American exhibitors won five Council Medals, including one to McCormick for his reaper, one to Charles Goodyear for his India rubber display, and one to Gail Borden, later more associated with canned evaporated milk, for his "patent meat biscuit."

The London *Times* generously editorialized that "Great Britain has received more useful ideas and more ingenious inventions from the United States, through the exhibition, than from all other sources."[9] And the London *Observer,* attempting to draw a political moral from the Yankee triumph, asserted:

> Our cousins across the Atlantic cut many degrees closer to the ground than we do in seeking for markets. Their industrial system unfettered by ancient usage, and by the pomp and magnificence which our social institutions countenance, is essentially democratic in its tendencies. They produce for the masses, and for wholesale consumption. No Government of favoritism raises any manufacturers to a preeminence. . . . Everything is

entrusted to the ingenuity of individuals, who look for their reward to public demand alone. With an immense command of raw produce, they do not, like many other countries, skip over the wants of the many, and rush to supply the luxuries of the few.[10]

It was sweet to Americans to gain praise from the British who, after all, had invented the Industrial Revolution. They were busily competing with that empire in the other markets of the world as well. In the age of Manifest Destiny, many Americans made little distinction between the territory within the nation's boundaries and the lands that lay just beyond, and that were "manifestly destined" to fall into American hands. Yankee and Middle Atlantic mechanics and manufacturers had for years treated the American South as an underdeveloped, semitropical land that neither built its own machines nor trained its own mechanics, and therefore needed northern help in establishing a technological infrastructure. The steam engines that ran Louisiana sugar mills in 1848 were built in northern manufactories, bought with money loaned by northern bankers, sent south in northern ships underwritten by northern insurance companies, and set up and operated with northern mechanics.

Just as Southerners saw the trans-American Caribbean islands as natural extensions of their slave economy, so northern businesspeople saw them as ripe for commercial exploitation. In 1851 the *Scientific American* noted in an extensive report that "Cuba is almost wholly supplied with machinists from the United States." The journal explained that

> there is in nearly every plantation in Cuba, a sugar mill driven by steam engines, built usually in New York or Boston. In these mills it is necessary to have some one well acquainted with machinery. The Cubans are not qualified for the situation, and the planters are forced to secure machinists from our country. Accordingly during the month of October, some hundred machinists of Boston leave for Cuba. The sugar crop commences soon after their arrival, and they are busily employed till June or July. Each man has a confidential negro whom he can leave in the charge of the mill, besides having as many assistants as he wishes. He is obliged to work but little, simply overseeing and directing.
>
> For this service he receives from six to ten ounces a month, varying from one hundred to one hundred and fifty dollars a month.
>
> In June, the crop is manufactured, and as the weather grows warm and unhealthy, the machinists return home, and spend the summer in a

> more pleasant climate. During the absence of the overseers, the mills are closed and repaired, and when the delegation again return in October, they find everything prepared for the commencement of the new season.[11]

Whether it was the "confidential negro" who repaired and serviced the machinery during the off-season is not clear, but the assumption that Cubans were "not qualified for the situation" must have been based on grounds other than that of technical competence.

In most of Latin America, North American engineers and machinists were operating in territories outside the formal boundaries of the American empire, despite the occasional filibustering expeditions in the nineteenth century or punitive raids by U.S. Marines in the twentieth. The *Baltimore Sun* noticed in 1850 that

> the bark George and Henry, bound from this port for the west coast of South America, carries out with her some matters of more than ordinary freight. Among her cargo is the machinery for two complete flour mills; one of them was built by Mr. Alfred Duvall, and the other by Mr. Thomas J. Matthews. All the iron work was made by Wells and Miller. The mills are to be run by water, no steam being used. Messrs. William Wiker and Chas. Thomas, millers, from Baltimore, also go out, with three millers from the Brandywine Mills, in Delaware, whose names we could not learn. One of the mills is for Delano, Ferral & Co., at Conception, Chile.[12]

Other parts of Latin America, like that contiguous to the Louisiana Purchase, fared differently. One-third of Mexico was annexed as war booty or by purchase as an outgrowth of the Mexican War, and in the 1870s and 1880s American mining and railroad engineers fanned out over the rest of Mexico in the employ of American capital. Here as elsewhere, the power of American technology, its engineering, and its machinery proved to be a powerful solvent of local control and national sovereignty.

The American adventure with "imperialism," however, is usually identified with the course and outcome of the Spanish-American War of 1898. The results of Admiral George Dewey's naval victory in the Philippines and the land campaign, highlighted by Teddy Roosevelt's "charge" up San Juan Hill, in Cuba, garnered a harvest of new fields for American technology, and capital, to exploit. The *Scientific American* knew exactly what to expect, and from whom. Editorializing on "Our New Possessions as a Field for Engineers," the journal advised:

> We have been asked to state our opinions as to the possibilities of our newly acquired possessions as a field of employment for engineers, both civil and mechanical. There is in our midst a large and rapidly increasing body of young men, graduates from technical schools and colleges, with more or less practical experiences in the shop or in the field, who think they see in Cuba, Puerto Rico, the Hawaiian Islands, and the Philippines an immediate field of employment of a more remunerative kind, and with opportunities for more rapid promotion, than are possible at home. The expectation is based upon the conviction that our possession or control of these islands will be followed by an immediate and extensive development of their natural resources, in the course of which the services of the civil and mechanical engineer will be in active command. . . .
>
> The location and construction of a railroad system in Cuba alone will call for the services of a very considerable force of engineers, and the rebuilding of sugar mills, the installing of electric light and power plants, the development of mines and other material resources of the island will present many excellent openings for young men in electrical and mechanical engineering.
>
> For the present, however, we would advise those who are contemplating a trip to one or the other of our possessions to stay at home and watch the course of events, meanwhile keeping in touch as far as possible, with such companies as may be formed for the exploitation of the West Indies and South Pacific possessions.[13]

Indeed, both civilian engineers and those with the U.S. Army Corps of Engineers did move quickly into the new territories, exploiting the resources of the interior regions and rebuilding, or initiating, the engineering infrastructure of utilities to make Manila and Havana modern provincial capitals. Sewers, safe water supplies, port facilities and warehouses, railroads and paved streets—all became progressive proof and justification of the benefits of American rule.

Repeated failures to build a canal across Mesoamerica were finally capped by success when the United States, after encouraging the secession of northern parts of Colombia and the setting up of the new nation of Panama, built the Panama Canal to link the Atlantic and Pacific Oceans. Opened in

The use of allegorical figures to represent the technological sublime continued into the twentieth century, investing machines and projects with mystical importance (while at the same time obfuscating their social and political meanings). Here, in "The Thirteenth Labor of Hercules," by Perham W. Nahl, the Isthmus of Panama is parted so that a small ship can proceed through the Panama Canal to the Panama Pacific International Exposition of 1915, for which the poster was originally produced.

1915, the project was celebrated with a Panama-Pacific Exposition in San Francisco. The fair was the occasion for the first coast-to-coast telephone call, called forth the establishment of West Coast sections of such organizations as the American Association for the Advancement of Science and the national engineering societies, confirmed the renaissance of the city after its disastrous

1906 earthquake and fire, and included a great engineering congress, marred only by the last-minute withdrawal of delegates from those European nations lately locked in the slaughter of the First World War. Most of all, however, the fair and the new interoceanic water link looked forward to a new American political and commercial presence in Asia.

Japan had long ago been "opened up" by Commodore Matthew C. Perry and the U.S. Navy. After the Civil War, the American engineer Benjamin Smith Lyman, who had trained at Harvard, the School of Mines in Paris, and the Royal Academy at Freiburg, was hired by the Japanese Kaitakushi (colonial office) to aid in the development of Hokkaido. In 1873 he made preliminary reports on possible oil resources and then went on to survey and map nearly the entire country, noting coal deposits, potential waterpower sites, and even customs and folklore. Lyman left in 1879 but the next year the Japanese government hired Col. Joseph U. Crawford, of the Pennsylvania Railroad, to build the first railway on Hokkaido, running 45 miles from newly developed coal mines to the harbor, and equipped it with rolling stock from the Baldwin Locomotive Works.

Overall development of Hokkaido was under the direction of Horace Capron, who had arrived in Japan in 1871 to be commissioner and adviser to the Kaitakushi. A former U.S. commissioner of agriculture serving under President Ulysses S. Grant, Capron was particularly concerned with agricultural development. He imported both nursery and livestock and set up model farms on which American tools and practices could be showcased. He introduced the strong Illinois breaking plow to cut through virgin prairie soil and in 1874 the first wheat was harvested by reaper on the island. N. W. Holt, of Dayton, Ohio, and an assistant built gristmills to handle the grain and sawmills, both steam- and water-powered. The latter was operated by water from a large, newly constructed dam, into the reservoir for which a new canal brought logs as well as water. The steam sawmill was capable of sawing 12,000 feet of boards a day and during its first season turned out about a million feet. In part this was to provide lumber, shingles, and sashes for the wooden dwellings that American technical advisers hoped the Japanese would begin to build instead of traditional "paper" houses. Overall, Americans provided a large amount of technical assistance to the administrators of the Meji restoration in Japan.

Two of Colonel Crawford's assistants in building the railway on Hokkaido were Japanese engineers trained at Rensselaer Polytechnic Institute in

New York. At the same time that the graduates of America's institutes of technology and colleges of engineering were seeking practical experience on overseas projects (much as they had gone west in previous decades), these same schools were playing host to engineering students from abroad. Opinion was divided on the wisdom of this. Some argued that these students would return to their countries and take over jobs that would otherwise have gone to American engineers. Others, however, argued that they were more likely to return home, gain positions of responsibility with their governments, and hire their old professors and classmates to undertake the actual work of technological development.

The class of '96 at Lehigh University may have been representative. Most seem to have worked with railroads or factories in the Middle Atlantic states,

This print from the turn of the century depicted the grand Higashi-Honganji Temple in Kyoto, Japan. Perhaps to balance the respect for tradition with an appreciation of the modern, the Shichijyo tram station was also shown in an insert. American engineers played an important role in the transfer of Western technology to Asia. Print in possession of the author.

By the early twentieth century, service on developments in foreign climes was considered an almost obligatory part of the training of the young American engineer. Tom Read and Harry Hazelton, the heroes of *The Young Engineers Series* of juvenile novels by H. Irving Hancock, went directly from high school to "practice and training in a local engineering office" before going "forth into the world to stand or fall as engineers." H. Irving Hancock, *The Young Engineers in Mexico* (Akron, 1913), book cover.

but some gained experience by working abroad. George Pomeroy Bartholomew, for example, fought in Cuba during the war and then, as he said, "looked around for more nations in trouble." He went to the Transvaal and worked in the mines until, on the day the Boer War broke out, he "left with the last load of white men for the coast."[14] Frederick Rawson ("Babe") Bartles "got the canal fever" and left for Panama in 1905, after working for railroads. In 1906 he was working on a reservoir for the city of Colon and claimed that two of his friends from the class of '96 were there as well.

The Panama Canal, the opening of which was celebrated in 1915, became not only a critical link in American shipping routes but also a symbol of the miracles American engineering practice (and armed force) could accomplish in what imperialists everywhere considered the "underdeveloped" world. Courtesy Museum of American History, Smithsonian Institution.

Moriz Bernstein went first to Nicaragua, where he worked on several dams and turned down an offered position as "provincial supervisor for one of the Philippine provinces." Robert Parsons Howell spent two years after the Spanish-American War in the employ of the Cuba Company, "locating and constructing on their railroad through the centre of the island." Victor Witmer Kline spent two years after graduation on the New York canals, then spent two more on the U.S. Deep Waterways' Survey and on the Nicaragua Canal Commission. Clifford Sherron MacCalla went to Sydney, Australia, to install streetcars and powerhouses, and James Gordon Mason, two years after graduating, traveled to the Transvaal to work for the Robinson Deep Gold Mining Co.

Rafael de la Mora, a native of Mexico, was a member of the Lehigh class of '96 who took three jobs in the United States to get practical experience before returning home. There he set up a consulting firm with another

Lehigh graduate named Lopez de Lara and accepted his brother Manuel, also from Lehigh, into the firm as well. George Homer Ruggles was working on the Panama Canal in 1906 while John Augustus Thomson reported in 1906 that he had been working in Mexico for nine years, "spent principally in acquiring a varied experience in mining, milling and general engineering work, such as is needed in frontier life."

Davis Sanno Williams was perhaps the most traveled of his class. He arrived in the Philippines at the end of the war. He "put up tons of English and United States steel in bridges and markets for the benefit of the little brown brethren; built sea walls, roads and all kinds of things as assistant city engineer of Manila, and had a good private practice besides." Hearing of railroad work in China, he went there "and became a locater, constructor, military commander, etc., etc., for the Canton-Hankow Railway. And here," he reported, "I had the time of my life for two and a half turbulent years, dodging pirates, sitting up at nights for tigers (which we never saw), scrapping with natives, and doing the heavy diplomatic with officials over numerous cups of tea." Though he believed that "there are lots of things to be done in China, and few people to do them," he returned briefly to the Philippines to build roads, then went back to "China, my true love." He claimed that "during my travels I have knocked against lots of 'Lehighers' in out of the way places. Manila has a few, Hong Kong, and Shanghai also, and the first Chink I met in Canton introduced himself as a Lehigh man. A fat, provincial looking old boy, most Chinese of all Chinamen, Lehigh, '80, and anti-American to the core, and, I think, a leader in the present trouble. But," he concluded, "generally the old boys are all right. Here's to them, and to you."[15]

Some quite prominent American engineers based a large part of their careers on overseas work. J. A. L. Waddell, a prominent bridge engineer, had been born in Ontario, Canada, in 1854. After studying in New York at the Rensselaer Polytechnic Institute and graduating with the class of 1875, he returned to Canada to work on various engineering projects. He soon returned to the United States to do bridge work and in 1878 joined the faculty at Rensselaer. In 1882 he took the post of professor of civil engineering at the Japanese Imperial University in Tokyo. While there he published *A System of Iron Railroad Bridges for Japan* (1885), in which he criticized the work in Japan of English engineers and the use of English bridges. The following year he returned to the United States, having been decorated by the emperor with the order of the Rising Sun, and resumed bridge work here. His biographer

J. A. L. Waddell, born in Canada in 1854, became not only the leading American bridge engineer of his time but one of engineering's most celebrated cosmopolitans. This formal photograph shows him wearing the many decorations he received from grateful governments around the world for the engineering advice and the structures he provided. J. A. L. Waddell, *Memoirs and Addresses of Two Decades,* ed. Frank W. Skinner (Easton, 1928), frontispiece.

declared that "many . . . bridges were designed by him throughout the United States, Canada, and Mexico, and for Cuba, New Zealand, Russia, and Japan." A truly cosmopolitan figure, Waddell belonged to the national engineering societies of Spain, Peru, and China; was knighted five times, in Japan, Russia, China, and Italy; and was a correspondent of the Academy of Sciences in Paris and of the Royal Academy of Sciences and Arts in Barcelona. Late in life he received yet another decoration from the emperor of Japan "as a recognition of his services in training, during thirty-five years, a large number of young Japanese engineers in his office."[16]

For Chinese students studying in the United States, Waddell had a message as much political as technical. China, he told them in 1924, was in a

"Slough of Despond" that only they could exorcise. Students returning to China should form a secret society to overthrow the corrupt government in power and establish a progressive, honest, and efficient replacement. Only after this could the new government borrow the funds needed to undertake the great engineering works that it required. Waddell even had a name for the society, the "Purer China Association." Without this regeneration, he warned, European powers would partition the country because it was rich in resources and the rest of the resources in the world had already been expropriated:

> Owing to the phenomenal development of Japan and to the wonderful progress she has made in all lines of engineering, there is not much now that I can do professionally in her interests; but it is otherwise in the case of China, for in that country I can see numerous ways in which I could be of great service professionally—not only in the development of new work, but also in the correction of numerous engineering mistakes made in the past by engineers of foreign countries whom China, unfortunately, had employed.[17]

"Young Men of China!" he exclaimed, "Harken to Me and Heed What I Say!"

Herbert Hoover, later to become president of the United States, was another peripatetic engineer at the turn of the century. Born in Iowa in 1874, he graduated, in geology, from the new Stanford University with its first class in 1895. After a brief apprenticeship in the gold mining regions of California and Nevada, he took a position with the British firm of Bewick, Moreing and Co. in 1897 and left for the towns of Coolgardie and Kalgoorlie in the gold region of Western Australia. A year later he was sent to China, where he was caught up in the Boxer Rebellion. Over the next years his itinerary took him back to Australia and to New Zealand, to South Africa, Burma, and Russia. In 1914, after seventeen years working primarily as an engineer, he had earned a large fortune and gained such prominence that he could devote himself thereafter to public service, first in Belgium Relief, then in the Department of Commerce, and finally in the White House.

In 1891, seven years before the Spanish-American War established the United States firmly in the Far East, the prominent engineer Octave Chanute looked beyond the maturing domestic railroad network and confidently asserted that "there is still a good deal for the railway builders and organizers to do, and foreign fields may yet be opened to the energy of Europeans and North Americans, should some of the Asiatic nations, like China, for instance, enter upon an epoch of railroad construction, or have the good fortune, like India, to fall into strong hands."[18] Looking back over half a century, Chanute could have pointed to many areas—such as California, Alaska, and the Philippines—that had indeed fallen into the strong hands of America. Other areas—such as Canada, Cuba, and the southern two-thirds of Mexico, which survived after its war with the United States—never did become part of the political domain of the nation. In some ways, it hardly mattered. Despite competition from other engineers and capital, particularly British, Americans poured into these areas to open mines, build factories, and create the railroads and port facilities necessary to exploit them.

While its period of classic "imperialism" may have been short-lived, attempts to extend American hegemony abroad through technology continued. In his inaugural address on January 20, 1949, President Harry S Truman advocated four points that would inform and guide his stridently anti-communist foreign policy for the next four years. Point Four, as it came to be called, noted that much of the world was living in what the president called "conditions approaching misery." To combat that misery, and the attractiveness of communism that it presumably led to, he declared that "the United States is preeminent among nations in the development of industrial and scientific techniques" and proposed that "we should make available to peace-loving peoples the benefits of our store of technical knowledge in order to help them realize their aspirations for a better life. And in cooperation with other nations, we should foster capital investment in areas needing development." Although the American technology that the president had in mind was capital intensive and could not be acquired without massive capital investments, he insisted that the "old imperialism—exploitation for foreign profit—has no place in our plans. What we envisage is a program of development based on the concepts of democratic fair dealing."[19] It was, therefore, the overseas analog of his Fair Deal program at home.

After a century of increasingly successful American technological penetration of the rest of the world, Point Four was hardly a surprise. Truman's Fair Deal rhetoric and anti-communist motivations bore a striking family resemblance to the reform zeal and Manifest Destiny of the antebellum decades of the mid-nineteenth century. From the antebellum South, through the Far West, into Mexico and Cuba, and across the Pacific to Hawaii and the Philippines, underdeveloped resources, preindustrial societies, less powerful governments, and people of color proved a powerful magnet to a maturing American technological base that at the same time was challenging the "workshop of the world" within its own boundaries.

IV

TECHNOLOGY AND HEGEMONY

9
THE COMING OF SCIENCE AND SYSTEMS

SCIENCE AND TECHNOLOGY had never been completely separate, and no tool or technique had ever existed in complete isolation from others. Nevertheless, as the nineteenth century turned into the twentieth, more and more technologies became transformed by science, and the centralizing effects of modern life gathered them into systems for more effective management. The way in which the nation dealt with natural resources and work, the government of cities, and even the generation of technological change itself became heavily infused with the methods, results, and spirit of the sciences.

As a result, many technical processes and techniques that had been traditional and widely practiced became instead rationalized and concentrated in fewer hands. Activities that in the nineteenth century had been widely available to Americans tended in the twentieth to be increasingly open only to those who were scientifically trained and formally permitted to enter them. Knowledge of the sciences, available primarily through colleges (which were largely confined to white, Anglo-Saxon, middle- and upper-class males), became the device by which technologies were controlled by some and denied to others. A kind of "wise-man's burden" dictated that those who knew should define and direct the efforts of those who did not. Capital retained its control of the factors of production but increasingly exerted that control through the offices of men (seldom women) who claimed to be acting only in accord with natural laws. One major stream of American Progressivism was that searching for social justice and humane institutions. Another was an effort to rationalize and manage society in the interests of stability and order.

Although American technology was transformed by and served the latter urge, many Progressives were firm in their belief that social justice and humane institutions could only grow out of stability and order. In a period torn by class struggle, violence in the workplace, cultural diversity, and general social unrest, "science" seemed to hold the promise not only of efficiency but also of impartiality and even inevitability. Technology appeared to be an effective and uncontroversial substitute for politics. What the new system of scientific technology delivered, however, was a vast increase in goods and services at the cost of a diminished degree of freedom and meaning. The vague but powerful notion of "efficiency" became the watchword of the Progressive Era.

Theodore Roosevelt, who in 1898 had led a charge up San Juan Hill in Cuba, was in the White House three years later, leading charges there as well, against all manner of "special interests" and old inefficient ways of doing things. His presidential administration (1901–9) stood preeminently for the conservation of the nation's natural resources, and this activity was seen by the president as a key to "the larger question of national efficiency." The involvement of the federal government in the discovery and exploitation of the nation's natural resources was nothing new in the administration of Roosevelt. A Scots chemist visiting America in 1850 commented that "so far the encouragement given to positive and experimental science by the general Government, as well as by the Legislatures of the several States, has been exceedingly creditable to all, and has been evidently dictated by an enlightened desire to develop, as early as possible, the boundless natural resources of the broad regions they govern. The States," he continued, "have nearly all employed scientific men to make geological and other surveys, more or less complete, and the general Government has done the same for the territories."[1]

Before the Civil War, western surveys and exploration had largely been in the hands of the military, but after 1865 more and more civilian responsibility was exercised. Such expeditions as Clarence King's 1867 geological survey of the fortieth parallel, Lt. George M. Wheeler's 1869 reconnaissance

The conservation movement in the United States began in those federal bureaus that had responsibility for natural resources. The diminishing New England fisheries led to the establishment of the U.S. Fish Commission, which carried on research in support of the fishing industry. This scene of workers in the commission's laboratory shows microscopes on the tables and specimens on the floor. *Harper's New Monthly Magazine,* 291 (July 1874), 214.

of Nevada and Utah, and John Wesley Powell's 1870 geological and topographical survey of the Colorado River of the West greatly extended knowledge of just what resources the West had to offer. By 1886 the federal government had spent some $68 million on such efforts. In 1879 much of this work was consolidated into the new U.S. Geological Survey.

Meanwhile, new conservation agencies were beginning to take shape. In 1871 the Congress set up the Fish Commission to look into the disappearing commercial fishery off New England and placed it under the ichthyologist Spencer F. Baird, secretary of the Smithsonian Institution. In 1896 the Department of Agriculture's old Division of Economic Ornithology and Mammalogy became the new Division of the Biological Survey, which was to carry out fundamental work in natural history to better understand the fast dwindling ranks of the continent's flora and fauna. It was the Forest Service, however, that came to lead the conservation forces in the federal government.

Set up in the 1880s under the German-trained Bernhard E. Fernow, the agency husbanded its small appropriation and capital of political support by concentrating on basic research, which, if it did little immediate good, at least avoided immediate conflict. In 1891 Congress authorized the president to withdraw the disappearing forestlands of the country from government sale, and within two years 17.5 million acres had been set aside, but with no real provision made for their management. Legislation of 1897, following a report from the National Academy of Sciences, laid down the groundwork for a management plan that would include research, and Gifford Pinchot was induced to take over the Forest Service (in 1898) from Fernow. Yale University's new School of Forestry, financed in large part by Pinchot family funds, provided a trained cadre of foresters and gave the service a firm academic base. When Roosevelt became president in 1901, after the assassination of William McKinley, the ground was laid for a broadly conceived government attack on the problem of disappearing natural resources.

Theodore Roosevelt was probably the first president since John Quincy Adams to have a good grasp on what science was and a faith in what it could accomplish in the public interest. He quickly made conservation a key to his administration: in no other area could "disinterested" scientific fact displace political "prejudice" with such startling and beneficial results as in the field of resource policy. There was opposition, of course, from western members of Congress who saw conservation as a hindrance to economic growth, from small exploiters who lacked the resources and stability to invest science in long-range prosperity, and from preservationists like John Muir who realized that "scientific" destruction of the wilderness was destruction still.

The conservation scientists and engineers, operating especially from the Inland Waterways Commission appointed by Roosevelt in 1907, prevailed upon the president to call a conference of the governors of the several states to meet at the White House, on May 13–15, 1908. This was the first of what has since become an annual affair. Besides the governors, the president invited representatives from business, labor, the Congress, and a host of scientific and technical societies. Opening the convocation, the Reverend Dr. Edward Everett Hale prayed: "Father, for this we have come together. Thou hast made for us the Paradise through which these rivers flow. Now give us the strength of Thy Holy Spirit that we may go into this garden of Thine and bring forth fruit in Thy service."[2]

Perhaps Hale misunderstood the purpose of the meeting. For many generations Americans had been bringing forth the fruit with such success that the garden was fast becoming emptied of its bounty. Roosevelt in his welcoming address came nearer to the mark. Conservation, he said, "is the chief material question that confronts us, second only—and second always—to the great fundamental questions of morality."[3] In fact, as he believed, the two were not separable. Waste was immoral, and efficiency was the very soul of morality.

The danger of running out of resources seemed a real one to the assembled delegates. Roosevelt warned that "the natural resources of our country are in danger of exhaustion if we permit old wasteful methods of exploiting them longer to continue." The president of the American Institute of Electrical Engineers told the delegates that under what he called "present wasteful conditions," the nation's forests would be gone in twenty-five years, its best grades of iron ore would be gone in seventy-five years (annual tonnage used had risen from 5 million in 1870 to more than 55 million in 1907), the coal supply might last another century, but the soil was so badly violated that yields per acre were already going down. "Perhaps no other body of men," he added, "come quite so close in contact with the problems involved in this historic Conference on the Conservation of our Natural Resources as Engineers," and speaking for them, he pledged "their hearty and disinterested cooperation in this great patriotic movement."[4]

One of the delegates present was Charles S. Howe, president of the Case School of Applied Science in Cleveland, Ohio, and a representative of the Society for the Promotion of Engineering Education. Reporting back to his organization, he called the meeting "one of the most notable ever held in the country" and pointed out the wisdom of having engineers in attendance. "The engineer," he reasoned, "adapts the forces of nature to the use of men and this adaptation should be done both economically and efficiently. It is not enough to show that a certain force can be made to work when a machine transforms raw into finished product. The work must be done efficiently." Similarly, the "work of conservation is the work of the engineer." "I believe," he continued, "the engineers of the country are capable of solving these problems, and that if they are given the necessary governmental and private aid that the problem of the conservation of our natural resources will be solved." This task, he concluded, "must be accomplished through new inventions. This means that the engineer of the future must be able to do more than the simple engineering work which comes to him from day to day. He must be so thoroughly trained in the principles of science and applied mechanics that he will be able to discover new processes and accomplish old results in new and more economical ways."[5]

The excitement of these engineering leaders was well founded. Members of their profession were subject to potentially conflicting allegiances. First, as professionals, engineers sought to serve the public good. Second, they of necessity had to serve their employers, and these, in turn, were increasingly large and bureaucratic corporations. Third, they had to do so in terms that served their own interests. Engineers, like everyone else, sought influence, prestige, and affluence. For once—because the future of the nation so obviously and unambiguously depended on the conservation of its natural resources, because it was in the interests of the largest corporate resource companies to raise the cost of entrance into their industry as well as guarantee the future viability of their own activities, and finally because conservation made a heavy call upon scientific and engineering expertise and thus raised its reputation and remuneration—all three goals appeared to coincide.

The widening of the conservation movement after the White House Conference, from a scientific and engineering reform into a moral crusade, picked up necessary popular support but also worried many engineers. One, deploring popular notions "drawn from general statements, which are often the product of imagination," told the American Institute of Mining Engineers that "we may justly feel some resentment at the harsh criticism which is now being so generally aimed by the press and the public at technical men. And this is partly true likewise," he continued, "of the strictures so indiscriminately passed upon the corporations which are instrumental in developing the country's natural wealth."[6]

Denunciations of waste were easily turned to condemnation of the wasters, but this engineer preferred to point to the progress that had been made. By way of example he cited recent advances in the design of blast furnaces that cut fuel consumption from 37 hundredweight of coke per ton of pig iron in 1875 to only 22 hundredweight per ton within a few years. It was in this sense that engineers believed inventions would solve the problem of conservation. While some engineers took a narrow view of their responsibilities, arguing, for example, that "it is neither our part nor within our power, as mining and metallurgical engineers, to reconstruct society or renovate the world," others made an exception when the interests of capital were concerned. One chemical engineer in 1909 urged his colleagues to get involved in conservation so that they, too, might help in showing "how the wholesome influence of conservation can be applied so as to broaden and extend the scope of the industry, to maintain and add to its remunerative character and to give it stability and promise of permanence."[7] Not very far away lurked the image of society as a finely balanced, well-oiled, and efficient machine. In 1916 an

A 1912 cover of *Scientific American* promised "wealth from waste," celebrating the power of industrial research to create new products from formerly wasted by-products. The scene also powerfully demonstrates that the new, twentieth-century venue for engineering conquest was the laboratory rather than the distant frontier. *Scientific American,* 106 (June 15, 1912), cover.

editorial in the influential journal *Engineering News* declared that "while the politicians have used the conservation movement for political purposes, engineers, who better than any other class appreciate the vast importance of real conservation to the public welfare, have continued to agitate the subject."[8] While the conservation crusade faded, like other parts of the Progressive Era, the more efficient utilization of natural resources continued during the 1920s to be the goal of a number of key governmental scientific and engineering bureaus and became again a main feature of the administration of the next Roosevelt to become president.

While the conservation movement promised to rationalize the exploitation of raw materials, another factor of production, work, was reformed by the movement for Scientific Management. Lockouts, strikes, pitched battles between workers and the well-armed private armies of employers, and growing political successes of the Socialist party all pointed to the fact that class struggle was a reality in America, no matter how un-American it might seem to some. The capitalist desire to discipline labor found a powerful expression in the work of the mechanical engineer Frederick Winslow Taylor. Once again, the danger and inefficiency of "politics" were to be avoided by the "neutral" adoption of science. As Taylor testified before a congressional committee in 1912, "the great revolution that takes place in the mental attitude of the two parties [i.e., capital and labor] under scientific management is that both sides take their eyes off of the division of the surplus as the all-important matter, and together turn their attention toward increasing the size of the surplus until this surplus becomes so large that it is unnecessary to quarrel over how it shall be divided."[9]

Born in Germantown, Pennsylvania, in 1856, Taylor graduated as a mechanical engineer from the Stevens Institute of Technology in 1883, the year before he became chief engineer of the Midvale Steel Works. His background of comfortable wealth, poor health, and apparently compulsive behavior all have tempted scholars to find deep psychological bases for his later system of scientific management, but its wide appeal during and after his lifetime was due in large measure to its reinforcement of a national inclination to try to solve social problems through technical means.

While working at Midvale Steel, and later as a private consultant, Taylor worked out, applied, and tirelessly proselytized for his system of scientific management. First presented in addresses to the American Society of Mechanical Engineers, which he served as president in 1906, Taylor's ideas on the subject were brought together in 1911 in a slim book entitled *The Principles of Scientific Management.* The book was intended to point out, first, "the great loss which the whole country is suffering through inefficiency in almost all of our daily acts." Second, it was to "try to convince the reader that the remedy for this inefficiency lies in systematic management, rather than in searching for some unusual or extraordinary man," and finally, "to prove that the best management is a true science, resting upon clearly defined laws, rules, and principles, as a foundation." In a sweeping claim that was to justify the efficiency craze that followed, he ended by asserting that "the fundamental principles of scientific management are applicable to all kinds of human activities, from our simplest individual acts to the work of our great corporations."[10]

Taylor's system grew out of his practical experience in machine shops as well as his class biases and, perhaps, psychological horror of things out of control. The typical machine shop of his time would have been poorly lit and noisy, with long rows of machine tools taking their power from leather belts attached to overhead shafts. Tools and piles of materials lay strewn about and workers carried out their tasks in their own way, hallowed by tradition and learned by emulation. Orders arrived in the office, were transferred to the shop, and in the fullness of time (a time also established by tradition and maintained by worker control of the shop floor) the finished product would appear. At times, the foreman would act as an independent contractor, producing the wanted goods without supervision from management. Taylor believed that "among the various methods and implements used in each element of each trade there is always one method and one implement which is quicker and better than any of the rest. And," he continued, "this one best method and best implement can only be discovered or developed through a scientific study and analysis of all the methods and implements in use, together with accurate, minute, motion and time study. This involves the gradual substitution of science for rule of thumb throughout the mechanic arts."[11]

With clipboards and stopwatches, and later with motion-picture cameras as well, the new efficiency engineers recorded the traditional methods of work, broke them down into uniform bits, and reconstituted them into new patterns that became the work processes enforced by management. As Taylor wrote, "the managers assume . . . the burden of gathering together all of the traditional knowledge which in the past has been possessed by the workmen and then of classifying, tabulating, and reducing all this knowledge to rules, laws, and formulae which are immensely helpful to the workmen in doing their daily work."[12]

Workers were not always grateful for this help. Years of patiently acquired skill could be lost overnight, leaving workers with nothing to sell but their time and muscle. When Taylorism was really successful, the resulting work process was rationalized to the point where skill, initiative, and control

The notions of conservation and efficiency converged in a desire to control the environment for maximum production. Even the environment indoors became something that could be controlled through technology. In this case, the Carrier air-conditioning company promised to "manufacture" weather so that it could make "every day a good day." Carrier Engineering Corporation, *The Story of Manufactured Weather* (n.p., 1919), title page.

were all reduced to their lowest possible level, and workers became interchangeable parts in the productive machine, contributing time and effort to processes they neither controlled nor understood. For many, the prospect of a small share of the increased productivity in the form of higher piece-rate wages was poor compensation for a work experience drained of initiative, satisfaction, and meaning.

In 1911, the same year in which Taylor's *Principles of Scientific Management* was published, the Society to Promote the Science of Management was established, called, after his death, the Taylor Society. Curricula in "efficiency engineering" or "industrial engineering," or any one of dozens of other titles were established at schools throughout the country. Henry L. Gantt, one of Taylor's leading pupils, developed his famous "flowcharts," which allowed management to track the progress of production, and was active in improving efficiency in such World War I agencies as the Shipping Board and the Ordnance Bureau. After his death, his charts found favor in the new Soviet Union, where the sudden need to plan a chaotic, sprawling, and rapidly industrializing economy was all but overwhelming.

Another apostle of Taylor was the self-taught former bricklayer Frank Gilbreth. He developed a system of breaking action down into eighteen basic movements, which he called "therbligs" (Gilbreth spelled backwards) and pioneered the use of the camera to record tasks to be studied. The father of twelve children, his life was humorously chronicled in the 1948 best seller *Cheaper by the Dozen.* His wife, Lillian Gilbreth, earned a doctorate in psychology at Brown University and worked with her husband until his untimely death, whereupon she attempted to carry on their engineering practice. Such was the prejudice against a woman in engineering that her very considerable talents and experience could find acceptable outlet only in the "female" sphere of housework, where her "kitchen efficient" became one of the best-known examples of the application of scientific management to the home.

When the respected British observer Lord Bryce described America at the turn of the century, he called its cities the most conspicuous failure of the nation. From the standpoint of the middle class, power in American cities had tended to concentrate at opposite ends of the social spectrum: votes, and therefore political power, in the masses of working-class citizens, many of them foreign born; money, and therefore economic power, in the hands of a few capitalists who controlled the industrial base of the cities. Since, in theory, those with the votes should control the government, it was apparently necessary for corruption to bring together what constitutions and charters had set asunder. Specifically, business leaders and the new professional groups who served their interests found that American cities were being run in a most unbusinesslike way. They were, in a word, grossly inefficient.

Much of the business of a city concerned its engineering infrastructure: streets, sewers, bridges, streetcar lines, water supply, electrical supply. These were in the hands of mayors and city councils that were themselves in the hands of political parties widely seen as corrupt and incompetent, headed by "bosses" who mediated between the working class, ethnic masses who had the votes, and the capitalists who had the money and controlled the means of production. For Progressive reformers, the idea of replacing this "political" leadership with technical expertise appeared to be a logical extension of the general movement for efficiency.

Engineers had passed through three stages in their relationship to American cities. In the colonial period and well into the life of the new nation, engineers, even if they were available, were largely ignored. Water supplies came from individually dug wells, and streets, even if laid out by surveyors, were unpaved. Sewers were nonexistent: pit-privies provided for human waste and storm water was simply allowed to run off. Wharfs and public buildings were built by local skill along traditional lines, and gas, electricity, and public transport were unknown. As problems mounted along with growing populations and technical advances, engineers were often contracted with to perform certain limited and well-defined tasks, such as providing a new waterworks or digging a canal. When the task was completed, the engineer moved on to another city and another job.

Eventually, some cities became large enough to need the services of resident engineers on a continuing basis. The career of Rudolph Hering illustrates this third stage of relationship. He returned to the United States in 1868 after studying engineering in Dresden, Germany. His first job upon arrival was carrying a chain on the survey then being made of Prospect Park in Brooklyn: he had been trained specifically in bridge work but had been advised to take any job he could get. The Prospect Park job had attracted people of considerable stature, including the chief engineer, Joseph P. Davis, later city engineer of Boston and after that an engineer with the Bell Telephone Co. Fredrick Law Olmsted and Calvert Vaux, fresh from their triumphal work on Central Park in Manhattan, were the landscape architects.

Hering next went to Philadelphia, where his father was a prominent and well-acquainted physician, and got a job in charge of a survey party working on the new Fairmount Park. Family connections then got him work as astronomer with an expedition sent out to draw the boundaries of Yellowstone National Park, but after a year he returned to Philadelphia where the city engineer put him, at last, in charge of the Girard Street bridge. In 1875 he was transferred to the city sewerage department as an outdoor assistant, in charge of the construction of main sewers and bridges, where a major part of his work was keeping contractors honest. He spent the rest of his distinguished career in city employment, on one occasion traveling to Europe at the expense of the federal government to study sewerage problems and solutions, and became widely known as an able, honest, and devoted engineer and public servant.

Such individuals as Hering did much to raise the stature of city engineers and to convince cities, on the one hand, that engineers were useful to have on the payroll, and engineers, on the other, that city service was legitimate and professional employment. Nevertheless, many remained unconvinced. The *Engineering News* editorialized in 1892 that the lot of the independent consulting engineer was the happiest of all. At the same time, they warned that "our strong advice to every young engineer is to avoid working for cities if he can, as the meanest and most uncertain kind of professional employment, with some honorable exceptions. After that avoid working for states, and after that working for the general government. It is better as a rule to accept a considerably lower salary from a respectable corporation than to work for either of these three."[13] The cause of this unhappy advice was the prevalence of political interference with engineers, and the pressure on them in many cities to subvert their professional opinions to fit the demands of often corrupt political expedience.

Unfortunately, and perhaps more important, it often worked the other way around as well. Morris L. Cooke, a leading mechanical engineer and reformer of the profession, made many enemies within mechanical engineering circles with his charges that cities, seeking advice on engineering matters, were at the mercy of engineers working for those "respectable corporations" that *Engineering News* so admired. "Some months ago," he wrote in 1915 while working for the city of Philadelphia, "I wanted to retain an engineer fully posted on the details of a certain sub-division of railroad operation. It was extremely difficult to find a man without recognized affiliations which would preclude his retention. Again I am informed that there are no asphalt experts who do not receive retainers from the manufacturers." Outraged fellow-engineers met his charges by denouncing him as being "unworthy of a place in the profession," but the problem was both real and systemic.[14]

The warnings of *Engineering News* and the accusations of Cooke were indicative of a fundamental revolution in American cities. The character of cities was rapidly changing, as were the power alignments in the American economy. In the context of the Progressive Era, the engineer came to the city in a fundamentally new role—as the instrument of redemption of civic virtue. Municipal reform was an important part of the Progressive movement and its character drew heavily upon the tone and direction given it by engineers.

The country's founding generation had established a government based on the notion of balanced powers. By the end of the Gilded Age, however, some power had become so disproportionately large that it could no longer be realistically balanced. Almost unimaginable accumulations of capital worked through increasingly complex and arcane business systems to shape the nation into a corporate state that could be little influenced by farmers and urban workers. With political power formally vested, through the vote, in these very people and economic power concentrated in the hands of quite another class, corruption and incompetence were hallmarks of many an American city. At the same time, a rising urban middle class, including the new and growing professions of teacher, accountant, librarian, and engineer among them, sought to work out the dynamics of service in a modern, industrial society.

Together, corporate leaders and the new professionals sought to wrest control of cities from the working class and the political bosses who claimed to represent them. One way of doing this was to make more elective officers run at large, so that neighborhood power was diluted by the funds and celebrity needed to poll votes citywide. Another technique was to try to make the city more "businesslike" by turning over the supervision of city services to paid professionals. No individual better epitomized this effort than did the city manager.

According to Leonard D. White, who made a study of the subject in 1927, "the origin of the council-manager plan is imbedded in the revolution of the civic and business interests of the American city, aided and abetted by various forward-looking groups, against the waste, extravagance, and sometimes corruption which characterized 'politician' government of the last century." The "constant growth of the physical size of the city and the constantly greater complexity of its problems made 'amateur' government more and more of an anomaly."[15] Indeed, White's assessment of the larger implications of the city-manager movement was perceptive:

> Governmental problems have become intricate and ever more insistent. They call for solution with the aid of science, not with the wisdom of a ward politician. The amazing mobility of the American people leaves no community a law unto itself; each and all are responsible for their own

> good government to the larger whole of which they are a part. What the whole world is witnessing is the emergence of government by experts, by men and women who are trained technicians highly specialized to perform some service by scientific methods. It is indeed a fair question, whether we shall not be forced to reinterpret American government as a means for utilizing the services of experts in the performance of ends democratically defined.[16]

Just as natural resource policy and the industrial work process must be rationalized by "experts" following scientific law, not popular will, so were cities to be governed by the "experts" in the name of efficiency and businesslike procedures.

The nation's first city manager was the engineer Charles Ashburner, born in India in 1870 to the wife of a British army officer. He was educated in England, France, and Germany, where he took an engineering degree from Heidelberg, and then came to the United States. Here he held, in rapid succession, jobs as a civilian employee of the U.S. Army Corps of Engineers working on the James River, with the highway departments of both the state of Virginia and the federal government, as an engineer with railroad and electric companies, as an independent contractor and consultant, and (perhaps most significantly) as engineer in charge of a company town in Virginia.

His career as a city manager grew directly out of advice he gave to the city of Staunton, Virginia. The city had a leaky dam that a contractor had claimed would cost $4,000 to repair. When a friend of Ashburner's on the city council asked his advice, Ashburner said the job could be done for $737, which turned out to be within a few cents of the actual price. Impressed, the city hired him as general manager at a salary of $2,000. He stayed at Staunton from 1908 to 1911, then moved on to the city of Richmond, Virginia, though not as manager. From 1914 to 1918 he was manager of Springfield, Ohio; from 1918 to 1923 he was manager of Norfolk, Virginia; and in 1923 he became manager of Stockton, California.

Ashburner was not a politician but a builder. In each city he managed, he set about creating or expanding the engineering infrastructure of commerce. In Staunton he specialized in streets and substructures; in Norfolk he laid out a modern system of streets, a great rail-water terminal, a grain elevator, and a new water system; and in Stockton, he built an elaborate system of flood control, a new city hall, and most important, he pushed through a deepwater canal to San Francisco Bay. When interviewed by White, he exclaimed with great vigor: "By God, I go into a town to build! When I can't build, I get out!"[17]

The career of Ashburner illustrates several important strengths and limitations of the city manager scheme of government. First, it concentrated heavily and appeared to work best on those problems of engineering infrastructure in which engineering expertise was a necessary component. On matters more reasonably considered to be those of social infrastructure—law enforcement, welfare, and education, for example—the technocratic approach proved less satisfactory. Second, since one motive for adopting the manager system was usually to avoid or circumvent "politics," it worked best in middle-size cities in which dissident political opinion was either absent or marginalized. Thus in large cities such as New York and Chicago, large and well-organized working-class movements were able to articulate and defend alternative political agendas, which put a larger value on political compromise at the expense of merely technocratic expertise.

In the smaller cities, however, Ashburner's example spread rapidly. By 1914, just six years after he went to work in Staunton, there were already forty city managers in the nation and they had just formed a professional association to represent their interests. At their annual meeting in 1917 it was remarked that between 80 and 90 percent of the managers attending had been originally trained as engineers. Their training in physical infrastructures and their presumed political neutrality (gained in large part from their study and use of science) both recommended them to cities seeking efficiency supported by consensus. A careful count by White in 1927 showed that of a total of 863 city managers then employed in the United States, 398, a group nearly four times larger than any other defined by occupation, were engineers. The second largest was defined as "business or executive classes" with 103. Only 16 lawyers were managers, and 2 were officers in the U.S. Army. Of the 398 engineers, 162 had been trained in civil engineering, 77 were unclassified, 21 were electrical engineers, and 5 mining. Thirty-four were listed as "city engineer," and one, surprisingly, was credited with being "owner of the city water department."

There are no overall statistics to indicate how well these engineers did as city managers. Col. Clarence O. Sherrill, an army engineer who had been a military adviser to President Calvin Coolidge, was made manager of Cincinnati, Ohio. In 1927 it was said, with obvious understatement, that in that city the "public welfare situation is unfortunate and it is not being given adequate attention." It was charged that while Ashburner was manager of Norfolk, Virginia, from 1918 to 1923 he did not follow closely "the work of even so important a department as the police," and there was evidence that his police administration had not been "as successful as his development of city plant." The "City Efficient," advocated by some engineers, proved to be much more complicated than merely sewers and bridges.[18]

Yet so rapidly were cities growing, and so costly were the new technical amenities like streetcars and electric lights, that even engineers unconnected with city government were caught up in their problems. The Cleveland Engineering Society, for example, long had a tradition of activism, and during the administrations of such Progressive mayors as Tom Johnson and Newton D. Baker, standing committees on smoke abatement, railroad crossings, the city building code, public safety, and planning gave unofficial but valuable aid to government. In 1917 the retiring president of the society told its members that "we have an administration [in Cleveland] elected by the people; we have a Mayor, a Council, etc., but I want to tell you that it needs the assistance of the members of the Cleveland Engineering Society to properly run this city."[19] The sentiment was widespread among engineers. The president of the American Society of Civil Engineers, addressing himself to "The Modern City and the Engineer's Relation to It," told his fellows in 1925 "the engineer recognized more than other citizens, and by force of example, by precept, and by teaching, he must show all men, that the days of waste are passing and the era of conservation has begun. With the waste must go all petty and partisan politics with so-called 'log-rolling' and in its stead must be substituted constructive statesmanship to go hand in hand with the principles of conservation."[20] The engineering of society continued to be an attractive dream for many engineers.

Among those technical activities that felt the quickening hand of science during these years, none produced such dramatic results as the application of something like the scientific method to the very problem of technological change itself. Sometime during the eighteenth century a new social type, "the inventor," emerged both in Great Britain and in America. In the United States such persons became the epitome of the national hero: those who by native drive and intelligence, unfettered by prejudice and tradition, created new devices to better the lives of the common people. The "flash of genius," the lightbulb going on over one's head, or the awakening in the middle of the night with the long-sought solution to a technical problem, became the popular notion of how technology moved forward. Inspiration struck seemingly at random and on a schedule of its own. The government encouraged this unpredictable event with the protection of letters patent, but otherwise little could be done.

By the mid-nineteenth century, however, the increasing complexity of some technologies and the growing power of scientific theory over natural phenomena were both leading to a closer interaction between science and technology. In part, this was also because of a widening gulf between the practitioners of each. The age of Jackson in America was a time when many social roles were becoming specialized and professionalized. To some extent, the emergence of institutions to bridge the gap between science and technology (both were new words that began to gain currency during these same years) was a response to a recent divorce between those who did and those who were presumed to know. Against the prejudice behind Dr. Charles T. Jackson's 1851 dictum that "no true man of science will ever disgrace himself by asking for a patent," Alonzo Potter in 1841 proposed the compromise: "On the one hand, science has furnished principles for the arts to apply: on the other hand, the arts have proposed problems for science to resolve; and this mutual aid and dependence have been the means of carrying both forward at a rate continually accelerated."[21]

Mutual dependence became particularly important for manufacturers and analytical chemists. E. I. Du Pont, before accepting and paying for a shipload of saltpeter, sometimes brought samples to a Philadelphia chemist for testing. In Cincinnati in the 1830s, a bell founder discovered that he had bought a shipment of adulterated copper pigs and was saved from great financial loss only through the intervention of a professor of chemistry at the Medical College of Ohio. At Harvard College, Daniel Treadwell, the practical inventor who was the first Rumford Professor and Lecturer on the Application of Science to the Useful Arts, anticipated his twentieth-century colleagues. His biographer has recorded that since Treadwell's teaching duties left him some free time, he used that to undertake experiments into the making of superior cannon. Working within the military bureaus, at institutes of technology, and in such private settings as the waterworks at Lowell, engineers during the nineteenth century used scientific methods and practical problems to create what Edwin T. Layton has called an engineering "mirror-image twin" of science.

In 1805 Oliver Evans, always ahead of his time, went as far as to suggest a regular program of federal government support for research:

> If government would, at the expense of the community, employ ingenious persons, in every art and science, to make with care every experiment that might possibly lead to the extension of our knowledge of principles, carefully recording the experiments and results so that they might be fully relied on, and leaving readers to draw their own inferences, the money would be well expended; for it would tend greatly to aid the progress of improvement in the arts and sciences.[22]

If government had not yet accepted the responsibility for subsidizing research of interest to manufacturers, certain industries, notably the electrical, were finding science more and more of a necessity in their operations. The journal *Scientific American* editorialized in 1894:

> Those who pursue electricity, by the inductive method—that of practical experiment and generalization therefrom—have lately entered the complaint that electricity has become nine-tenths mathematics. To many real experimental discoverers, the higher branches of mathematics constitute a trackless maze. . . . [Instead, it suggested, inventors should look to "experimental chemical research."] Much of such research has been, is and will yet be, pervaded with what may be called the *random* element. All experimental attempts to penetrate the unknown necessarily partake of this element. It is only when facts have so multiplied as to render *generalization* possible that *systematic* research begins.[23]

By the mid-1870s Thomas Alva Edison, the quintessential American "lone" inventor, had already surrounded himself with a research team, including scientists with doctorates from leading colleges, all working in a well-equipped laboratory with an attached library.

Powerful forces were also at work, however, to separate such fertile inventors from the products of their own genius. Typically, an inventor would devise and then patent a new or improved device. With this patent, capital could be sought for developing, manufacturing, and marketing it. Industrial capitalists like J. P. Morgan, however, needed to protect their sometimes very large investments in new devices by placing a controlling interest on the board of the corporation (sometimes newly formed) entrusted with its exploitation. For further protection, the original inventor was sometimes required to assign all future devices in the same field to the corporation and, at other times, was simply forced out of the company. When Morgan, for example, took over the Edison electrical interests and formed what became General Electric, Edison himself lost control of his own work. By 1889 the Edison General Electric Company contained not only the Edison interests but also the electric street traction inventions of Frank J. Sprague, a former Edison employee. In 1892 this group merged with another giant firm, Thomson-Houston, formed originally from the companies of those two inventors and in turn merged with other inventor-established firms such as that of Charles F. Brush, the Cleveland inventor of the principal arc-light system.

By 1900 most of the inventors who had been absorbed by General Electric were either dead or had drifted off to other activities, with the exception of Elihu Thomson, who continued to work in his laboratory at Lynn, Massachusetts. The mathematician Charles P. Steinmetz had recently joined the company but worked largely on his own problems. The financiers and attorneys who ran the company concentrated on using their patents to control the market for General Electric products. They hoped that when the lightning of inspiration struck some new inventor in their field, they could simply buy the rights to the new patent. It was a gamble, of course. The new device might instead come into the possession of their only large American rival, the Westinghouse Company, or form the basis of some entirely new firm.

It is not clear whose idea it was to create a research laboratory at General Electric, a sort of institutional equivalent of those individual geniuses whose inventions were responsible for the company's prosperity. Thomson and Steinmetz were both involved, and in the fall of 1900 Willis R. Whitney, an instructor in chemistry at the Massachusetts Institute of Technology, was approached about founding the laboratory. So new and uncertain was the venture that Whitney accepted only on a trial basis and refused to move from Cambridge to Schenectady until May 1904.

The first decade of the research laboratory at General Electric was a huge success—so much so that the laboratory's mission, unclear at first, was defined by the work itself. The laboratory's primary concern was to improve lighting, but important work was also done on vacuum tubes and X rays. In 1905 William D. Coolidge, an assistant professor of chemistry, also from MIT, was hired and almost immediately became Whitney's principal aide. Part of his work was to keep up with the rapid developments in Europe pertaining to new filaments for lightbulbs. On his recommendation, the company paid $760,000 for American rights to manufacture certain types. In 1908 Coolidge himself developed a process for drawing tungsten wire, and new bulbs using this filament were brought out in 1911. They were two and a half times more efficient than the best carbon lamps then manufactured by the company and 30 percent more efficient than the nonductile tungsten lamps then on the market. It is estimated that General Electric lost $1 million in obsolete stock and equipment through this innovation.

Irving Langmuir, who joined the staff in 1909, was perhaps the laboratory's most famous scientist. Although he concentrated on theoretical work only loosely related to the immediate needs of the company, his experiments led directly to the filling of incandescent lamps, which had previously been marketed in an evacuated state, with nitrogen, which further company development later replaced with argon. After the patent was taken out, the company spent another $194,473 on development, but the lamp that resulted was 25 percent more efficient than its predecessor.

Such successes gave the laboratory a firm place in the reputation and corporate structure of General Electric. The staff of 8 members in 1908 grew to 22 in the next year, and the number rose to 298 in 1917. Just how good corporate science could be at its best was demonstrated in 1932, when Langmuir won a Nobel prize for his work on surface chemistry. By 1960 the original laboratory had been joined by twenty-two other, separate facilities within the company.

By the eve of American entry into the First World War, some 375 industrial laboratories had been established in the country, and by 1931 that number had risen to more than 1,600. A growing number of companies had come to anticipate the formulation made in 1962 by S. W. Herwald, the vice president for research at the Westinghouse Electric Corp., who insisted that future corporate profits came from new products and processes, the economic benefits of which far outstripped the costs of the industrial research that lay behind them. Companies that might want to enjoy the benefits of research but were reluctant to invest in their own facilities, or that had need for one-time research for which their own laboratories were not well-equipped, had recourse to a host of competing institutions.

One of the oldest and best-established firms doing contract research and development was that of Arthur D. Little of Cambridge, Massachusetts.

Little had dropped out of MIT in 1885 and the following year set up a chemical laboratory to carry on routine analysis commercially. Like Edison, he had a flair for publicity and was widely believed to have successfully accepted the challenge to "make a silk purse out of a sow's ear," thus proving the power of modern science. Many firms came to Little for help, including General Motors, which hired him in 1911 to help it establish its own laboratories. In 1917 he built new laboratories on Memorial Drive in Cambridge, the now-famous Research Row. Some companies called on Little many times over the years: Owens-Illinois Glass, for one, consulted him eighty-two times. In 1927 Little helped the glass manufacturer standardize its plant and products by inventing a viscosimeter to gauge the viscosity of molten glass, which allowed Owens-Illinois to cut its manufacturing formulas from forty to six. Little, who was considered the "Nestor" of commercial research, set the standard.

It was important, he insisted, to do only justifiable science; to put the client's money into research, not brick and mortar; to keep the relationship confidential, although the client eventually got any patents that might result; to keep red tape and paperwork to a minimum; and to insist that all professional staff be available for any client. Little's concept of shifting research teams, chosen for relevant skills and organized temporarily around problems rather than disciplines, echoed the successful techniques pioneered by the U.S. Department of Agriculture years before.

Whereas Little worked for profit, other laboratories carried on industrial research on a nonprofit basis. One such laboratory was the Mellon Institute of Research, established finally at the University of Pittsburgh and the brainchild of the Canadian chemist Robert Kennedy Duncan. After traveling to Europe on a journalistic assignment for *Harper's* magazine to investigate industrial research on the Continent, Duncan settled at the University of Kansas and established his "industrial fellowships." His notion was that companies needing specific chemical research could finance graduate fellowships. The students receiving the money would do the research. In this way manufacturers would get their research, students would have their education paid for, and the host university would have an ongoing research program with students and facilities paid for by industry.

In 1911 Duncan moved to Pittsburgh, and his effort was endowed two years later by the Mellon family, financiers and philanthropists of Pittsburgh. Upon the death of Duncan, the directorship passed to Raymond F. Bacon, one

Just as World War I was sometimes called the chemists' war, the decades before World War II were called the Chemical Age. Depicted here as a priest elevating the host at the altar, the chemist was seen as performing miracles in the field of industrial technology. More ominously, trees, clouds, and birds have been replaced by smokestacks, smoke, and airplanes. A. Creesy Morrison, *Man in a Chemical World* (New York, 1937), 31. Reprinted by permission.

of the most active apostles of industrial research in these early years. A new building was put up in 1915, and two years later there were forty-two fellowships, sixty-four fellows, and contributions of $147,000. Bacon proudly proclaimed at this time that a study from 1911 to 1916 had shown that about 70 percent of the problems submitted by industry to the institute had been solved to the satisfaction of the donors.

Even without industrial fellowships engineering and applied science research began to flourish at institutions of higher education elsewhere in the country. Between 1893 and 1933 the number of engineering schools increased from 100 to 160, and the estimated number of students grew from about twelve or thirteen thousand to perhaps sixty-five thousand, including by this time a sprinkling of women. During these same years the number of graduate students in engineering, a better indicator of possible research activity, rose from practically none to about four thousand. In the early 1890s a master's program was started in electrical engineering at the University of Wisconsin, one of the first in the nation. Harvard's Lawrence Scientific School offered a master's degree in the same field beginning in 1893 and MIT gave one in civil engineering in 1894. Graduate programs in chemical, mechanical, sanitary, and electrical engineering followed at that school within the next decade.

In 1903 the University of Illinois established an engineering experiment station as a deliberate analog to the agricultural experiment stations found in every state and territory of the nation. Another was set up at Iowa State College in the winter of 1903–4, and others soon followed, including one at Purdue in 1917. An attempt on the eve of World War I to gain regular federal subsidy for these stations failed, as did a similar attempt during the Great Depression. Nevertheless, those established continued to provide something very like industrial research, often geared to the prevailing economic interests of the regions served by their respective universities.

In striking ways, the First World War became the last of the great Progressive crusades for America—in this case, to make the world safe for democracy. In April 1917, thirty-two months after Europe was engulfed in the most terrible conflict it had ever known, the United States entered the battle. With an army of 200,000 men, half the number of casualties suffered by Great Britain in the single battle of the Somme in 1916, but with an industrial base that could be greatly expanded and focused, America decisively shifted the balance of power and broke the long stalemate in the trenches.

Notions of conservation, efficiency, and research were quickly applied to the war effort. In 1915 the National Advisory Committee on Aeronautics was established to undertake the research needed to bring American aircraft up to European standards. That same year the secretary of the navy set up the Naval Consulting Board to advise him on new technology for the fleet. Chaired by

The need for munitions in World War I created a fertile field for the talents of engineers who championed the just-emerging field of scientific management. That same need brought women into machine shops where they had never before been welcomed, violating gendered notions of technological propriety. U.S. Department of Labor, *The New Position of Women in American Industry,* Bulletin of the Women's Bureau, No. 12 (Washington, 1920), facing 102.

Thomas Edison, it was filled out by such luminaries as Willis R. Whitney of General Electric and Elmer Sperry, the gyroscope inventor. In 1916 the astrophysicist George Ellery Hale, a political activist in the National Academy of Sciences, prevailed upon President Woodrow Wilson to set up the National Research Council under the auspices of the academy, to direct research in the national interest. Within the military, the Chemical Warfare

Service was organized to provide poison gases, and defenses against them, for the troops.

Engineers, too, organized. In 1916 the Engineering Foundation was established with an endowment from the Cleveland machine and telescope maker, Ambrose Swasey. Its entire first year's income, however, was turned over to the new National Research Council to finance the research of scientists. Indeed, the ambiguous distinctions between science and technology and between engineering research and scientific research, and the basic professional orientations of many active scientists and engineers led to no little conflict over who was to organize, lead, and eventually get credit for the mobilization of the nation's technical resources.

Despite the glamour and destructive power of such new weapons as the machine gun, the submarine, tanks, and poison gas, the war was sustained and ultimately won by young soldiers expending vast amounts of bullets and shells. The industrial infrastructure of the United States was tested and expanded by the effort to produce enough for itself and its allies. To get the personnel and material across the Atlantic, Congress chartered the new Emergency Fleet Corporation and charged it with building the needed ships. New shipyards were planned on the precepts of scientific management, and Gantt flowcharts were used to keep track of materials and components. Women poured into the munitions works in America, as they had in Great Britain and on the Continent, and with them came an infusion of labor-saving machinery.

A generation of knowledge on how to conserve and rationalize the factors of production in American industry, how to organize workers and machines into efficient combinations, and how to target applied scientific research to solve particular technological problems was now set loose through the war effort. Some engineers had long chafed at the realization that the nation's great corporations were organized essentially to produce profits rather than goods. Now, with the federal government picking up the bill for a virtually unlimited flow of products and with patriotism in the air, engineers in a host of industrial settings were unleashed and eager to show what they could do. The Great War, as it came to be called, showed the new twentieth century that death, too, could be mass-produced efficiently.

10

THE DECADE OF PROSPERITY AND CONSUMPTION

THERE WAS A CERTAIN EDGE of arrogance, and perhaps even aggression, about those who chose to defend technology in America during the 1920s. It was the golden period of High Modernism, and the machine aesthetic as well as the old "mechanistic" ideal deplored by Thomas Carlyle seemed on the verge of triumph. If the Enlightenment project was ever to be brought to a successful conclusion, now looked like the time. An article by that publicist for chemistry, Edwin E. Slosson, appeared in the magazine *Independent* in 1920 under the title "Back to Nature? Never! Forward to the Machine." Deploring such cultural tendencies as the movements to "jazz your music and cube your painting," he charged that "the cult of naturalism is now dominant everywhere. The call of the wild is drowning out the appeal of civilization. 'Back to barbarism!' is the slogan of the hour. Sink into savagery. Praise the country and denounce the city."[1]

It was, he believed, "a reactionary spirit, antagonistic to progress and destructive of civilization. Science and Christianity are at one in abhorring the natural man and calling upon the civilized man to fight and subdue him. The conquest of nature, not the imitation of nature, is the whole duty of man." Gradually, Slosson claimed, "[man] will substitute for the natural world an artificial world, molded nearer to his heart's desire. Man the Artifex will ultimately master Nature and reign supreme over his own creation until chaos shall come again. . . . In the ancient drama it was *deus ex machina* that came in at the end to solve the problems of the play. It is to the same supernatural agency, the divinity of machinery, that we must look for the salvation of society."[2]

The succeeding decade did indeed witness an accelerated recourse to labor-saving methods and an elaboration of what Thomas P. Hughes has called "Networks of Power." To these were added a dazzling array of machines and devices: automobiles, airplanes, radios, and a host of appliances for the home. It was a flood of consumer durables unprecedented in history. By the end of the 1920s, in the more sober phrases of the social sciences, President Herbert Hoover's Research Committee on Social Trends wrote that "the magnificent material portion of our culture has been developed by scientific discoveries and inventions applied to a rich natural heritage."[3] The flow of inventions was increasing, and its effects were spilling into other areas of culture. "Thus," wrote the sociologists W. F. Ogburn and S. C. Gilfillan, "woman suffrage was the outcome of a great number of forces and converging influences. Mass production, urbanization, birth control, the typewriter, education, the theory of natural rights, and many other factors contributed."[4] The thread of technology ran throughout social relations.

The scholars on President Hoover's committee took ample account of the technological explosion of the 1920s. The great danger was not, they felt, the number or nature of technological advances, but the problem of "cultural lag": "There is in our social organizations an institutional inertia, and in our social philosophies a tradition of rigidity. Unless there is a speeding up of social invention," they declared, "or a slowing down of mechanical invention, grave maladjustments are certain to result."[5] By the time the report of *Recent Social Trends* was finally issued in 1933, that adaptability was being sorely tested.

Throughout these rhetorical flourishes there was widespread agreement on the basic relationship between science, on the one hand, and the public, on the other. As the formula had it, scientific research discovered laws of nature that were then converted into technologies that were consumed by the nation's citizens. An early and defining statement of the formula was adopted as the official motto of the 1933 Chicago Century of Progress world's fair: "Science Finds—Industry Applies—Man Conforms."

Industry was seen as the link between science and the consumer, but during the decades after the close of World War I the federal government

The motto of the 1933 Chicago Century of Progress world's fair, "Science Finds—Industry Applies—Man Conforms," was illustrated by a heroic sculpture in the Hall of Science, showing a robot pushing an apparently reluctant woman and man into the future. *The Official Pictures of a Century of Progress Exposition,* Chicago 1933, intro. by James Weber Lynn (Chicago, 1933).

played an increasingly active role in the relationship. The power and resources of the state were mobilized to stimulate the production of science, facilitate its transfer to industrial control and embodiment, and create social mechanisms and facilities for maximizing public consumption of these technologies. As historian A. Hunter Dupree has written, the federal research establishment had been substantially completed in the years before the war. The new scientific bureaus, he noted, "renewed the ability of the government to conduct its own business in a society dominated by a complex technology that increasingly depended on research for guidance." In addition, he wrote, in "the name of the general welfare, they sought answers to those problems that industry needed to have solved but were unable or unwilling to answer for itself. . . . Without the efforts of these [government] bureaus," he concluded, "the use of science would have penetrated more slowly into technology."[6]

The growth of federal subsidies to scientific research is easily documented. In his study *The Growth of the Federal Government* (1934), Carroll H. Wooddy calculated that between 1915 and 1930 "science and research" expenditures rose 323 percent, compared with an increase in all civil functions of only 237.2 percent. The budget of the Bureau of Standards, for example, went up from $661,100 in 1915 to $2,759,200 in 1930. During these same years its number of employees rose from 233 to 1,055 and the number of cooperative research assistants from none to ninety-seven. The funds for the National Advisory Committee for Aeronautics (NACA) went up from $3,900 in 1915 (its first year of operation) to $979,700 in 1930. A decade later, the National Resources Committee explained this increasing subsidy to science by pointing out that the government had long since passed the point of merely collecting data to better inform legislation. "The government today," it wrote, "assumes responsibility for scientific studies which deal with many general problems, such as the improvement of agriculture, the conservation of natural resources, and the development and maintenance of physical standards. . . . In all these extended researches the Government is serving the double purpose of directing its own operations and supplying the people with important scientific findings which they need for their private purposes."[7]

While the federal government was spending $78.5 million for research (on social science, statistics, surveying, and mapping, as well as natural science), private industry was spending in the neighborhood of $300 million for science, employing some 70,000 researchers in more than 2,200 laboratories. This "continuous and increasing application of science by industry is contributing significantly to the high standard of American living," according to one governmental report in 1940. "Viewed in this light, industrial research is a major national resource. . . . Industrial research as a national resource capable of contributing to the public welfare," it concluded, "should be fostered."[8]

Many of the government agencies listed by Woody as primarily engaged in research were, in fact, virtual adjuncts of the industries they served. In 1910 the Bureau of Mines was created to investigate the safety of miners and appliances designed to prevent accidents, possible improvements in the condition under which mining operations were carried on, the treatment of ores and other mineral substances, the use of explosives and electricity, and "other inquiries and technologic investigations pertinent to said industries."

"Since the mining and metallurgical industries are country-wide in extent and varied in nature and in methods of control," one student of the bureau pointed out in 1922, "the Bureau of Mines has entered into many cooperative agreements for the conduct of work within its field." Reflecting the pluralism of science and technology within the American system, as well as a willingness to cross jurisdictional boundaries to aid industry, a wide range of agencies were involved. "They have included," it was reported, "other branches of the national government, many state and local governments, educational institutions, scientific societies, and private agencies wishing to pursue a basic investigation of importance to the whole industry. It brings into this cooperative work, the viewpoint of the engineer and metallurgist."[9]

Another prime example of government research for industrial betterment was the work of the National Advisory Committee for Aeronautics. Established during the period of preparedness in 1915, the NACA (which was absorbed into the National Aeronautics and Space Administration in 1958) pioneered in the use of grants and contracts to achieve its statutory mandate to "supervise and direct the scientific study of the problems of flight, with a view to their practical solution" and to "direct and conduct research and experiments in aeronautics." Eventually operating three of its own large research and development facilities, the NACA saw as one of its major purposes the close coordination of federally funded research, the needs and capacities of the aircraft industry, and the wishes of the air arms of the military services.

Jerome C. Hunsaker, in 1955 chair of the NACA, recalled that in 1916

> the executive committee invited engine manufacturers to attend a meeting on June 18 in Dr. Walcott's office at the Smithsonian Institution to discuss the problem of obtaining more powerful and more reliable engines and to bring about a better understanding between builders and users. Representatives of the military services were in attendance, and although it is to be doubted that many problems were solved, unquestionable good was done by bringing them into sharp focus. Another benefit from the meeting was an arrangement whereby the Society of Automotive Engineers became active in providing assistance in the solution of aircraft problems.[10]

It was an early example of what would later be called the military-industrial complex, a cooperation that was institutionalized by the NACA in a series of annual get-togethers for the mutual information of government, military, and industrial leaders in aeronautics.

Early in 1917, before the United States entered into World War I, the War and Navy Departments complained to the NACA about the high cost of airplanes, a condition that the military blamed on high royalty fees charged by holders of individual patents. Under the good offices and at the recommendation of the committee, the Manufacturers Aircraft Association was formed to regulate the licensing of patents within the industry. It was what Hunsaker was later to call a "happy solution" to the problem of an important, but underregulated, technology.

Perhaps the best example of government aid to industrial efforts to create and utilize new technologies during the 1920s was the National Bureau of Standards (NBS). In 1901, responding in part to the establishment of national standards laboratories in Germany and Great Britain, the United States created a bureau of standards within the Department of Commerce. Its organic act charged it with the making and keeping of standards (the foot, gallon, and pound, for example) and the determination of physical constants and properties of materials. It also set standards for goods purchased by the government and helped industry set dimensional standards. Total funding for the bureau reached $4.1 million in 1931.

In the decades between the wars, the NBS created two devices that tied its work closely to the needs of American industry. One was the Visiting Committee, provided for in the organic act and composed of "prominent men in the various interests involved and not in the employ of the government, who are to visit the Bureau at least once a year and report to the Secretary of Commerce [the engineer Herbert Hoover during much of the 1920s] upon the efficiency of its scientific work and the condition of its equipment." The Visiting Committee—made up of such important figures as Albert Colby, secretary of the Association of American Steel Manufacturers; Samuel W. Stratton, president of the Massachusetts Institute of Technology; Charles F. Kettering, director of research and vice president at General Motors; Frank B. Jewett, vice president in charge of research for the American Telephone & Telegraph Company; and Crawford H. Greenwalt, president of E. I. Du Pont de Nemours & Company—kept the bureau keyed to the corporate and educational needs of the country and its work consonant with those needs. In 1925 there were, in addition to this committee, no fewer than eighty-two advisory committees, "all made up of representatives from the industries, appointed by the industries, or by technical, engineering, or scientific societies."[11]

Besides this intricate web of committees, the bureau also hosted a number of research associates. These were "stationed at the Bureau by trade associations and others, working on fundamental problems on which the industries want specific answers. These associates," which numbered sixty-one by 1925 and represented thirty-six organizations, "have the use of the laboratories, and they work with the experts of the Bureau under the Bureau's direction."[12] In 1925, the five eminent members of the Visiting Committee, the sixty-one research associates, and the members of the eighty-two advisory committees represented a massive effort to knit together the work of the bureau and the industries that mediated American technology. It was part of a deliberate and constantly elaborating effort to move science out of the laboratory and into the ongoing activities of industry.

The begetting of science and its transfer to industrial practice still did not ensure the social acceptance of the resulting technological innovations. Wooddy, in his study of the growth of government, asserted that between 1915 and 1932 "the change of greatest apparent significance was the tremendous expansion of activities designed to control or promote commerce and industry and transport on sea and land."[13] Aeronautics and radio were two areas he specifically singled out as benefiting from increased government aid.

Among the agencies listed by Wooddy in this category were the Patent Office, the Aeronautics Branch of the Department of Commerce, the Federal

Radio Commission, and the Federal Oil Conservation Board. In comparison with the 237.2 percent rise in funding between 1915 and 1930 for all civil functions, and the 323.4 percent for science and research, the increase for commerce was a prodigious 809.1 percent.

Between 1914 and 1927 the number of aircraft manufacturers in the United States grew from sixteen to sixty-one and the number of airplanes produced each year to 1,888. By 1929 some eighty-four domestic air transport routes had been established. To help rationalize this activity, the Air Commerce Act of 1926 established the Aeronautics Branch of the Department of Commerce. Appropriations rose to $6.4 million in 1930, but what was more important, this sum was expended on a dozen activities ranging from the licensing and inspection of aircraft to the installation and maintenance of airways. Along with the independent NACA, charged with research, and the Aeronautical Board, which was a joint army-navy coordinating agency, the Aeronautics Branch made an indispensable contribution to the widening utilization of aircraft technology.

The government also promoted air transportation by having the army begin, in 1918, to fly a scheduled mail service between New York and Washington, D.C. The Post Office began service between New York and Cleveland the next year, then extended service to Chicago and, in 1920, to Omaha. Regular around-the-clock flights between New York and San Francisco began in 1924. Meanwhile air shows and races, and "barnstorming" entrepreneural pilots, helped sustain what Joseph Corn has called "The Winged Gospel": an almost religious belief in the spiritual as well as the commercial benfit of aviation. In May 1927 Charles A. Lindbergh flew solo from New York to Paris. The sensation of that flight in the *Spirit of St. Louis* was explainable in part by the fact that it neatly wedded two seemingly disparate parts of the American myth: a belief in the power and accomplishment of the "lone" individual, and a commitment to the social construction of a modern, industrial technology.

In yet another field, the Federal Radio Commission sought to untangle the chaos of underregulated wireless transmission. On November 2, 1920, KDKA in Pittsburgh made the first radio "broadcast": two thousand listeners learned of Warren G. Harding's presidential victory. After the breakdown of earlier attempts at governmental regulation, the Radio Act of 1927 established the Federal Radio Commission to, among other things, classify stations, mark out zones for them to serve, prevent interference, assign frequencies, determine power to be used, and regulate broadcasting apparatus.

The need was great. Between 1922 and 1925 the number of families with radios in America grew from 0.2 to 10.1 percent of the population, and the value of radios produced rose from $5 million to $100 million between 1922 and 1926. At the same time, 222 manufacturers of radio equipment were also active as broadcasters. The number of stations broadcasting rose from eight in 1920–21 to 700 in 1928, by which time 40 million Americans were listeners. In 1927 the new Radio Commission accepted as fact the proposition that advertising should continue to be the major support of broadcasting, and that regulation would have to be depended on to prevent the abuse of private privileges, publicly bestowed.

When the commission began its work, 732 stations had already been licensed and the problems entailed in setting this considerable house in order were worked out with the advice and cooperation of the manufacturers and broadcasters. A 1928 amendment to the Radio Act, tightening restrictions, was met with the call for a "conference with radio engineers and representatives of the trade and industry in order to work out allocations. . . . The principles governing the new allocations were embodied in General Order No. 40, issued August 28, 1928."[14] Writing in his memoirs thirty years later, former President and Secretary of Commerce Herbert Hoover still believed that the evils of mid-twentieth-century broadcasting might "be much reduced by resuming the annual conferences of the early twenties."[15]

It seems likely that few Americans noticed or cared greatly about these govenmental and corporate accomodations. The social reality for most was a diminishing of rural isolation, particularly for women, and the forging, through commercial programming, of common cultural icons and contexts. Radio personalities and shows for this generation created a new national culture: it seemed that everyone tuned in to Amos 'n' Andy, the Goldbergs, Jack Benny, and a host of other characters who became better known to some than many of their neighbors. In a pattern renewed a generation later by television, the mixture of fictional lives and consumer products instructed the nation in attitudes, aspirations, and styles of living.

Between the wars, the government accepted a responsibility not only to help foster basic science but also to guide it to industrial acceptance and finally create both the rules and the playing field for public acceptance. The burgeoning of radio use and air travel, automobiles and home appliances,

Herbert Hoover, the "Great Engineer," became a symbol of the 1920s. In his cutaway coat, speaking into radio microphones, the president became a spokesperson for technological progress, clothed in the trappings of a conservative ideology. Herbert Hoover, *The New Day: Campaign Speeches of Herbert Hoover 1928* (Stanford, 1928), frontispiece.

would have been impossible without the regulatory activities and engineering infrastructure provided by governments at all levels, from local to national. In a nation increasingly under the domination of large, liberal corporations that were effectively harnessing science and technology to achieve their goals of growth, stability, and market share, governments served as the focus of efforts to socialize the administration and expense of rationalizing the technological environment. Almost exclusively, governments serviced rather than challenged this corporate hegemony. The cornucopia of consumer goods developed (with or without the aid of science), produced, and made available to the American public by these corporations was truly astonishing.

No device better exemplified this process than the automobile. The first of these built in the United States had been crafted in the 1890s, but in 1908 the industry took on a familiar look with the organization of General Motors and the introduction, by Henry Ford, of his Model T. Just two years later there were already nearly half a million cars registered in the country, and in 1915, 895,930 new cars were manufactured.

Historian James Flink has codified the many reasons that Americans accepted the automobile so readily: a lack of prejudice against identical devices, a somewhat broad distribution of income, a lack of tariff barriers, and a belief that cars were cleaner, safer, cheaper, and more dependable than horses, especially in cities. The car was a useful status symbol in a country that claimed to ignore matters of class, lineage, and birthplace but placed a high value on material things. It reified deeply rooted values of individuality, privatism, free choice, and control over one's life. It was a perfect example of the nation's habit of trying to replace politics with technology (Henry Ford had bragged that "we will solve the city problem by leaving the city"). In sum, to use Flink's words, "automobility permitted a restructuring of American society, via technical efficiency along lines dictated by traditional cultural values."[16]

Many key innovations were necessary to make this public acceptance possible. In 1913 Ford had begun the first of his famous assembly lines, the last link in the chain of production technology innovations stretching from the American system of manufacture to mass production. Along with such techniques as stamping sheet metal and electric arc welding, borrowed from the bicycle industry, Ford's moving assembly line revolutionized the making of cars—and by imitation, many other devices in many other industries.

In a piece written in his name for inclusion in the thirteenth (1926) edition of the *Encyclopaedia Britannica,* Ford defined mass production as "the focussing upon a manufacturing project of the principles of power, accuracy, economy, system, continuity, speed, and repetition. To interpret these principles, through studies of operation and machine development and their co-ordination, is the conspicuous task of management. The normal result," he concluded, "is a productive organization that delivers in continuous quantities a useful commodity of standard material, workmanship and design at minimum cost."[17]

As numerous examples were to prove over the years, many useless as well as useful commodities could be made this same way, and "minimum cost"

By the 1920s the "Good Roads movement" had succeeded in convincing governments that the bicycle, and then the automobile, had made road improvement a political, as well as social, necessity. Courtesy National Museum of American History, Smithsonian Institution.

referred to the producer, not necessarily the consumer. The litany of "power, accuracy, economy, system, continuity, speed and repetition," it turns out, can chill as well as thrill the heart. The key to the whole process, he claimed, was "simplicity," which was underlaid by three more principles: "(a) the planned orderly and continuous progression of the commodity through the shop; (b) the delivery of work instead of leaving it to the workman's initiative to find it; (c) an analysis of operations into their constituent parts."[18]

A second need was for some way of enabling the public to afford the mass of cars produced by Ford and others. In this case it was General Motors that provided the innovation. In 1919 the General Motors Acceptance Corporation (GMAC) was introduced, and with it the birth of modern consumer installment credit. Soon even children could buy their pedal cars "on time" and, indeed, were encouraged to do so.

And finally, automobiles needed what were called "good roads" to operate on. In the 1890s the nation's bicycle enthusiasts had inaugurated a "Good Roads" movement to force the upgrading of roads, an engineering infrastructure that had since colonial times been a local responsibility. Now, as the *Engineering News-Record* put it in 1917, "to Henry Ford belongs the honor and glory of having furnished the only good roads argument which hundreds of thousands of people in the backwoods sections of the country are capable of absorbing"—the Flivver.[19]

When the federal Bureau of Public Roads took the first official census of roads in the country, it found 153,662 miles of rural highways, only one-twelfth of which were surfaced. Then in 1916 the principle of local control was effectively destroyed, when the Federal Aid Road Act offered federal monies to aid road construction in any state that would set up a state division of highways and draw up a five-year plan for trunk-route development. Road building rapidly became one of the main forms of capital improvement engaged in by governments at all levels. In 1925 alone, slightly more than a billion dollars was spent on roads. The idea of toll roads was revived, and in the West, where roads ran for many miles past few people, a gasoline tax was first instituted and then written into the several state constitutions. Over the years, the powerful "highway lobby" of trucking companies, teamster and auto unions, building contractors, oil companies, the auto industry, insurance

The coincidence of the push for better roads and the trend toward the use of reinforced concrete in the fabrication of highway bridges has left a legacy of handsome structures across the American landscape, like this patented, open-spandrel arch bridge in California. Postcard from the author's collection.

companies (sometimes attached to so-called auto clubs), public works officials, and others proved a formidable force for the favoring of the motor vehicle over other forms of transportation in the nation.

The federal government had been financing and coordinating the nation's highways since 1916. As one engineer pointed out, however, "the needs of a larger volume of intensely mobile traffic which desires to move from one place to another with the least resistance compatible with the physical conditions of the territory through which it must pass," demanded not just more roads but a new type.[20] Early extensions of the system had left cities and towns strung out along highways like beads on a string. Even highways designated as federal, like U.S. 1 running north and south along the Atlantic Coast, passed down the main streets and through the downtowns of innumerable urban areas.

One of the earliest roads to provide a "freeway" through such areas traversed the highly urbanized section of New Jersey adjacent to New York City. There Route 25, which passed from the Holland Tunnel through Elizabeth, was threatening to strangle on its own congestion. Traffic in Newark between 1912 and 1922 increased an average of 340 percent, but was up as high as 1,183 percent on some streets. Vehicles passing through Rahway every twenty-four hours rose from 12,000 to 44,000 between 1921 and 1926. As one engineer wrote, "neither citizens who desired to shop, shopkeepers who desired to sell, or manufacturers who desired to move materials, could do business except at great expense and delay."[21] And all because of travelers who did not want to be in Rahway at all! By 1931 the first eight miles out of the Holland Tunnel were provided with a partly raised, partly lowered freeway, provided with ramps for access and egress. After World War II, the building of freeways became a major federal and state undertaking.

The social and cultural impact of the automobile can hardly be measured. By 1923 California already had more than a million registered cars, many of them, of course, in southern California. As if to prove Henry Ford a prophet, by 1925 Angelenos boasted that light cars in the possession of the working class allowed them to commute from suburban homes to workplaces downtown, or in other parts of the sprawling city. Surely it is no accident that as the movie industry grew up in this same environment, the aggressive and deliberate smashing and dismantling of automobiles became a standard feature of silent comedies, and that audiences in the darkened movie palaces roared their approval. Such scenes revealed a basic ambivalence about this machine, so newly necessary but at the same time so profoundly upsetting of settled habits and so costly in terms of personal danger.

The car was hailed as a savior of family values, as the patriarch assembled the family for a Sunday drive. At the same time, some deplored the fact that the family was together in a car at all, and not in church. It seemed also to undermine the family as children sped off to entertain themselves (and each other) out of the parental gaze. A young woman who once had invited a beau to her house, entertained him in her parlor, and treated him to her cookies and lemonade, now accepted his offer of a date, got into his car, and accepted his offer of refreshment. Little wonder then that Beth Bailey, the author of a history of courtship in America, *From Front Porch to Back Seat* (1988), pointed out the shift in power that this change of venue implied. The electrical self-starter, developed in 1911 by Charles Kettering of General Motors, was advertised as making driving more available to women, and indeed "chauffeur" soon became a part of the job description of the middle-class housewife and mother.

A study of attitudes toward automobiles in the South discovered at least one aspect of the way race as well as gender shaped the impact of this technology. Although all races and classes were found to be ambivalent about the coming of the automobile, African-American Southerners saw in it the virtue of allowing them to avoid the humiliating Jim Crow accommodations on public street cars. Indeed, the story persisted that southern politicians, fearing the implied equality of a single highway system with both black and white having the same rules of right-of-way, toyed with the idea of a dual, segregated system. Since highways cost more than drinking fountains, however, fiscal reality prevailed.

While the car was in the garage (like *chauffeur* a French word, first used in English only in 1902), the middle-class house, too, was filling with new technologies. Ruth Schwartz Cowan has shown that, from the earliest years of the Industrial Revolution in America, new technologies reshaped the lives of housewives in all parts of the country. As food, fuel, and clothing were commodified, traditional men's work (chopping wood, taking the grain to the custom mill for grinding into flour) was reduced around the home, and women's work (cooking, for example) was increased. Commercial fine-grade flour and the cast-iron cooking stove replaced not only the course meal and fireplace hearth, but also recipes, baking techniques, and culinary expectations.

In an age of consumer durables (many of them household appliances), it was hoped that women would find a technological route to "Freedom!" that would avoid the messy political need to renegotiate power relationships within marriage. Note that the enabling power that appears, transcendent in the sky, is "GE," the corporate deity. General Electric Co., *Freedom!* Kitchen Appliance Catalog (n.p., June 1933), front cover.

The outpouring of home appliances, though often hailed as "labor-saving," ironically made, in Cowan's phrase, "more work for mother." The overwhelming evidence is that although the number of hours spent in housework by women employed outside the home did fall, the hours spent by those who were "solely homemakers" actually rose by a few hours between 1926 and 1966 (to about fifty-five hours per week). The statistics for home laundry work are typical: from 1925 to 1964, after the introduction of the electric

In parallel with the chemist-priest invoking industrial research pictured in chapter 9, the chemist here (still in his vestments) reinforces the spheres of masculinity and femininity and conjures up a single-family, detached home in which the strong verticals are still trees and the domestic hearth, clouds prevail over smoke, and birds are still seen, rather than airplanes. A. Cressy Morrison, *Man in a Chemical World* (New York, 1937), 159. Reproduced by permission.

washer, the automatic washer, the automatic dryer, commercial soap powders, and wash-and-wear clothing, the hours devoted to laundry actually went up from just over five hours a week to just over six. What had changed dramatically, however, was the shift from a back-breaking task to a relatively easy one, and the increase in cleanliness, with concomitant benefits for child health.

Not only was the middle-class urban home mechanized through the introduction of new machines, there were concerted attempts to make it more efficient as well, through the introduction of scientific management. Ellen Swallow Richards, a chemist on the faculty at MIT, had spent many years testing new appliances and encouraging a more "scientific" attitude toward housework. With the publication of Frederick Winslow Taylor's classic *Principles of Scientific Management* in 1911, women's clubs and others across the nation sought to raise the prestige and psychological well-being of housewives by transforming them into "domestic engineers."

A leader of the movement was the efficiency expert Lillian Gilbreth, who became closely identified with the "kitchen efficient" movement. For years she had been a partner with her husband, the scientific management pioneer Frank Gilbreth, in engineering consulting work. Now, in her advocacy of the "kitchen efficient," like so many of her co-workers in the field, she emphasized the psychological benefits that a feeling of being in "control" of housework would bring to housewives who worked efficiently.

The effort to apply scientific principles to the running of domestic workplaces proved to be both an opportunity and a dead-end for professional women. The establishment of the American Home Economics Association in 1908 formalized an effort to both raise the status of housewives and improve the conditions of housework through research, education, and the application of the newest science and technology to the home. Richards, who became the first president of the organization, declared in 1910 that "the work of homemaking in this scientific age must be worked out on engineering principles and with the cooperation of trained men and trained women. . . . Tomorrow, if not today," she concluded, "the woman who is to be really mistress of her house must be an engineer, so far as to be able to understand the use of machines."[22] Nutritious recipes were worked out, the more efficient placement of kitchen cabinets and counters was planned, time and motion studies were made of how women ironed, and dustpans with long handles were recommended.

In the end, either because an efficient home appeared to be a poor haven from increasingly efficient workplaces for men, or because "labor-saving" devices and techniques threatened to truly liberate women from at least some aspects of patriarchy, the typical middle-class American home remained a "separate sphere" where women worked outside the logic of either Taylorism or Fordism. Despite some success in more conveniently designed kitchens (with floor plans to cut down on wasted steps, for example), the effort to Taylorize homes was a failure. The exquisitely detailed division of labor in a Ford automobile plant could not even be approximated in the average home, in which one person, the "lady of the house," had to do all the work herself.

As Ruth Cowan has pointed out, the effects of even the partial industrialization of the home were nearly opposite those experienced by factory workers. The managerial component of housework grew smaller, not larger, as servants disappeared from the home. Instead of housework being drained of emotional content, it was infused with portentous meaning. Instead of an increased division of labor, the housewife became a Jill of all trades. The revolutionary potential of household mechanization was thwarted by social norms that needed the market for consumer goods that women represented, encouraged them to look on appliances as evidence of material progress and personal fulfillment, but ultimately wished to preserve the traditional gender distinctions and limitations on women.

One of the signal features of the early twentieth century was the increasing commodification of leisure, and no device better epitomized this trend than the motion pictures. Developed late in the nineteenth century by Thomas Edison and others, the "movies" were, in the words of historian George Basalla, an art form at once "in essence mechanical, industrial, and commercial."[23] Like the contemporary phonograph and the mechanically produced visual arts of the earlier nineteenth century, the movies raised questions about the nature and role of art in modern life. Produced first for the urban working class, movies were at their best when showing machines in rapid motion: car chases, locomotives, airplanes, streetcars—anything moving fast.

The nickelodeons of the early days were eventually replaced as the technology of motion pictures evolved and they became the entertainment and educational property of a wide stratum of American society. By the early 1930s, the movie palace was not only a place frequented by masses of Americans in search of a good time and practical information on how to live the good life, but a mechanical marvel in itself. Often constructed on fantasy themes (Egyptian, Spanish, and so on), cinemas showed moving pictures that were accompanied by organ music, light shows orchestrated from intricate banks of switches and rheostats, "starlit" skies intermittently covered by drifting clouds, and eventually by sound itself and were increasingly "air-conditioned." It was a totalizing mechanical experience that shaped the way patrons thought about and lived their lives.

"Science" was never portrayed in a more masculine guise than this engineer, with his mustache, pipe, necktie, and sleeves rolled up to reveal muscular arms. "Science in the Home" represented the male invasion of a territory that in the nineteenth century had been assigned unambiguously to women. Technology could, when people chose, dissolve as well as reinforce gendered boundaries. *Scientific American,* 106 (April 13, 1912), cover.

The flood of household machines, the automobiles, and airplanes, and perhaps even more the drama of assembly lines, scientific management, and mass production, struck the intellectuals of Europe with what Thomas P. Hughes has called "the second discovery of America." A Europe destroyed morally as well as physically by "the Great War" of 1914–18 saw in *Fordismo* and *Taylorismo* a way out of its morass of failed traditions. Powerful, clean, scientific, modern, and, at least in terms of its product, democratic, American technology attracted both attention and emulation.

The Modernist movement in art, literature, and architecture had its adherents in America as well, of course. Many American intellectuals and artists, however, expressed profound unease at the direction of modern life. In his failed stage play *Dynamo* (1928), Eugene O'Neill examined the ultimately fatal anguish of the son of a minister who fell in love with the daughter of an engineer. Already Willa Cather, in her first novel, *Alexander's Bridge* (1912), chose as a protagonist the most virile and admired embodiment of American masculinity, an engineer, and found him as flawed and dangerous as the bridge that was to have been the capstone of his career. As though in answer, Sherwood Anderson's odd book *Perhaps Women* (1931) suggested that because of a mysterious, secret "inner life" that was immune to the damage caused by modern technology, perhaps women alone had the ability to save American men from the emasculating effects of their own machines.

Modernism made a sharp theoretical distinction between high and low art, but in actual fact they tended to blend. The emerging profession of industrial design began to shape the form of manufactured products. In the decade ahead, "streamline moderne" became a distinct style for everything from ash trays to sports stadia. Art deco, while not originally an American style, put its imprint on a generation of buildings and interior decoration, costume jewelry and radio cabinets. Streamlined design may not have always facilitated the swift passage of pencil sharpeners through the resistant air, but it undoubtedly helped the mind slip with minimum friction from the nineteenth century to the twentieth.

As always, class and race made a difference as to how and which technologies were used. In rural areas of the nation, such basic technologies (to middle class city dwellers) as running water and electricity, indoor toilets and telephones, remained beyond the grasp of many. Often electricity was simply not available; other times husbands controlled expenditures and insisted that the barn, not the kitchen, receive whatever new technology was available and affordable. In the popular culture of the time, rural folks were often the butt of unkind jokes, in part at least because their lack of the latest technology made them seem less than modern.

As usual, African-Americans were less likely still, in the country at least, to enjoy the benefits of modern technology. Sara Brooks, who was born in the rural South in 1911, later recalled that her father, who owned his own farm, had a Ford that he used for running errands into town. Over the years he also had a wagon and various buggies, but the family had no bathtub, electric lights, or even cooking stove. Yet Brooks titled one chapter of her autobiograpy "The Things He Knew How to Do, He Did Em," and the book is filled with references to traditional tools and technologies: sacks, a well, a dipper, bed, dishes, tongs, anvil, plow, matches, curn, hat, "spider," hoe, buckets, and much more.[24] Her parents made and mended, as well as bought, an array of technologies sufficient to sustain life, a family, and a large measure of both independence and dignity, even in the racist environment of time and place.

If only a few Americans actually jazzed their music and cubed their paintings, as Edwin Slosson feared, many millions of others, if they could, embraced modernity in the most intimate and profound aspects of their lives—the way they worked and the way they lived. Despite the inequalities of race, gender, and class, between urban and rural environments or religious and ethnic proscriptions, modern advertising, installment credit, and government-sponsored infrastructures made it possible for a growing number of people in the nation to "jazz and cube" the very fabric of their lives.

II

DEPRESSION: STUDY AND SUBSIDY

THE GREAT CHICAGO Century of Progress exhibition, held in 1933 to celebrate the one hundredth birthday of the city, picked as its motto an extraordinary message to accompany the new technological marvels on display: "Science Finds—Industry Applies—Man Conforms." The first two decades of this century had seen the flowering of industrial research, the third had witnessed an unprecedented outpouring of consumer goods (particularly consumer durables), and now the fourth decade was experiencing the twin scourges of depression at home and fascism abroad. With the passage of time it is difficult to see how the motto, which was certainly meant to be proscriptive as well as descriptive, can have been seen as positive, let alone something to celebrate.

The 1920s had seen a concerted effort to embrace what was considered a new machine civilization in the country: one founded on the rock of science and dedicated to human improvement through increased production and leisure. When in 1928 the historian Charles A. Beard edited a volume entitled *Whither Mankind: A Panorama of Modern Civilization,* he included chapters from some of the best-known intellectuals of America and Europe. Bertrand Russell wrote on science, Sidney and Beatrice Webb on labor, Lewis Mumford on the arts, and John Dewey on philosophy. In his introduction, Beard denied that what he called "machine civilization" was necessarily destructive of aesthetic, religious, or humanistic values and asserted instead that "these ancient forces will become powerful in the modern age just in the proportion that men and women accept the inevitability of science and the machine, understand the nature of the civilization in which they work, and turn their faces resolutely to the future."[1]

It sounded a good deal like "man conforming," but still the book brought a hot response from a defensive group of scientists and engineers who convinced Beard to edit a companion volume giving their own answer to the question *Whither Mankind.* The new volume, published in 1930, was titled *Toward Civilization* and contained chapters by Elmer Sperry on invention, Robert Millikan on science, and Lillian Gilbreth on work and leisure, among others. In the introduction, Beard expressed his pleasure that those who had "solved the problem of production" were "now deeply concerned about the next stage—a wider distribution of wealth and a nobler use of riches and leisure."[2] They took credit for a society that was, however, already in serious jeopardy.

An even more aggressive claim for technology's triumph was made by Howard Scott, founder of the Technocracy movement. Soon after the First World War the maverick American economist, Thorstein Veblen, had gathered a study group at the New School in New York to investigate the nature and purpose of a possible "soviet of technicians." The group broke up in 1921, but Scott, one of its members, continued to be interested in the problem, forming during the winter of 1932–33 an energy survey of the United States in rooms provided by Columbia University's Department of Industrial Engineering. In an article entitled "A Rendezvous with Destiny," Scott articulated his belief that what he called "political government" was incapable of dealing with the crisis of a crumbling capitalism. Only when that government had collapsed, and another made up of engineers and other technicians had taken power, would the problem of production and distribution be solved. God, he claimed, "in His kindness is on the side of the greatest technology. . . . This generation of Americans has the technology, the men, the materials, and the machinery" for the accomplishment of "a new civilization."[3]

Revelations about the shady past of Scott himself, and the fear that the scheme could be used "as an instrument of oppression and exploitation" (it seemed to some to be, under Scott's leadership, fascist), led to a rapid decline

"Chemical Industry, Upheld by Pure Science, Sustains the Production of Man's Necessities." Ideologies of gender, race, and nature reinforce the belief that technology flows from science, and human happiness from industrial productivity. A. Cressy Morrison, *Man in a Chemical World* (New York, 1937), frontispiece. Reproduced by permission.

of the movement, but throughout the 1930s in at least some parts of the country, the gray cars and trucks of Technocracy, with their red and white yin/yang logo, were a common and striking sight. The old dreams of using technology to avoid politics and of having a government run by expert technicians died hard.

If the fantasies of technocrats were exaggerated, the real impact of new technologies could hardly be. President Herbert Hoover's Research Committee on Social Trends produced a two-volume report, issued some months after

Hoover's electoral defeat in 1932, which contained chapters on population, the arts, women, labor, the law, and many other topics. In his chapter, "The Influence of Invention and Discovery," the pioneer sociologist of technology, W. F. Ogburn, declared that "science and technology are the most dynamic elements of our material culture." To demonstrate the widespread impact of invention, he listed no fewer than 150 of what he called the "effects of the radio telegraph and telephone and of radio broadcasting."[4]

One notable result of technological change was what was widely referred to as the "saving" of labor. One scholar made an intensive study of *Mechanization in Industry,* published in 1934, which showed that between 1899 and 1929, American manufacturing industry had increased its installed horsepower per wage-earner by 130.3 percent, ranging from a mere 22.6 percent in food and kindred products to 257.9 percent in nonferrous metals and their products. A survey of 101 executives produced the estimate that 25 percent of the resulting disemployed were skilled workers, 43 percent were semiskilled, and 32 percent were common laborers. The author admitted that the mechanization did "result in a substantial period of unemployment for the men displaced and frequently necessitated their taking employment at a lower wage." He could not, however "find convincing the evidence for theoretical arguments sometimes advanced to demonstrate an inherent tendency for mechanization to create an ever larger permanent body of unemployed."[5]

Fortune magazine was more blunt. Noting that after the passage of new restrictive immigration laws in the early 1920s "the hunkies thinned out of the employment offices, engineers came into the plants." The replacement of workers by machines had, in its opinion, created a dilemma for American society: "from the purely productive point of view, a part of the human race is already obsolete and a further part is obsolescent. But from the consuming point of view, no human being is obsolete: on the contrary, an ever-increasing human consumption is not only desirable but necessary."[6] Much of the prosperity of the 1920s had been traceable to the desire of consumers, spurred on by a newly aggressive advertising industry and enabled by a new wave of consumer credit, to buy the products of industrial research and mass production. Automobiles, radios, home appliances, and a host of other products were purchased with past or anticipated wages.

But unemployment, not wages, was the signal result of the depression that descended on the nation in October 1929. During the following year the

Massive unemployment was a signal feature of the Great Depression, and after a decade of boasts that technology had transformed production with new labor-saving devices, it was difficult to avoid the suspicion that perhaps the nation had had too much of a good thing. Here unemployed workers kill time around an idle technology. Still photo from the film *Valley Town* (1940), directed by Willard Van Dyke. Courtesy The Museum of Modern Art Film Stills Archive.

ranks of the jobless grew from 429,000 to 2.9 million. In 1931 that army had swelled to 7.1 million and in 1933 it numbered 11.8 million. The disaster was of staggering proportions and reached beyond the general level of factory workers. In 1932 the unemployed counted among themselves some 112,000 engineers. In May of that year 1,400 of the 4,500 engineers in Cleveland, Ohio, were out of work, and an attempt was made to resettle them on 1,200 acres of farmland outside the city where they could "cultivate truck gardens, raising crops for their own use or for barter."[7] By September the Philadelphia Technical Service Committee had registered a thousand engineers in need of relief, and by December two thousand of the twenty thousand chemists and chemical engineers within a 50-mile radius of New York City were out of work. One mechanical engineer, forty-one years old, married with seven children, found himself in 1934 digging ditches for the Civil Works Administration. Discussing his fate anonymously in a popular magazine, he wrote that "the biggest thing to learn is to bring your mind down to the ditch." In a moment of populist irreverence, he noted that he worked with "men of all trades; men who have lost their businesses; salesmen, engineers and executives. . . . Strangely enough, we find no politicians, soldiers or clergymen in that ditch!"[8]

Perhaps taking his cue from New Deal farm policies, the eminent bridge-designer David B. Steinman claimed that "the only way to improve the economic situation of the engineer is to increase the demand for his services and reduce the oversupply of engineers." Projecting that there would be 230,000 unemployed engineers by the year 1960, Steinman asked "is it not time for the engineering profession to do what the medical profession did in 1904,—undertake a program of rigorous selective limitation for education and for admission to practice?"[9]

The supreme artistic statement of labor and the machine was the 1935 film *Modern Times,* written, directed, and starred in by the "Little Tramp," Charlie Chaplin. In it the hero was torn by the modern dilemma: either he was unemployed or he could find work only in a factory where the assembly line and speed-up created a frenzied monotony that drove him mad. The scenes of mass unemployment, and the corrosive effects of that on people, drew their power in part from the fact that they were taken from the everyday lives of millions of Americans. The apotheosis of the modern industrial worker, however, was symbolized by the employed Charlie's actual descent into the very gears of the machine.

At least three times during the New Deal, government agencies attempted to study the effects of technology on society. In *Technology on the Farm,* issued in 1940 by the Department of Agriculture, the authors warned that "it is impossible to forecast the nature of inventions, whims, fashions, and movements that will affect agriculture," but insisted that "it is possible

In his 1936 film *Modern Times,* Charlie Chaplin created the depression's single most powerful image of humanity literally caught up in its own technology. Courtesy The Museum of Modern Art Film Stills Archive.

to forecast that there will be inventions, whims, fashions, and movement." "It would be useless," they concluded, "for us to try to curb this march of technology, for we know that it gives jobs as well as takes them away. Our task, rather, is to study ways to equalize the advantages brought by technology and to help plan a more stable economy."[10]

The Temporary National Economic Committee (TNEC), established by the Congress in 1938, took 1,753 pages of testimony and exhibits on the subject of technology and the concentration of power, and published monographs on both patents and the role of technology in the nation's economy. By the time it issued its reports in 1941, however, the chance of any real reforms had passed. "Dr. Win the War" had replaced "Dr. New Deal," as Roosevelt himself said, and the TNEC satisfied itself with such palliative suggestions as dismissal wage contracts and job retraining programs.

In the 1937 report *Technological Trends and National Policy,* issued by the National Resources Committee, an attempt was made to extrapolate from current research which new technologies were likely to be available twenty-five years in the future. While admitting that "there is as yet no science capable of predicting the social effects of inventions," the committee maintained that "invention is a great disturber and it is fair to say that the greatest general cause of change in our modern civilization is invention."[11]

With hindsight it is obvious that many of the guesses by some of the nation's most eminent scientists were wildly off the mark: General Electric Laboratory's William D. Coolidge, for example, flatly denied that it would ever be possible to unlock the power of the atom. E. B. Wilson of the Harvard School of Public Health laughed at the thought that the sociologist Ogburn "has been 'hipped' on the devastation that the mechanical cotton picker would bring to the South," but pointed out that even the ardent New Dealer Secretary of Agriculture Henry A. Wallace was known to "sneer at the proposition pointing out the very obvious difficulties of a mechanical cotton picker which would be satisfactory—difficulties which Ogburn with his ignorance of the problems of machinery doesn't see."[12]

Ironically, at the very time of Wilson's amusement, the Rust Brothers were perfecting their successful cotton picker, first patented in 1928. The "problems of machinery" had indeed been formidable, but field trials of the Rust machine in 1935 demonstrated conclusively that it would work. "It will," they claimed, "do the work of 50 to 100 men. Thrown on the market in the manner of past inventions, it would mean, in the share-cropped country, that 75% of the labor population would be thrown out of employment. We are not willing that this should happen. How can we prevent it?"[13] The brothers realized that others were working on similar machines, and that a unilateral suppression of their device would do little good. They considered giving market control to the Southern Tenant Farmers Union, leasing machines only to growers who would promise to provide for workers, and were enthusiastic about sending machines to the Soviet Union, where, as a matter of policy, unemployment was not allowed. Finally, though without effect, they called for "Federal and state aid in working out a program for painlessly absorbing the picker into the South's economy." During the late 1930s and

The coming of the gasoline tractor provided farmers with relatively inexpensive flexible, portable power and transformed agriculture into an industrial enterprise. Still photo from the Pare Lorentz film, *The Plow That Broke the Plains* (1936). Courtesy The Museum of Modern Art Film Stills Archive.

early 1940s, many large farmers in such states as Oklahoma and Texas purchased these machines, and in the years after the war the hand-picking of cotton was everywhere abandoned, throwing many thousands (mostly African-Americans) out of work.

During the 1930s the single most important and disruptive device in agriculture was the small, flexible, rubber-tired, gasoline-powered tractor. The Department of Agriculture estimated that there were 1.6 million tractors in use on American farms in 1939, almost twice the number of 1930. Their use varied by region: "the highest degree of adoption is in the small-grain areas, the Corn Belt, and the specialized sections like the dairy, truck, and orchard areas of the Eastern and Western States. In the Southern and Eastern States, small farms and low incomes have not favored the purchase of tractors, but some large speciality farms have been mechanized."[14]

In the Corn Belt of the nation, the tractor was a key part of a larger revolution in the growing of that major crop. The first step was often the purchase of a tractor, and this usually then necessitated the purchase of a whole range of ancillary machinery since that previously pulled by horses proved inefficient when connected to the tractor. Also by the 1930s hybrid corn, developed through careful genetic selection at various state agricultural experiment stations, was finding increasing popularity among middle-western farmers. Finally, by the period 1927–30, some ten thousand corn-picking machines were being sold annually. This number dipped in the early years of the depression, but rose again to fourteen thousand sales in 1937. In

1938, 35 percent of the Iowa corn crop was machine picked and in Illinois, 43 percent. The following year twenty thousand mechanical pickers were counted in Iowa. It is important to note that in this, as in so many similar cases, the three components of the changed method of handling corn were closely intertwined and reinforcing. The mechanical picker was not feasible until the development of the tractor. The requirements of the picker—that the crop should ripen all at once, the stalks be strong and straight, the ears at the same height, all made the use of hybrid seed more attractive, and even necessary.

The use of cotton-picking machines by southern growers had, of course, been preceded by the purchase of tractors. Indeed, it was not uncommon for large Southern plantation owners to turn their sharecroppers (often whole families of African-Americans) off their farms, collect government allotment checks from the Agricultural Adjustment Agency for not planting cotton, and use that money to buy tractors. Between 1930 and 1940 in Oklahoma, for example, the number of tractors increased 75 percent. John Steinbeck, in his disturbing novel of the great farm migration *Grapes of Wrath* (1939), described in moving terms what a new tractor on the farm could mean for sharecroppers and tenant farmers:

> Across the dooryard the tractor cut, and the hard foot-beaten ground was seeded field, and the tractor cut through again; the uncut space was ten feet wide. And back he came. The iron guard bit into the house-corner, crumbled the wall, and wrenched the little house from its foundation so that it fell sideways, crushed like a bug. . . . The tractor cut a straight line on, and the ground vibrated with its thunder. The tenant man stared after it, with his rifle in his hand. His wife was beside him, and the quiet children behind. And all of them stared after the tractor.[15]

The tragic dust bowl of the decade could be, and was, traced back to an inappropriate imposition of a too-powerful agricultural technology on the Great Plains. In 1936 the classic documentary *The Plow That Broke the Plains,* made for the Farm Security Administration by Pare Lorentz, drew a clear and powerful analogy between the recent war on the Western Front in France and the war that farmers had made upon the land which sustained them. A combination of heavy machinery and speculative land practices stripped the protective sod from thousands of square miles of land, so that when drought came, as it must, the land dried up, sending a steady stream of "Okies" heading west for another chance. It represented an official awareness of the interplay between technology and the environment that was not seen again until the 1970s.

Lorentz was forced to add a new and upbeat ending to his pessimistic film: one that purported to show a host of New Deal agencies working to make things right again. The insistence that what technology had taken away it also promised to give back proved a persistent and politically conservative argument. The century-old notion that technological change created as many jobs as it destroyed, coupled with the new faith that science was the best and most reliable source of new technologies, led to a number of schemes to deliberately increase innovation. Ralph E. Flanders, then president of the American Society of Mechanical Engineers and later to be a U.S. senator, insisted in 1935 that it was better to try to recover the status quo than to reform it, since jobs did more to uplift workers than welfare. Paradoxically, it seems, he believed that jobs could best be created through the continued application of "labor-saving" machinery. It was the sort of panacea that led Secretary of Agriculture Wallace to remark that, in his opinion, three-quarters of all engineers before the depression had been social and economic conservatives, who even in the face of the worst depression in the nation's history believed that if only nothing new was done, the good old days would soon be back.

The problems and new possibilities of the early New Deal years did lead to some innovative thinking in the technical community, however. In 1933 Karl T. Compton, president of the Massachusetts Institute of Technology and director of President Franklin D. Roosevelt's newly appointed Science Advisory Board, introduced, along with Alfred D. Flinn of the Engineering Foundation, the $16 million Recovery Program for Science Progress. The program went through several drafts, but in essence, was to provide federal funds for support, on a project-by-project basis, of research proposals that promised to address the creation of new technologies, and through them new industries. In this way, it was hoped, the program would be self-liquidating. An attempt, dating back to the nineteenth century and most recently put forward in 1916, to provide federal funding on a regular basis for state engineering experiment stations again failed to pass the Congress.

The Department of Agriculture had better luck in 1935 when it secured passage of the Bankhead-Jones Act authorizing the secretary to "conduct scientific, technical, economic, and other research into basic laws and principles and processes relating to the improvement of the quality of, and the development of, new and improved methods of production of, distribution of, and new and extended uses and markets for, agricultural commodities and byproducts and manufactures thereof." The National Bureau of Standards was not so fortunate. Its plan for increased funding for what might be called "mission-oriented" research fell afoul of bickering and turf fights within the technical community. Director of the bureau Lyman J. Briggs had envisioned support for a kind of research that lay between basic, as carried on in the universities, and applied, as undertaken by industry. It must, he said, be "quite fundamental in character, but have some distant practical objective."[16]

Without direct support from the technical community, Senator Jennings Randolph (Democrat of West Virginia) introduced a series of bills to create something like a scientific and technological advisory committee with broad powers to test new inventions and processes, grant support for research and development, and hold and administer patents on devices developed with federal funds. It, too, failed in the Congress. All of these schemes contemplated a larger and more active role for the federal government in stimulating and assessing technological progress through government subsidies to research and development. The widespread acceptance of that new role, however, came only with the Second World War, and the projection of that experience into the postwar years.

Part of the problem was that science, even among those who thought it a good source of new technologies, was assumed to have a relatively long payoff period. Ogburn, in making his official study for *Technological Trends,* thought it took about twenty-five years for research to appear in the market as technological innovation. In the midst of the worst economic disaster ever to strike the nation, with its failed farms, bankrupt factories, and massive unemployment, a quarter of a century seemed too long to wait. Instead, the government proved much more willing to pour funds into an upgrading of the technological infrastructure of the nation. Public works were a well-known and understood instrument for the spending of public monies and the employment of masses of workers.

Boulder Dam, later renamed Hoover Dam, stands as a prime example of the public works undertaken during the depression. The arched gravity dam had been a product of western needs and the increasing sophistication of engineers in the use of reinforced concrete. As late as 1931 the largest such structure undertaken was the San Gabriel Dam in California, which was planned to rise 500 feet above its foundations. Boulder Dam, which was eventually to rise 726 feet above bedrock, was a daring engineering project, unprecedented in dimension. The great Colorado River, which drained all or part of one-twelfth of the continental United States, overflowed its banks in 1905–7, causing extensive damage to the Imperial Valley of California and creating the present Salton Sea.

Such disasters caused the Reclamation Service to recommend, as early as 1918, that a dam of unprecedented size be built in Boulder Canyon. Legislation authorizing the project was passed in 1928 and signed by President Calvin Coolidge. In 1930 word was sent down to expedite plans for the dam because of the employment the job would provide, and the following year a contract was awarded to the Six Companies, Inc., of which Henry J. Kaiser, of World War II Liberty ship fame, was a member firm. The stated purpose of the dam well reflected the multipurpose philosophy that guided the Reclamation Service: flood control, improvement of navigation and regulation of water flow, storage and delivery of stored water for reclamation of arid lands, and the generation of electrical power. The work was pushed forward with vigor: the first cement was poured in June of 1933 and the last in May of 1935. The first electric generator was put into full operation in 1936, and by 1939 installed capacity reached 704,800 kilowatts, making it the largest powerplant of its kind in the world, a distinction it had until the Grand Coulee project, authorized in 1933, was completed in 1949.

The statistics on construction were staggering: 5 million barrels of cement, 18 million pounds of structural steel, 21 million pounds of gates and valves, 840 miles of pipe, 5,250 workers employed at one time. The Bureau of Standards, which had to check the quality of supplies purchased, was able to build a new hydraulics laboratory to test the characteristics of high dams. New techniques were worked out on site. For example, it would have taken a century for the mass of cement to cool on its own, producing damaging cracks and warping. A plan that was developed for pouring the concrete into huge blocks, into which were set pipes carrying ice water, solved both problems.

A more elaborate undertaking was the long-contemplated Tennessee Valley project. Authorized in 1933, the Tennessee Valley Authority (known as

The Hoover Dam, shown here under construction in 1934, was a massive engineering enterprise that epitomized the federal government's commitment to using subsidized, large-scale technology to create an infrastructure that would stimulate the economy and improve people's lives. Courtesy Bureau of Reclamation, U.S. Department of the Interior.

the TVA) was assigned the task of taming the floodwaters of the Tennessee River and its tributaries in such a way as to restore the social, economic, and environmental health of the region. A series of dams were to be built that would impound water at flood time, releasing it gradually to produce electrical power. At the same time, such measures as contour plowing and reforestation were to rebuild the natural environment, while electrification and model villages were to restore the health of the social body. Part of the motivation of the TVA's backers was to create a yardstick for measuring the electrical service provided by private power interests. Like the latter, however, the authority had to deal with economies of scale and sought to increase consumer use of electricity.

The Electric Home and Farm Authority was established to provide low-interest loans to people for "appliance purchases and worked with manufacturers to get low-cost appliances onto the market."[17] As scholars have noted, the whole effort of the TVA expressed "a faith that electricity will exorcise social disorder and environmental disruption, eliminate political conflict and personal alienation, and restore ecological balance and a communication of man with nature."[18] Economic subsidies and political policies, some southern agrarians had argued, had worked to centralize both people and economic power. The TVA's Industrial Division was designed to use the power of electricity to encourage smaller and more scattered industrial development. It was an ideological commitment to social engineering, which was not saved from criticism by the fact that it was to be guaranteed by physical engineering. The presidentially appointed Mississippi Valley Committee, made up of water experts in the Progressive tradition, sought to plan for an even larger TVA-type development of the entire Mississippi basin, but Roosevelt failed to support its report.

One member of that committee, however, the Progressive engineer Morris L. Cooke, was appointed to head up the new Rural Electrification Administration (REA) created in 1935. In that year fewer than 10 percent of American farms were served by electricity from central power systems, although a small number had their own diesel or gasoline-driven generators. The problem had been that private power companies were reluctant to extend lines for long distances to serve only a few customers. The REA was designed to encourage the formation of farmer-owned and -controlled electric cooperatives that would receive government loan guarantees to build the needed lines. As with the TVA, there was an initial fear that farmers would

The push for rural electrification in America during the 1930s played upon the theme of labor saving, especially for women. This advertisement by General Electric, however, raises more questions than it answers. The text reads "Slavery is the cornerstone of civilization," and although the conflation of slavery and technology was at least a century old, the invocation of the Hindu goddess Vishnu (worker and preserver of the world) was uncommon. Most of the tasks to be performed were gendered feminine and the electrical "slave" was also pictured as female, suggesting that technology could revolutionize the work process without disturbing traditional gender roles. *Fortune,* 7 (April 1933), 36.

not use enough power to make the effort worthwhile, and a campaign was launched to stimulate consumption. This included demonstration workers, pamphlets, 4-H projects for children, meetings, and a demonstration farm equipment tour, called the "REA Circus," which featured a tent seating a thousand and "a dozen equipment sideshows, furnished by manufacturers." A film entitled *Power and the Land* (1940) was circulated to convince farmers not only that electricity would be useful in their work, but that cooperating to build the lines was traditionally American (like a barn-raising) rather than a dangerously Bolshevik activity.

By the end of the decade it was discovered that this campaign had been so successful that farmers actually used more kilowatt-hours per person than did city-dwellers. Forty to 50 percent of that power went to irrigation, 30 percent was used for household appliances, and 20 to 30 percent mainly for lighting barnyards and outbuildings, but also for chick-brooders, milking-machines, and similar miscellaneous devices. Indeed, it was notorious that modern electrical technology found its way into the barn before it appeared in the kitchen. Undervaluing the work that women did, it seemed an obviously good idea to "invest" the electricity in production rather than "consumption." By 1940 some 78.5 percent of farms on the Pacific Coast had been electrified, but only 10 percent in the West South Central region. The average for all American farms was 29 percent, and twenty-five thousand new farms a month were being added to the total.

By 1939, new world's fairs in New York and San Francisco again held out the promise of better things for better living through science, to paraphrase a contemporary advertising slogan of the Du Pont company. By this time large corporations had long since given up on showing how things were made and concentrated on how they were to be used. A decade of experience, however, had shown that the motto of the Century of Progress exposition, "Science Finds—Industry Applies—Man Conforms," had a sinister as well as a utopian message. The massive economic disaster of the Great Depression, especially with its devastatingly high levels of unemployment, had sparked an unprecedented public debate over the sources and effects of technological change, as well as the role of science in stimulating that change.

The slogan itself appeared true enough. Schemes put forward by the Science Advisory Board and other agencies to increase government investment in scientific research were always based on the proposition that it would pay off in new technologies, which would form the base for new industries and thereby create more jobs and profits. It was still true, also, that "industry applied." Federal initiatives to bring new technology directly to the people were few and severely limited. The full potential of such experiments as the TVA and the REA was never allowed to demonstrate itself. Suggestions like that of Senator Randolph to establish some sort of federal agency for the stimulation of applied science in the public interest never even received a full hearing, let alone the backing of the president. Industrial capital remained the principal link between science and technology.

And the notion that "man conforms" was also universally accepted. This meant, by the end of the decade, more than simply that people seemed to want to buy the appliances that appeared on the market in such profusion. Much of the destructive nature of the depression was still understood in terms of cultural lag—that is, that society was lagging in its ability to deal with the material changes wrought by science and technology. An antiquated economic system could not distribute the increased production of a mechanized industry and agriculture. An antiquated governmental system could not adequately protect its citizens from the impacts of new technologies, nor could it provide fast enough the engineering infrastructure necessary to use them most effectively. All of the major government reports and most of the public debate on the subject agreed that scientific advancement and technological change were inevitable and that society had no option but to change to accommodate these. The popular economist Stuart Chase, in a book entitled *Men and Machines* (1929), put the case in plain terms: "we can drift with the tide as at present. We can officially adopt some simple formula like 'government by business,' or 'state socialism,' and thus attempt to run a dreadnought with a donkey engine. Or we can face the full implications of the machine, relying on no formulas because none adequate have been created, with nothing to guide us but our naked intelligence and a will to conquer."[19] The Department of Agriculture was more practical and less utopian. "The basic problem," it said, was to "provide employment and security to the displaced and underprivileged persons most adversely affected by technology."[20] It was a humane but meliorist approach, one that aptly characterized the New Deal's approach to the entire Great Depression.

During the 1930s, the full and disastrous impact of the technological changes that had accumulated during the preceding generation staggered the American institutions of welfare capitalism and representative democracy. Overseas, the planned economies of Communist Russia and Nazi Germany seemed to some to suggest that only a strong, centralized, command-and-control government with a regimented society could harness the full power of modern technology for the good of the state, if not of all its people. Such an extreme break with the nation's basic political ideology was never seriously contemplated, however. If the logic of mass production, the rapid mechanization of agriculture, and an increasing demand for consumer durables was imposing a collectivist imperative on American society, it was a development noted but not seriously challenged by the New Deal administrations of Franklin D. Roosevelt.

12

WARS AND THE "AMERICAN CENTURY"

THE DECADE OF THE 1930S, characterized by the triumph of fascism abroad and the failure of capitalism to revive the economy at home, led to numerous schemes for reform and revolution. Except in a few areas, such as big dam building in the West and rural electrification, suggestions for enlisting the federal government directly in the planned stimulation of technology to solve the depression were rejected. The threat of social dissolution, sufficiently blunted by New Deal measures, was not strong enough to overcome the private interests of large business organizations, protected as they were by a still-powerful ideological commitment to the idea of free enterprise. American technology, therefore, remained the instrument of corporate rather than public policy.

With the coming of world war to Europe in September of 1939, however, a new public purpose of immense proportions appeared on the horizon. The common defense provided what the general welfare had failed to produce—a consensus that technology should be shaped and mobilized by the government to accomplish a great public purpose, even at the risk of private profit and privilege. As it was worked out, not surprisingly, the decision to prepare for and then win World War II with as little disruption as possible to the prewar economic and social status quo meant that public purpose was designed so as to largely enhance rather than threaten private arrangements. For over a generation, the decisions on how to fight the war gave shape to the nation in peace.

The three decades following the end of the war seemed to provide that "American Century" of which some had dreamed. A generation witnessed the nation's unparalleled growth, prosperity, and influence around the world. The new technologies of the interwar years had been rationalized by the war, and wartime research and development added new ones. Just as important, after nearly two decades of New Deal and defense planning the federal government was accustomed to playing an even more active role at home and abroad in creating and subsidizing a host of initiatives, from the Marshall Plan for the capitalist reconstruction of Europe to urban renewal at home. Of particular significance was the spread of transportation and communications infrastructures such as interstate highways and television networks. This marriage of the influences of Henry Ford and John Maynard Keynes appeared to produce those sturdy progeny of corporate hegemony, consumer prosperity, and American world dominance.

The growing military threat of Nazi Germany worried Vannevar Bush in 1939 and 1940. A professor of electrical engineering and vice president at the Massachusetts Institute of Technology (MIT), Bush had been elected the president of the Carnegie Institution, an influential philanthropic foundation that was actively supporting science from its headquarters in Washington. Once in the nation's capital, Bush was appointed to the National Advisory Committee on Aeronautics (NACA) and a year later was made its chair. When wearing any one of a number of these hats, he could see that the United States was not, in his opinion, adequately responding to the German menace, particularly its large and technologically advanced air force.

Bush's response was to gather together a high-level group of scientists representing all the important constituencies of American science. Frank B. Jewett, recently made president of the National Academy of Sciences, was a physicist who directed the Bell Laboratories and served as a vice president of AT&T. James B. Conant was a chemist and president of Harvard University. Karl T. Compton was president of MIT. All these men had strong ties with the most powerful corporate, educational, and philanthropic patrons of science in the nation. In the spring of 1940 they succeeded in having President Franklin D. Roosevelt appoint them to the newly formed National Defense Research Council (NDRC), where they were joined by others, notably representatives from the army, navy, and Patent Office. The latter positions were ex officio. Bush and his colleagues from Harvard, Bell Labs, and MIT were in agreement that they could not rely on the vaunted Yankee ingenuity of their fellow citizens for the new "instrumentalities of war" that they hoped to produce. A separate National Inventors Council was established to handle the inventions suggested by "people unknown to us," as Bush put it. Real progress was to be made through the granting of contracts to institutions and individuals with proven records of research excellence.

When the NDRC was, in turn, made a branch of the newly established Office of Scientific Research and Development (OSRD) in 1941, Bush was moved up to chair the OSRD, and it continued the policy of buying new technology from the largest institutions with the best reputations. Thus it is no surprise that through 1944 the OSRD spent $337 million (out of a total of $1.8 billion for the government as a whole, excluding the Manhattan District) on research and development, and that among nonindustrial agencies, MIT received more of these funds than any other, followed by the California Institute of Technology, Harvard, Columbia, the University of California, and Johns Hopkins University, in that order. In 1964 these institutions were still among the top twenty universities in terms of defense research and development, and MIT was still at the top of the list. Among industrial contractors, Western Electric (the manufacturing arm of AT&T), the Research Construction Corporation (formed specifically to build facilities at MIT), General Electric, RCA, Westinghouse, Remington Rand, and Eastman Kodak were the favored contractors, in that order.

The overall result of this policy was to make the rich richer and the poor poorer. Although such corporations as General Motors and Boeing turned out vast quantities of more or less traditional materials of war, the research and development of the OSRD pointed toward the creation of what has been called "an electronic environment for war." Such devices as radar and proximity fuses, and especially the first generation of computers, developed to facilitate operations research and related activities, all gave the major OSRD contractors a firm grounding in the key technologies of the future.

The least-known but most expensive new technology to emerge from the war was, of course, the atomic bomb. Originally suggested by a group of émigré European scientists, the idea for such a weapon was first investigated by a special committee headed by Bush and then turned over to the Army Corps of Engineers for development. The corps, in turn, called in industrial firms, especially the Du Pont Company, to manage the project. All told, a

The thermonuclear (hydrogen) bombs that were designed at national laboratories were key elements in an arms race, which, ironically, was run as much against America's own research and development capabilities as against any credible Soviet threat. Reinforced by scientific, political, and corporate interests, these weapons seemed to grow, to a remarkable degree, out of their own logic. Courtesy Lawrence Livermore National Laboratory.

small army of scientists and engineers spent perhaps $3 billion producing the test weapon at Trinity and the two bombs that devastated Hiroshima and Nagasaki in 1945.

The massing of resources from industry and educational institutions under military necessity became paradigmatic for the future of American technology. An amalgam of experience (such as the building of the Tennessee Valley Authority [TVA] projects in the 1930s), current practice (as with the grants and contracts being used so successfully by the OSRD), and innovation (the free flow of Pentagon funds directed to narrowly military ends), the Manhattan Project became the godparent of the space program of the 1960s and a host of lesser efforts in Big Science. A half century later the University of California still operated national laboratories inherited or spun off from the Manhattan District, and they in turn spawned and supported two generations of military technologists who were responsible for, among other projects, the hydrogen bomb and President Ronald Reagan's "Star Wars," the Strategic Defense Initiative.

After the innumerable cries of wolf that heralded so many false "new technological ages" over the years, the genuine discontinuity of nuclear technology seems hardly to have been appreciated. The idea that it had given rise to "just another bomb" gave way to the declaration of a new Atomic Age, but one miraculously without cost or danger (except in the hands of America's enemies). Energy too cheap to meter, nuclear-powered aircraft and artificial hearts, and myriad other miracles seemed to beckon just over the horizon. The setting up of the Atomic Energy Commission (AEC) in 1946 deliberately placed the powerful new technology in civilian hands, but it was the nuclear card in the growing cold war that powered the rapid development of the field. The Atomic Age became one of anticipated fantasy and postponed reality.

The newly established AEC began the long and expensive process of trying to create a civilian nuclear reactor program, which eventually, in 1953, resulted in the funding of the first nuclear power plant in Shippingport, Pennsylvania. Even more hopeful was the inauguration, in 1951, of Project Sherwood, aimed at developing the ability to harness controlled fusion. "If the present efforts are successful," wrote one AEC official, "man will have found the ultimate solution to one of his most pressing problems. He will

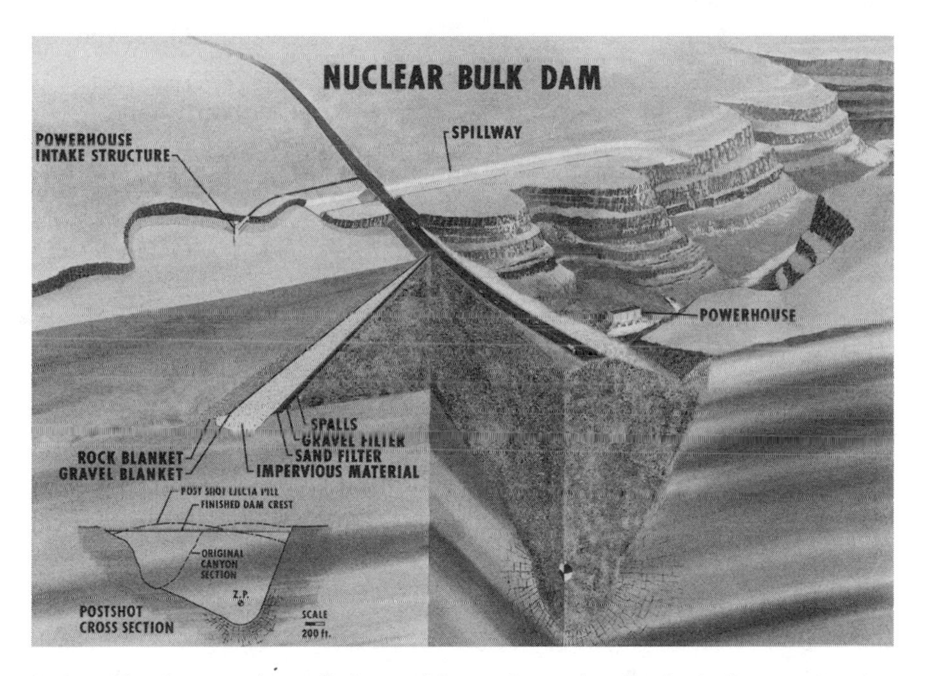

Project Plowshare sought to find peaceful uses for nuclear bombs in large engineering projects. This artist's conception of a "nuclear bulk dam" shows how such a structure might look after a nuclear blast had thrown up a "post-shot ejecta pile." Courtesy Lawrence Livermore National Laboratory.

have developed a new and practical source of energy which will meet his needs, not for just the next hundreds or thousands of years, but for as far into the future as he can see."[1] In 1957 the AEC undertook Project Plowshare, which sought to use nuclear blasts to excavate for civil engineering projects both above and below ground. The idea of using nuclear power to generate electricity or dig artificial harbors was always overshadowed, however, by the declared need to continue to expand and elaborate the nation's stockpile of atomic bombs. The most dramatic leap forward came with the development of the hydrogen bomb, the first of which was tested in 1954.

A heavy national commitment to ever-elaborating nuclear arms, as it turned out, was only one aspect of an arms race that was instituted during the cold war. During World War II the Army Air Corps had dreamed of and planned on achieving separate service status with, presumably, more independence and at least a third of the military budget. Separate status was justified on the basis that strategic bombing, rather than close support of ground troops, was the way to use airplanes in winning wars. The projected size of the planned service was based on the personnel who could be expected to enlist (a peacetime draft was not considered politically possible). So many personnel meant so many airplanes, which dictated a certain number of air bases on which to park them, and, most important, a likely enemy worthy of such a buildup. The wartime innovations of jet propulsion, radar, strategic nuclear weapons, and so forth made new technology supported by strong science an obvious need, and the RAND (Research and Development) Corporation was set up to keep the new air force (separately established in 1947) in the game despite its lack of research facilities of its own.

The German rockets, the V-2 especially, had presented a new and terrifying weapon to the arsenal of war, and before V-E Day, Project Paperclip was on the European landmass, trying to get back to America as many German plans, drawings, rockets, and rocket specialists as possible. The setting up of the Redstone Arsenal at Huntsville, Alabama, to accommodate the German scientists and engineers was the army's response to the interservice rivalry that developed over this new weapon, and it provided the capability that finally produced an American response to the Soviet launching of *Sputnik* in October 1957. One wag remarked that the Soviet success in launching the world's first satellite merely proved that the Soviet Union's Germans were better than America's Germans. Ironically, October 1957 was the same month in which the Ford Motor Company unveiled its notorious new car, the Edsel.

The early years of NASA's space program were sprinkled with spectacular mishaps. Usually referred to as "engineering failures," they contrasted dramatically with "scientific successes," that is, those launches that went according to plan. In 1965, this Atlas/Centaur launch failed when the vehicle's main stage cut off prematurely. Courtesy National Aeronautics and Space Administration.

The political embarrassment of *Sputnik* shook up the government's structure for missile development and science policy in general. Service rivalries and a palsied NACA program were replaced (or at least overshadowed) by the creation, in 1958, of the National Aeronautics and Space Administration (NASA) and of both a science adviser to the president and a President's Science Advisory Council (PSAC) for that person to chair. It was to be a brief but golden age for science. The National Defense Education Act provided funds especially for young people who chose to become engineers or scientists, and the National Science Foundation pushed a curricular reform, reaching down into the elementary schools, featuring the "new math." Not only could American schoolchildren not read; their technical competence was thought to be well behind that of their Soviet counterparts. Military demands mixed with the 1960 presidential-campaign claims by John F. Kennedy of a dangerous "missile gap" spurred a politically charged atmosphere ending, in 1961, with by-then-President Kennedy's declaration that this nation's space program goal was to place a person on the moon within the decade.

Whether Americans looked at nuclear weapons, missile delivery systems, or myriad other, often more conventional, weapons, two lessons seemed to stand out starkly. First, the arms race appeared not to be with the Soviet Union at all, but rather with themselves. Since every offensive or defensive weapons system could theoretically be neutralized by a new generation of its opposite, it was necessary to begin immediately to counter that possibility. Thus the real race was with America's own laboratories. Second, with so much of the nation's limited resources of monies and technical talent committed to the arms race (it was claimed that one year's increase in the NASA budget was sufficient to hire every Ph.D. engineer graduated that year), the opportunity cost to civilian technology was enormous. A massive arms race (and the huge national debt it created) privileged certain segments of the nation's economy, but cost others dearly. Claims that civilian spin-offs made the effort worthwhile were less than convincing. Such supposed civilian by-products of arms technology as Teflon and dried orange drink looked very much like the scorched rat that resulted from burning down the barn.

The research, development, production, and deployment of a large number of high-technology weapons systems had substantial advantages, however, for a relatively small but influential number of industries and people. In his farewell address to the nation in 1961, President Dwight D. Eisenhower warned that "in the councils of government we must guard against the acquisition of unwarranted influence, whether sought or unsought," by what he called the military-industrial complex. While we should hold "scientific research and discovery in respect," he added, "we must also be alert to the equal and opposite danger that public policy could itself become the captive of a scientific-technological elite." The country, he said, had "been compelled to create a permanent armaments industry of vast proportions." Also new was the scale and nature of the military establishment, which, Eisenhower said, "today bears little relation to that known by any of my predecessors in peacetime, or indeed by the fighting men of World War II or Korea." And even science itself had been transformed. "A Government contract," he charged, "becomes virtually a substitute for intellectual curiosity. For every old blackboard there are now hundreds of new electronic computers."[2]

Despite Eisenhower's doubts, one wartime technical effort undoubtedly sped up the development of what proved to be a key machine of the postwar years. The prehistory of the computer went back many years, at least to the nineteenth-century efforts of Charles Babbage to develop a calculating engine. More recently, Vannevar Bush, in the early 1930s while still at MIT, had developed a differential analyzer, made up of electrically driven mechanical components, designed to solve differential equations. With the coming of the war, the need to manipulate large numbers of figures to solve problems in such areas as ballistics led to the hiring of thousands of women who were known as "calculators." To save this labor and speed up the tedious process, larger machines were undertaken on the Bush-MIT model, but at the University of Pennsylvania a group of technicians began work on the first electronic digital computer. Pennsylvania's ENIAC was ready for use toward the end of 1945, too late to help with the war effort but marking an important step forward in the development of the computer. Significantly, not only was this wartime device not funded by the OSRD (it was funded rather by the Ordnance Department of the Army), but that agency actually refused to share important data with the development team.

The massive government involvement with technological development that characterized the war years, while by no means unprecedented, was vastly larger and more coordinated than anything that had gone before. The nation found itself in 1945 not only in possession of powerful new technologies, most notably the atomic bomb, but economically and diplomatically poised for an aggressive role as, in the euphemistic phrase of the time, "the leader of the Free World." In an atmosphere of cold war with the Soviet Union and following the dream of an "American Century," strong government deployment of that technology would surely lead to prosperity at home and commercial hegemony abroad.

In order to rebuild the shattered economy and physical infrastructure of Europe along safely capitalist lines, President Harry S Truman's Marshall Plan resulted in the spending of large sums of money in the European nations of the Western bloc. In Japan, too, the devastated technological base of a former enemy was rebuilt along American lines through American aid and direction.

In less-developed portions of the world, many of them newly emerging from colonial domination after the war, the threat of communist takeover seemed even more imminent. Poor and hopeless people, declared Truman, might turn to the Soviet Union out of desperation unless provided with decent housing, food, and opportunity by the West. Within the Truman Doctrine, Point Four, as it was explained in a speech on June 24, 1949, was designed to provide for the "technical assistance [which] is necessary to lay the ground-work for productive investment." Happily, "investment, in turn, brings with it technical assistance."

Truman's conception of the way in which American technology could trigger the development process (and, of course, bind the new economy to that of the United States) was thoroughgoing:

> The aid that is needed falls roughly into two categories. The first is the technical, scientific, and managerial knowledge necessary to economic development. This category includes not only medical and educational knowledge, and assistance and advice in such basic fields as sanitation, communications, road building, and government services, but also, and perhaps most important, assistance in the survey of resources and in planning for long-range economic development.
>
> The second category is production goods—machinery and equipment—and financial assistance in the creation of productive enterprises. The underdeveloped areas need capital for port and harbor development, roads and communications, irrigation and drainage projects, as well as for public utilities and the whole range of extractive processing, and manufacturing industries.[3]

Over the years the export of American technology, along with the social, economic, and political institutions and habits necessary to tie it to American interests, developed into a vast and continuing "foreign aid" program. Sometimes directly, sometimes indirectly through such international agencies as the World Bank (itself largely supported by American capital) and the International Monetary Fund, the nations of what came to be called the Third World were helped to build dams and nuclear power plants, cut down rain forests, import luxury consumer goods, adopt the "green revolution," and beef up military hardware. In the long run it proved easier to destroy local and indigenous cultures than to transplant our own, as Mark Twain realized in his 1889 novel *A Connecticut Yankee in King Arthur's Court* when he tried to make Sir Boss the benefactor of Arthurian England. Just as that experiment in democratic and technological uplift ended in an environmental disaster and military slaughter, so later did it prove easier for President Lyndon B. Johnson to devastate the country of Vietnam than to realize his dream of a Mekong Valley Authority to reproduce the miracle of the TVA in southeast Asia.

Part of American aid abroad was in the form of credits to buy American goods, particularly military weapons. Part of it, however, went into the building of the engineering infrastructure that was deemed necessary for modernization. The privately owned Bechtel Corporation, which in 1983 had revenues of $14.1 billion, was an example of the type of firm that spread American technology abroad. Formed as a small construction company in California in 1906, Bechtel had been one of the partners in the building of Boulder Dam in the 1930s, and during World War II had prospered mightily not only by building Liberty ships (as did their erstwhile associate, Henry J. Kaiser) but also engaging in other projects, such as building the infamous Canol oil pipeline to Alaska.

Having good relationships with Washington proved equally important when Bechtel became heavily involved in building oil production facilities in Saudi Arabia and when it made a major commitment to building nuclear power plants at home and abroad. Like Brown and Root, the Texas engineering firm that was favored by President Johnson and did so much construction in Vietnam during the war, Bechtel built political relationships as assiduously as it did refineries and power plants. During the administration of President Richard M. Nixon, George Schultz served as labor secretary and Casper Weinberger as director of the Office of Management and Budget. Later, Schultz became president of Bechtel and then secretary of state under President Ronald Reagan, while Weinberger became general counsel of Bechtel and then secretary of defense under Reagan. W. Kenneth Davis, a former member of the AEC, became head of Bechtel's nuclear-development operations and later was made deputy secretary of energy under Reagan, in charge of increasing the nation's nuclear industry. Throughout the postwar decades, foreign and domestic technological policies, both private and public, were closely coordinated.

At home, the restructuring of society moved more smoothly. The pent-up demand of the wartime years for consumer goods managed to soak up the expanded industrial capacity of the nation as well as employ the millions of troops brought home and demobilized, taking back "their" jobs from the thousands of women who had been recruited into heavy industry during the war but were now sent home to be housewives and mothers. During the war a fear of renewed depression after the war led to calls for a vast new public works program to keep the economy moving, and in fact something very like that was put in place.

The census of 1920 had shown that, for the first time, more Americans lived in cities than anywhere else. In 1970, the census showed that those same cities had lost population to the suburbs, where most Americans now lived. The millions of homes constructed after the war were much more than just dwellings, of course. By design they gave dimensionality to prevailing racial, gender, and class assumptions about the country. Through devices such as red-lining, neighborhood tracts were segregated by race and class and embodied the assumption of separate spheres for men and women, an assumption (verging on an insistence) that Father would leave the house to work and that Mother would spend her day caring for the home and family. The predominant single-family, detached dwellings, isolated in residentially zoned areas, were to prove awkward for the families and lifestyles of a future generation. By the mid-1980s, three-quarters of the available housing stock of the nation had been built since 1940: of these, two-thirds were single-family detached homes.

The vast tracts of speculative housing that ate up farmland around the nation's cities (in 1949 Levittown became the symbol and best-known example of the result) were dependent on a host of consumer durables, nearly all of which had been available before the war, to make them viable and attractive. Houses, built quickly and cheaply with the century-old technique of balloon-framing and filled with appliances, were connected with points of production, commerce, and sociability by the automobile, which now solidified its role as the symbol of the American Way of Life. Unavailable during the war, cars rolled off the assembly lines in vast numbers after 1945. Annual production of passenger cars reached 7.6 million in 1963, to which were added 1.5 million trucks and buses. During one week in 1966, U.S. motor vehicle

After World War II, pent-up consumer desires and accumulated savings came together in an orgy of spending for domestic technologies. In the mid-1950s, this housewife, dressed like a bride in her own living room, stands between a mural depicting the out-of-doors and a new air-conditioner, which, beneath the blinded window, mediates between the "natural" environment outside and the feminine space within. Photo from the Warcha Air Conditioning Company, ca. 1955. Courtesy National Museum of American History, Smithsonian Institution.

production reached 208,604 units, and in 1971 it was estimated that over 112 million motor vehicles were currently registered in the country, an increase of 3.6 million over the previous year. By the 1970s it was claimed that one out of every seven American paid workers gained her or his livelihood from the automobile.

In 1940, just before World War II, the first freeway in California, from Pasadena to Los Angeles, was opened. After the war, the Collier-Burns Act, passed by the California state legislature in 1947, expanded the 19 miles of freeway in the state to 300 miles ten years later. In 1959 new legislation created the California Freeway and Expressway System, which was to extend 12,500 miles by 1980. By 1968 just one of these freeways, the Santa Monica,

"It's Fun to Live in America." In 1947 Kiwanis International began a series of leaflets prepared in the "interest of the American Way of Life." The first in the series compared the number of cars, radios, and telephones, per capita in the United States, Britain, France, and the Soviet Union. Reprinted with permission of Kiwanis International.

was carrying 210,000 vehicles a day. A highway lobby successfully protected the state gasoline tax (which supplied 54 percent of the funds for the system) as a specifically highway, rather than transportation, fund. The state's dependence on the automobile, with its attendant smog problem (first identified in the early 1950s by A. J. Haagen-Smit, a chemist at the California Institute of Technology), continued as alternatives were neglected.

By the early 1950s the freeways of Los Angeles had come to symbolize the triumph of the private automobile in American life. Two-thirds of the surface area of the city was given over to the automobile, and the car culture had decisively shaped everything from courtship rituals to suburban lifestyles. Pictured is the Hollywood Freeway near downtown Los Angeles in 1953. Copyright © 1953 California Department of Transportation.

Nationwide, the Federal-Aid Highway Act of 1956 committed the country to spending $50 billion to construct 41,000 miles of what came to be called interstate highways. As so often was the case, the ostensible reason put forward for this system was national defense, but the interstate trucks that were busily taking over freight carriage from the railroads, and the oversized family sedans and station wagons taking Americans on long vacations, turned out to be the real beneficiaries.

The building of the nation's interstate highway system, and freeways in general, fitted nicely with yet another massive government project to rebuild the nation's infrastructure, the urban renewal program of the 1950s and 1960s. As the technologies of public transportation, especially streetcars and commuter trains, were driven out of business or curtailed by the advocates of

In 1971, the elaboration of the California freeway network was continuing apace. Here in northern California, as elsewhere, older neighborhoods were wiped out to allow for faster commutes from more distant suburbs, contributing to the sprawl that was taking over more and more prime agricultural land between established cities. Courtesy State of California, Department of Public Works.

the automobile, traffic in cities became increasingly congested. The spread of suburbs continued to encourage a flight of the white middle class from the cities proper, and the need of businesspeople to reenter and leave the city during rush hours led to the design and building of inner and outer beltways to "speed" traffic to downtown areas. More often than not, these wide and expensive swaths of freeway were built through African-American or ethnic working-class neighborhoods, where land values, the tax base, and anticipated citizen outcry were least. Because of these projects, combined with plans to tear down "blighted" parts of the city to make them available for modern and more lucrative development, square miles of historic structures were pulled down, and entire neighborhoods destroyed.

Together, the interstate highway system, urban renewal, and suburbanization combined to create a kind of government-aid development program for the country. Tax monies and a host of subventions, regulations, subsidies, incentives, tax loopholes, and abatements encouraged the creation of a national culture grounded in consumption rather than production, mobility rather than stability, homogeneity rather than diversity, and conformity rather than individuality.

The very viability of individual citizens appeared to be in question. The early republican ideal of the "independent producer" was all but destroyed with industrialization in the nineteenth century; now the very concept of most people as producers at all seemed to be in danger. In 1946, D. S. Harder, a vice president of the Ford Motor Company, in describing a new plant to produce engine blocks, used for the first time the new word *automation.* The word appeared in print the next year, in the pages of *American Machinist,* and soon the nation was talking about the new phenomenon. Harder himself had been referring only to transfer machines that fed parts to machine tools. The concept at its largest, however, seemed to promise, in the not too distant future, the ultimate conclusion of two hundred years of industrial mechanization: the workerless factory.

In 1948 Norbert Wiener published his book *Cybernation,* coining that term to encompass the interconnection of human nervous systems and mechanical control systems. His 1950 book *The Human Uses of Human Beings* predicted a rapid spread of automation and raised alarms about its social impact, including massive unemployment. Several strong labor unions, particularly the United Auto Workers in 1954, began to seek protections against automation in their bargaining sessions with management, and soon the specter of mass disemployment through automation caught the public's attention. Kurt Vonnegut's first novel, *Player Piano* (1952), painted a dismal picture of an American society governed by a technical elite, in which the major decisions were made by computer and most citizens either served in the army or worked in gangs as common laborers in something like the Works Projects Administration of the depression days.

The popular use of the term *automation* to mean anything significantly more automatic than previously existed obscured the technological basis of the change. In essence, three elements were needed to produce the effect: machines to do the actual production of goods, other machines to pass the workpiece between the production machines, and some control system to regulate the whole process. The last, of course, had most recently been augmented by the coming of the computer. One had been purchased by the Bureau of the Census in 1952, and in 1957–58 others were applied to process monitoring and control in both oil refineries and electrical generating plants. Within a few years the first computer-controlled production line was in operation at a Western Electric plant in North Carolina. In 1966 it was estimated that there were fifteen to twenty thousand computers in use in the United States, though not all, of course, were used in manufacturing.

Earlier types of control mechanisms were of fairly ancient vintage. Cams had long governed the motions of machine parts, as when the cam shaft of an automobile engine raises each exhaust and intake valve in turn. Punch cards went back at least to the eighteenth century and formed the critical information storage base for the Jacquard loom. The principle of feedback was embodied in the ball-governor used by James Watt to regulate the speed of his steam engines and was widely used in bimetallic thermostats. Now, coupled with the computer, these became even more powerful control devices.

Automatic production machines also went back to the eighteenth century. Oliver Evans's "automatic" flour mill of the 1780s was an outstanding example of the use of mechanism to replace human intervention in the productive process. By the 1870s, examples could be found in nearly every field of production. Transfer machines were used in the Waltham Watch Company factories as early as 1888, picking up workpieces at one machine and passing them along to another. In the 1930s, handling systems were widely used in the automobile industry.

While union leaders pointed with alarm at mass unemployment, industry representatives scoffed at "the poorly rationalized and undocumented claims" of unemployment, which amounted to what they called an "automation hysteria." The public, it was claimed, thought of automation as "simply *increased automaticity,*" a "phenomenon as old, as pervasive as mechanism itself."[4] If automation was not a break with the past, past measures could be expected to deal with it adequately. If it was truly a revolution, however, revolutionary social controls and adjustments were called for.

Indeed, one group of intellectuals not only believed that automation was revolutionary, but saw it as intimately connected with two other contemporary revolutions as well. The 1966 manifesto on the "Triple Revolution" defined the three: "The Cybernation Revolution: A new era of production has begun. . . . The cybernation revolution has been brought about by the combination of the computer and the automated self-regulating machine"; "The Weaponry Revolution: New forms of weaponry have been developed which cannot win wars but can obliterate civilization"; and "The Human Rights Revolution: A universal demand for full human rights is now clearly evident."[5] Thus, in America the conjunction of social movements for nuclear disarmament and an end to the Vietnam War, the Civil Rights movement, and a concern that increases in production efficiency should go toward alleviation of poverty, not an increase in unemployment, put the automation problem in a large context of social concern.

Not yet politically wounded by the anti–Vietnam War movement, President Lyndon B. Johnson appointed a National Commission on Technology, Automation, and Economic Progress. It, too, reported in 1966 and took a somewhat more melioristic view of the question. Finding that output per person-hour had increased at 2 percent per year from 1909 to 1947 and only 3.2 percent from 1947 to 1965, it admitted that "this is a substantial increase, but there has not been and there is no evidence that there will be in the decade ahead an acceleration in technological change more rapid than the growth of demand can offset, given adequate public policies." It did go on to recommend "a program of public service employment, providing, in effect, that the Government be an employer of last resort, providing work for the 'hard-core unemployed' in useful community enterprises," but, like its other policy recommendations, this fell victim to the president's increasing commitment to guns in Vietnam rather than butter at home.[6] In the end, neither anger in the streets nor the cool academic application of technocratic planning brought about the social changes needed to produce a just industrial policy for the nation. The major unions negotiated "a piece of the machine," as Harry Bridges of the West Coast Longshoremen's Union termed it, and the dramatic armies of angry unemployed failed to materialize.

Three elements combined to produce the automation scare of the 1950s and 1960s from these disparate precursors. First, the machines of production, transfer, and control were combined into powerful systems within single plants. Second, the computer greatly enhanced the control factor over products as disparate as gasoline in a refinery and paperwork in a banking branch. Third, the confluence of these at a time when the periodic American mood of reform was in full flow led to a critical appraisal of the phenomenon that made the concept, if not yet the reality, a fact for everyday consideration. As it turned out, the impact of the computer far outran the worst-case scenarios of those who looked only to the workerless factory.

In rural America a similar pattern of increased mechanization and tighter coordination of systems was asserting itself. For the quarter century after 1940, the number of farms in the United States fell steadily and the number of people living on farms continued to drop, while the size of farms grew. With the coming of the rubber-tired, all-purpose gasoline tractor and rural electrification in the 1930s, the Industrial Revolution truly came to the farm. After the war, irrigation spread, even for farms not considered to be in semiarid regions. So true was this that wells in farming areas quickly drew down the underground aquifers and cast doubt on the long-term viability of the practice. Large water projects continued to be built with tax monies, most notably the California Aqueduct system, which entailed the moving of more earth than the building of the Panama Canal. The California system was opened over its 444-mile length in 1971. Its completion not only opened up the western half of the San Joaquin Valley for cultivation but also enabled the urban Los Angeles basin to continue its runaway growth. An interstate highway was immediately built along the valley, parallel to the canal.

The equally old technologies for the use of chemicals on farms—such as fertilizers, pesticides, and herbicides—were also continued on an ever-enlarging scale. In 1965 the American Chemical Society established its Chemical Abstracts Service to keep track of all the chemicals mentioned in the technical literature. By 1978 these numbered more than four million, of which some 50,000 were thought to be commonly encountered in everyday life. That number did not include the 1,500 active ingredients found in pesticides alone, nor the 5,500 chemicals added to food after it left the farm, to improve appearance, nutritional value, or taste (all of which were often compromised by extensive processing) or to extend the commercial viability, including the shelf-life, of the products.

The years after World War II also saw a great increase in farm mechanization tied, it is important to note, to the use of not only all-purpose tractors but also a whole new line of specialized tractors to harvest specific crops. The cotton picker, developed during the 1930s by the Rust brothers, came into general use only after the war. Another important machine proved to be the tomato picker.

During the war, high prices for tomatoes and high wages for labor led several inventors to investigate the mechanical picking of this crop, two

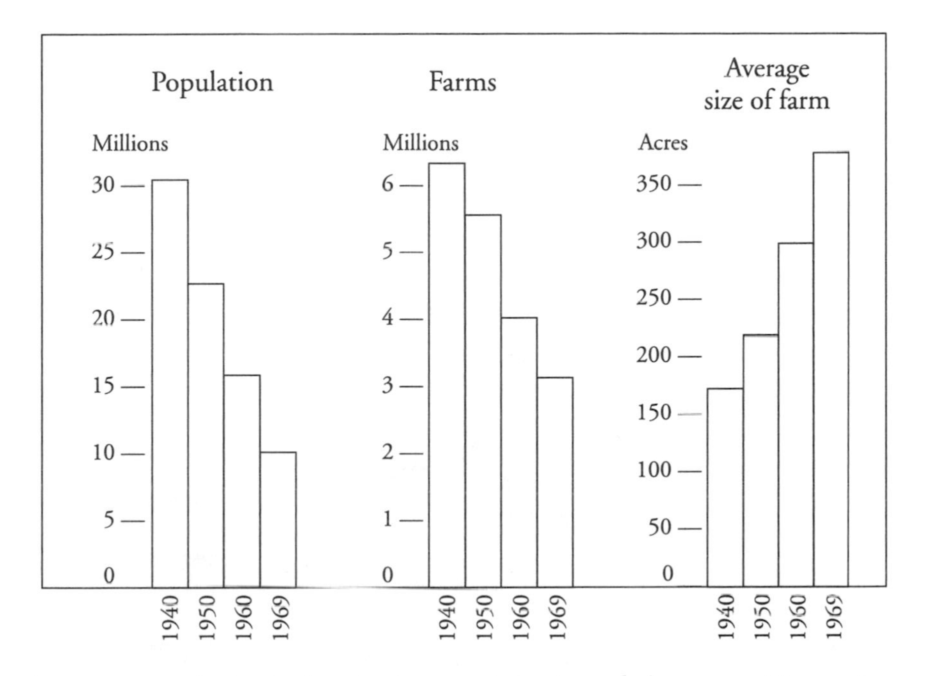

Between 1940 and 1969, the American farm population dropped, the number of farms sank as well, and the size of the average farm increased. New farm technologies were centrally involved in making larger farm size more efficient and profitable, as well as in allowing a shrinking farm population to support a burgeoning urban population. Redrawn from U.S. Department of Commerce, *Statistical Abstracts of the United States, 1970* (Washington, 1970), 581.

major drawbacks to which were that tomatoes ripened slowly over many weeks, producing not one but a number of crops, and that their tender skin and juicy interior made them vulnerable to mechanical manipulation and the intensive handling required for long-distance marketing. In 1949 G. C. Hanna, a professor of vegetable crops at the University of California, Davis, invited his colleague Coby Lorenzen, a professor of agricultural engineering, to join him in the project to develop a tomato-picking machine. Hanna had already made a critical conceptual breakthrough: if he could not invent a machine to pick the tomato, he would invent a tomato that could be picked by machine. The wartime farm labor shortage had been solved by a federal Bracero program to import Mexican nationals on contract to work in the nation's fields. Hanna was quoted as saying, however, that he had "seen nationality after nationality in the fields, and I felt that someday we might run out of nationalities to do our hard work."[7]

The great California Aqueduct snakes south from the Sacramento–San Joaquin Delta, through the arid west side of the Central Valley and over the Tehachapi Mountains to the Los Angeles Basin. Paralleled by Interstate 5 (following the base of the hills to the left), the aqueduct has made a massive environmental impact all along its route. Courtesy State of California, Department of Water Resources.

Hanna and Lorenzen produced their first successful machine (and the new tomato to go with it) in 1959. Sale of the picking machine, which was locally produced by a small manufacturer, was slow until the final ending of the Bracero program in 1964. The following year César Chávez brought his new National Farm Workers Association in to support an ongoing grape strike in Delano, California. The prospect of newly militant American farm workers quickly spurred an interest in a machine that would, at least in the tomato fields, greatly diminish their power and numbers. By 1966, there were 460 machines in the fields of California, already picking some 80 percent of the crop.

The tomato-picking machine provides a superb case history of the extent, scale, and interdependency of the industrial agribusiness that was constructed in America in the postwar years. A self-propelled machine was developed that cost $15,000 (in 1960) and carried a crew of sixteen: a driver, a supervisor, and fourteen sorters. Some owners claimed that women made the best sorters because men developed motion sickness, but others asserted that female sorters argued too much with each other. As the machine proceeded along the rows of crop, it cut off the plants and lifted them along a conveyer belt to the sorters. Clods of dirt, spoiled fruit, and vines were discarded by hand and the produce turned into large boxes, which were trucked to processing plants. The tomatoes themselves were bred to be ready for harvesting more or less all at once, to be picked green and ripen later, and to have tough skins to resist bruising. Taste was inadvertently largely bred out of them, but was later "enhanced" in the processing by adding large amounts of sugar and salt.

The inventors were publicly employed scientists and engineers working at the state university, and the machine itself was licensed for manufacture by the university regents. Processing firms provided an indirect subsidy to the machines by agreeing to accept a lower standard of fruit, and a chemical company developed a product that could be used to redden the fruit while in storage; its advertising slogan was "Etheral helps Nature do what Nature does naturally." Because the expensive machines could harvest large acreages of crop, tomato farms were consolidated into larger holdings.

Because a field could be harvested only once, it was important that the fruit mature at the same time, and therefore that the plants blossom and fruit set at the same time. This required carefully timed applications of water, fertilizers, pesticides, and so forth, and this new and specialized knowledge came increasingly from computer programs controlled by the owners and conveyed to farm managers, whose traditional knowledge of how to grow tomatoes was now less than useless. Finally, American society divided the benefits and costs of this increased efficiency in such a way that the growers (who would buy the machines) saved substantially on labor costs, while field workers, most of whom were Mexican-American, lost their livelihood and moved to the barrios of the large cities. It was all a telling example of the way in which after the war American agriculture came to be the most productive in the world if measured by output per worker (though not by output per acre). At the same time, the surviving farm owners had more heavily invested in technology and found it necessary to spend more and more on petroleum fuel and petroleum-based chemicals.

Not all mechanical amenities on the farm were totally new. At the end of the war only 5 percent of farmers in Alabama had telephones, and four years later two out of three farm families nationwide still did not have this basic technology. In 1949 Congress passed the Rural Telephone Act to create an analog to the successful Rural Electrification Administration (REA) of the New Deal years. Only three groups opposed the bill: the private power industry, which was fighting a rearguard action against the REA; the Bell system of telephone companies; and the independent telephone companies. By 1964 slightly over three-quarters of farms had telephones, but only in part because of the success of this extended REA effort. It was also in part because there were 2.2 million fewer farms in the country in 1964 than there had been in 1950. Not surprisingly, the figures varied by region and by race. In rural Louisiana, as late as 1980, 29 percent of African-American families (but only 11 percent of white families) lacked telephones. At the same time, 55 percent of African-American families (as opposed to 13 percent of white families) were without air-conditioning and 26 percent of African-American families (compared to 6 percent of white families) were without motor vehicles.

The coming of telephones to rural America was only one small part of an elaborating system of communications that, in conjunction with the transportation network, was bringing Americans into closer contact, if not with each other, with national centers of political, social, economic, and cultural authority. Again, the hegemony extended abroad as well as at home. In 1941 Henry Luce, the founder of both *Time* and *Life* magazines, crafted the phrase "American Century": "The world of the 20th Century," he proclaimed, "if it is to come to life in any nobility of health and vigor, must be to a significant degree an American Century." To play this role, America would have to "send out through the world its technical and artistic skills. Engineers, scientists, doctors, movie men, makers of entertainment, developers of airplanes, builders of roads, teachers, educators. Throughout the world, these skills, this training, this leadership is needed and will be eagerly welcomed," he assured his readers.[8] It was the field of communications that Luce knew best.

Three large communication systems were perfected after the war. First was the domestic network of commercial radio stations, which had been put into place essentially during the 1920s. With the addition of FM (frequency modulation) technology, available on the eve of World War II, this has merely extended itself along well-known lines of development. Second was a worldwide network of communications, made up partly of American investment overseas, partly of government-owned propaganda efforts such as Radio Free Europe, the Voice of America, and, more recently, Radio Marti, aimed directly at Cuba. The worldwide distribution of Hollywood motion pictures should be added to this and, especially after the introduction of communication satellites (starting with *Telstar* in 1962), the saturation of the globe with American television. Satellites, along with other innovations such as fiber optics, make it possible to dial Australia directly, to have a near-instantaneous connection and voice communication as good as "across the room." Third was the U.S. military communications system used to exert command and control at any distance, providing a new "control of the air" over such war zones as the Middle East. These three are, of course, closely integrated, and many of the same personalities and corporations turn up in all three. Not surprisingly, all three also tend to display and enforce a coherent message of American political, economic, and cultural hegemony.

The spread of television provided a striking example of the cultural, technological, and economic nexus of communications development. First shown commercially on a limited scale in the 1930s, the television became the indispensable consumer durable appliance of the postwar years. Commercially, it developed very much like the radio, with individual broadcasting stations organized into large networks and financed through advertising. Technically, it steadily improved, the screen growing larger, transistors replacing tubes, and black and white giving way to color. Socially, it briefly drew people together as crowds gathered in front of furniture stores to watch the flickering blue image and friends "came over" to admire and enjoy the first set on the block. Then, as sets spread to nearly all American households and multiple sets into various private rooms of the home, it helped to make strangers of people in the same family even as it brought images of strangers over many thousands of miles into the home. It was the phenomenon of the radio, played out now for the next generation.

Until the radio, as Raymond Williams has pointed out, electronic communications technologies provided "specific messages for specific people." Now television powerfully extended radio's ability to send "various messages to a general public."[9] Despite the fact that this new technology proved to be an inferior means of re-creating sound and images (which were not half as good, for example, as the sound and pictures in a motion picture theater), it was the ability of television to bring those images into the private home that fit it so well to postwar American social forms. Isolation was not, then, an unfortunate consequence but a valued purpose of the medium.

Raymond Williams coined the phrase "mobile privatization" to characterize what was at the same time a "mobile and home-centered way of living."[10] He was writing particularly of television, but the phrase serves as well to describe American society as a whole from the end of World War II through the upheavals of the movement for reform during the late 1960s and early 1970s. Together, the car culture and housing tracts, the automation and agribusiness, the city planning and arms race marked a high point of the systems-based technologies we associate with modernity. In this severely abbreviated American Century, the United States played out the logic of the Enlightenment Project through the technologies of mass production and private consumption that had been developed and perfected by a generation of prosperity, depression, and finally, again, world war.

13

CHALLENGE AND CHANGE IN A POSTMODERN WORLD

AT THE BEGINNING of the twenty-first century, American technology appears to be going through a sea change in response to mounting contradictions within its own structure and operation. The megalith of modern technology—highly structured, centralized, rationalized, homogenized, and universally applied—has been criticized from both ends and, of course, defended in the middle. During the 1970s, a concerted critique was mounted from the left, emphasizing smaller scale, environmental concern, and human purpose. During the 1990s, industrial capitalism itself appeared to be groping toward a restructuring of production, one that would restore flexibility, efficiency, and quality in the face of international challenges. At the center there has been a reification—a commitment to continued elaboration as the only possible alternative to backing down from a posture of technological advance that seems at once natural, even inevitable, but at the same time strangely fragile and threatened.

In 1966 the social critic Lewis Mumford unleashed a withering criticism of the American automobile, but his words stood for modern technology in general:

> Some of the critics have dared to say that the Sacred Cow of the American Way of Life is overfed and bloated; that the daily milk she supplies is

> poisonous; that the pasturage this species requires wastes acres of land that could be used for more significant human purposes; and that the vast herds of sacred cows, allowed to roam everywhere, like their Hindu counterparts, are trampling down the vegetation, depleting wild life, and turning both urban and rural areas into a single smudgy wasteland.[1]

Typically, Mumford used his criticism to call for a reformation of American life, to be based on a rethinking of priorities. There was, he wrote, a

> need for a conception of what constitutes a valid human life, and how much of life will be left if we go on ever more rapidly in the present direction. What has to be challenged is an economy that is based not on organic needs, historic experience, human aptitudes, ecological complexity and variety, but upon a system of empty abstractions: money, power, speed, quantity, progress, vanguardism, expansion. . . . In short, the crimes and the misdemeanors of the motor car manufacturers are significant, not because they are exceptional but because they are typical.[2]

Mumford's critique ran across a wide front. It explicitly lay the main charge at the door of advanced industrial capitalism, but just as clearly indicted American culture in general and its technology in particular. Since both culture and technology are socially constructed, of course, the three are inseparably linked. It would have been unrealistic, therefore, to expect that the multifaceted reform movement of the 1970s would fail to include technology on its list of what needed to be changed about America.

A resurgence of muckraking literature in the 1960s dramatically exposed some of the worst failures of the technology of industrial capitalism after midcentury. Ralph Nader's *Unsafe at Any Speed* (1965), a review of which triggered Mumford's jeremiad, detailed the willingness of Detroit automakers to knowingly produce unsafe automobiles and thus presented a classic portrait of corporate greed and irresponsibility. Rachel Carson's *Silent Spring* (1962) revealed the ecological devastation that was the underside of the nation's infatuation with chemical pesticides. In both instances, the response of the accused—in the one case Detroit, especially General Motors; in the other the powerful network of chemists, chemical corporations, the Department of Agriculture, and their allies—merely drove home the message that technology was above criticism, that progress was not to be questioned, and that those agencies empowered by that technology would go to extraordinary lengths to protect their interests.

Newton N. Minow, the youthful appointee to the Federal Communications Commission, called television a "vast wasteland," an attack that raised an outcry but little reform. More generally, Michael Harrington, in his book *The Other America* (1962), revealed that the one-third of a nation of which Franklin D. Roosevelt had spoken was still ill-clad, ill-housed, and ill-fed. Betty Friedan, in her classic *Feminine Mystique* (1963), helped revivify the feminist movement by demonstrating that women's place in America was in many ways less free, not more so, nearly a half century after the franchise. If technology had created or exacerbated a host of serious new problems, it had also failed to significantly solve older, persistent ones as well.

Finally, the Vietnam War drove home the point that America's massive and sophisticated technological capability, so grotesquely out of scale when compared to that of the invaded Third World nation, could wreak sickening devastation but seemed powerless to win minds and hearts. President Lyndon B. Johnson wanted a Mekong Valley Authority to reproduce the economic miracle that had transformed the Tennessee Valley thirty years before, but instead he provided napalm and carpet bombing. Just so had Twain's Connecticut Yankee, Sir Boss, found that he could destroy the people and culture of Arthurian England, but could not through his technology bring them quickly into a modern world of industrial capitalism and representative government.

All these signs pointed toward a technology that was unpalatable, dangerous, and often ineffective at home, and at the same time inappropriate and often vastly destructive abroad. The same social forces that drafted young Americans to go halfway around the world and use napalm and Agent Orange against the people and environment of a small and weak country also poisoned the workplaces—farms, homes, and shop floors—of America itself. The scientific and engineering research and development that the military-industrial complex channeled into the work of death stole resources from the civilian needs of a country struggling with industrial decline, social inequality, persistent poverty, and environmental degradation.

Before the problem of the Vietnam War could be solved, the nation became aware of the extent and persistence of its environmental crisis. Those involved in the conservation crusade of the Progressive Era had worried

Images of the earth as a beautiful, fragile, and precious whole reinforced the resistance of the environmental movement in the 1970s to further elaborations of projects for global engineering. Released by NASA in 1972, the photo was taken from Apollo 17, the final lunar landing flight of the Apollo program. Courtesy National Aeronautics and Space Administration.

almost exclusively about the disappearance of natural resources rather than where to put the waste that was inevitably produced when those resources were used. Rachel Carson's decimated population of robins laid low from DDT proved to be only the tip of the iceberg as a series of disasters, from the Santa Barbara oil spill through Love Canal to Three Mile Island, reminded Americans that technological successes bred problems of a comparable or even greater magnitude.

And while the environmental movement was still gaining momentum, the energy crisis of the mid-1970s hit. Over the past century, two obvious and significant trends had reached critical proportions: energy consumption per capita had risen sharply, and there had been a shift in source away from renewable fuels (wind, water, and muscle) to nonrenewable ones (coal, oil, and natural gas). When the Organization of Petroleum Exporting Countries embargoed oil exports in 1974, the true dimensions of our dependence on petroleum was brought home to most Americans in the form of higher prices and long lines at the service station.

This constellation of challenges called forth an outpouring of analysis and action that was seen by the defenders of technological advance as a repudiation of progress and, indeed, the entire Enlightenment Project to better the lot of humankind through direct and deliberate action. The hoary specter of Luddism (that often mythologized episode of machine-breaking in Regency England) was raised once more and connected with any questioning of the social uses of science, the priorities of American scientific and technological policy, and the reading of Tarot cards by hippies. Not surprisingly, some of the most Gothic warnings against the questioning of technological progress came from political, military, corporate, and technical leaders of the nation who, quite naturally and rightly, saw their hegemonic sway attacked (though hardly seriously threatened). Ronald Reagan, then governor of California, asserted that trees were the major source of environmental pollution, and President Richard M. Nixon celebrated the high per capita energy use by Americans as sure evidence of their superior civilization.

On both the state and federal levels, however, as well as in myriad private initiatives, there was experimentation with a new (if not terribly radical) approach to technology. The National Environmental Protection Act (1970) laid down the idea of environmental impact reports—studies in advance of development that would try to anticipate negative changes in the environment that might reasonably be expected to result from such development. Already in 1967, Representative Emilio Daddario (Democrat of Connecticut) had introduced legislation to establish what came to be the congressional Office of Technology Assessment to provide an analogous service. The National Academy of Sciences (NAS), along with the academies of engineering and public administration, was asked to report on the feasibility of carrying out the two missions that Daddario envisioned: to anticipate the social and environmental effects of new or projected technologies and also to identify potential technologies that needed support and encouragement to be brought forward. The academy ratified the possibility of assessing technology, but articulated four preconditions that revealed both the potentially radical power of the process and the limits beyond which it would not be allowed to probe.

First, the NAS maintained that technological change would continue. "The choice," said the panel, "is between technological advance that proceeds without adequate consideration of its consequences and technological change that is influenced by a deeper concern for the interaction between man's tools and the human environment in which they do their work." Second, they admitted that "our panel starts from the conviction that the advances of technology have yielded and still yield benefits that, on the whole, vastly outweigh all the injuries they have caused and continue to cause." Third, it maintained that "technology as such is not the subject of this report, much less the subject of this panel's indictment. Our subject, indeed, is human behavior and institutions, and our purpose is not to conceive ways to curb or restrain or otherwise 'fix' technology but rather to conceive ways to discover and repair the deficiencies in the process and institutions by which our society puts the tools of science and technology to work."[3]

Finally, given all the foregoing, the panel insisted that "it is therefore crucial that any new mechanism we propose foster a climate that elicits the cooperation of business with its activities. Such a climate cannot be maintained if the relationship of the assessment entity to the business firm is that of policeman to suspect." In a final effort to contain the potential of technology assessment to radically alter the way in which we think about the subject and the way in which public policy deals with it, the NAS panel admitted that it had not looked at all at technologies for the military. In short, criticism of technology was not to be allowed to prevent any technology from being implemented (progress was to continue), and both business and the military were to be exempt from criticism, let alone constraint. As President Nixon said, "technology assessment" was all right as long as it did not become "technology harassment." A congressional Office of Technology Assessment was finally established in 1972.

A concept closely related to technology assessment was that of "appropriate technology." The large-scale foreign aid programs undertaken after World War II by the American government, as well as by the Soviet Union and other major powers, had for years been supplemented by development projects financed by the World Bank and the International Monetary Fund. Some projects, like the Soviet-supported Aswan Dam built across the Nile River in Egypt, had obvious and immediate side effects, such as the submersion of ancient monuments, the spread of water-related diseases, and the diminution of the fisheries off the mouth of the river. On a more diffuse scale, new agricultural techniques like those advocated by participants in the green revolution were slowly seen to raise more problems than they solved. In the latter case, dramatic increases in yield per acre of rice, for instance, were purchased at the price of destroying a peasant culture and making farmers dependent on large inputs of outside resources: bank credit, chemical fertilizers, complex machinery, and the skills to manipulate all these. Stories of large diesel tractors sent to India for use by farmers still using stick plows pulled by oxen began to raise the possibility that some technologies, effective in certain environments, were useless or even disastrous in others. That is, for any given place and purpose, some technologies might be appropriate, while others might be inappropriate.

Public awareness was galvanized by the publication, by the British author E. F. Schumacher in 1973, of his book *Small Is Beautiful: Economics as if People Mattered.* Schumacher concentrated mainly on advocacy of an "intermediate technology" for developing nations: machines and tools better than they had already but not so vastly much more powerful and complicated that they destroyed indigenous cultures, devastated the local environment, stifled local initiative, and would soon be abandoned in any case by people who had neither the training nor the money to maintain them properly. Green revolutions and nuclear power plants, jet fighters and great dams were logical extensions of Western technological progress and expanded the power and prestige of local elites, but often did little good for common peoples who most needed potable water, latrines, literacy, and family planning.

The notion of appropriate technology, however, also had obvious applications to those nations, like the United States, that could be defined as "overdeveloped" rather than "underdeveloped." The election of Jerry Brown as governor of California in 1978 opened a window of opportunity in that state. Working with his state architect, Sim Van Der Ryn, Brown established an Office of Appropriate Technology to stimulate publicity and programs in that area, and the legislature pressed additional funds on the University of California to institute an appropriate technology research and demonstration program. A special fleet of state vehicles was converted to burn a gasoline-alcohol fuel mixture, and Van Der Ryn designed state office buildings using both passive and active solar energy. The state constitution was amended, by popular vote, so that a portion of the Gasoline Tax Fund could be diverted to

When Jerry Brown became governor of California, he set up the Office of Appropriate Technology (OAT) to foster technological alternatives for both public and private projects. Before being abolished by his Republican successor, George Dukemajian, OAT made significant advances in the support of solar, wind, light rail, and other technologies. The gentle and humorous logo of OAT betrayed the style that earned Brown the title of Governor "Moonbeam." Photo in possession of the author.

other forms of transportation; light rail systems and bicycle paths were built; and the state became the largest producer in the country of electricity from wind power.

Some of this same technological initiative was evident on the national scene. A National Center for Appropriate Technology was established in 1977 but, significantly, was situated safely in Montana, not inside the Washington beltway. The Army Corps of Engineers was ordered to survey existing dams in the nation to discover which of them might be retrofitted for low-head hydroelectric production. Legislation was passed that required electric power companies to buy electricity from any provider at prices reflecting the cost of new production facilities (gas, coal, or nuclear fired) rather than prevailing prices, which reflected older plants and cheaper hydroelectric facilities. An initiative to develop solar power was both hopeful and instructive.

The idea of solar power was an old one. In 1805 Oliver Evans had advocated solar power to raise steam for his engines. "I am fully of the opinion," he wrote, "that the time will come when water will be raised in great quantities by the heat of the sun at a very small expense, for various purposes; but," he warned, "the expense of such inventions cannot, in many instances, be borne by those who have the mental powers to design; at least it is highly imprudent for them to risk it. In such cases," he concluded, "aid from government becomes necessary."[4] In 1974 the Congress appropriated $77 million for research and development on and demonstration of solar power generation and an additional $60 million to demonstrate the feasibil-

Although the appropriate technology "movement" of the 1970s disappeared, the technologies themselves persisted and developed. In the 1990s, a California firm building advanced-design windmills for electrical production, in part a legacy of the Brown years, won the right to construct a windfarm on the Queen's lands near Bridgend, Wales. Courtesy Wind Harvest Company.

One "high-tech" answer to environmental concerns and threats of energy shortages was the development of fusion technologies. In an attempt to "harness fusion as a safe, economical, and inexhaustible energy source," the Department of Energy spent more than $176 million on Nova, the world's most powerful laser, the target chamber of which is shown here. Courtesy Lawrence Livermore National Laboratory.

ity of solar heating and cooling of buildings. At the same time, the heavily subsidized (though still not competitive) nuclear energy program received $2.2 billion, a fair measure of how closely wedded the nation remained to powerfully entrenched subsidies and technologies. In 1977 the first funds were released to the new Solar Energy Research Institute, located at Golden, Colorado, and run by the Department of Energy (DOE), which ran the nuclear program as well.

Significantly, it proved extremely difficult for the DOE to break its habit of massively supporting only huge programs emanating from the military-industrial complex. One of the major proposals for solar development came from the Boeing Aerospace Company, which wanted to orbit a mirrored satellite (perhaps 25 square miles in area) 23,000 miles above the earth to focus and reflect the sun's rays down to a "power tower" below. The tower would then reflect the rays to mirrors on the ground, which would use the intense heat to raise steam for the generation of electricity in a nearby power plant. The resulting electricity would, of course, then be distributed by an existing electrical utility to its customers along high-tension lines.

The proposal demonstrated the persistence of the example of nuclear technologies and the utilities industry. As the magazine *Science* commented in 1977, "Solar energy is democratic. It falls on everyone and can be put to use by individuals and small groups of people. The public enthusiasm for solar is perhaps as much a reflection of this unusual accessibility as it is a vote for the environmental kindness and inherent renewability of energy from the sun." As the journal also noted, however, "despite the diffuse nature of the resource, the [DOE's] research program has emphasized large central stations to produce solar electricity in some distant future and has largely ignored small solar devices for producing on-site power—an approach one critic describes as 'creating solar technologies in the image of nuclear power.'" The government program, in short, was bent on developing a solar technology that would perpetuate Taylorist and Fordist techniques and structures. As *Science* concluded, "The massive engineering projects designed by aerospace companies which dominate much of the program seem to have in mind the existing utility industry—rather than individuals or communities—as the ultimate consumer of solar energy equipment."[5]

By 1990 the DOE was spending $12.2 billion a year, but very little of that was going for research and development that might reduce the nation's reliance on oil as an energy source: $8.8 billion was still poured into its nuclear weapons program, and another $1.5 billion was expected to pay the first year's bill for an open-ended program to clean up the nuclear waste from weapons plants around the country. Another $1 billion was devoted to atom-smashing research, not including the estimated $11 billion cost of the projected Superconducting Super Collider. Clean coal, electric cars, and so forth accounted for less than $1 billion, of which $35.5 million was earmarked for research on solar-voltaic cells to convert sunlight directly into electricity, a technology first developed in 1954. Dr. Maria Telkes, on being given the first Achievement Award by the Society of Women Engineers in 1952 for her work on solar energy, remarked that the field had not been developed as fast as nuclear energy because, "you see, sunshine isn't lethal."[6] Her comment makes obvious that, contrary to the old saw, not everything that can be done will be done.

The demise of appropriate technology as a movement during the early 1980s suggests some of the dimensions of both its appeal and its danger. On one important level, of course, it was a potential threat to those economic and political interests with much invested in large-scale systems of modernist technologies, like the automobile, the agroindustrial farm, and the fossil-fuel or nuclear power plant. Technologies are never simply machines; they are also the nub of accumulated power. On another level, the masculine values of violence, conquest, and dominance that are built into so much of U.S. technology were equally threatened by the possibility of a more responsive, gentle, "feminine" set of technologies. A movement that celebrated the ideals of small and soft could hardly have recommended themselves to believers in more traditional masculine values. After the feminizing effect of losing the war in Vietnam, the decade of Reagan and Rambo, with its commitment to standing tall and being number one again, created a powerful cultural backlash against any attempt to redirect or even rethink American technology.

An attempt to redress the gender imbalance of the American engineering profession was also, in part, an implied criticism of the way things were being done. The debate over whether women in engineering somehow "softened" engineering itself, or whether they themselves were more likely to be masculinized by the experience, was only one aspect of the major effort, during the 1970s, to attract more women into what was the most male-dominated profession in the nation. Less than 1 percent of engineers were female, and a perceived national shortage of engineers was reinforced by the contemporary movement for women's rights to highlight the issue.

The Society of Women Engineers, which had been established in 1950 to support that corporal's guard of women who entered the profession during World War II, chose for itself two apparently conflicting goals. First, it sought to attract more young women into engineering schools. Second, it necessarily had to deal with the powerful and long-entrenched patriarchy of the profession. To the extent that it openly battled prejudice and sexism in engineering schools and practice, however, it risked frightening off prospective recruits. Over the next two decades, the proportion of women students in engineering schools rose to around 15 percent, but the larger possibilities inherent in even a partial degendering of American technology has thus far failed to develop. The profession has remained a bastion of white, male culture.

The search for new technological paradigms and initiatives of the 1970s revealed promising areas for research and development as well as suggesting new relationships among those social agencies that made the choices between and shaped technological developments. The political boundaries articulated by the NAS in its report on technology assessment, however, proved to be controlling. Technological change was not to be slackened, its overall beneficence was not to be doubted, corporate control was not to be threatened, and the economic and environmental effects of military technology were to be ignored. The cultural hegemony of industrial technology might be questioned, but it was not to be seriously challenged.

No system lasts unchanged, however, and dominant institutions maintain their hegemony by making concessions to external criticism from time to time. Safety belts became mandatory in cars, fleet mileage improvement was required by legislation, the supersonic transport was killed (or at least postponed), solar-powered calculators became available (if only from Japanese manufacturers), and at some sites around the country freeways were not built, dams were designed with fish-ladders, and toxic sites were cleaned up. More important, the contradictions internal to modern technology began to force many corporations to rethink the very Taylorist and Fordist bases of the system.

In particular, three important trends have developed over the past two decades. First, entrepreneurs have rediscovered the opportunities in smaller market shares that can be exploited by smaller, more flexible, and specialized technologies and firms. Second, the vastly increased mobility and speed of transportation and communication have encouraged the globalization of operations. Third, mass production, at least in some key heavy industries, is giving way to what has come to be called "lean" production.

Just as textile and shoe production fled the Northeast earlier in the century to seek a cheaper and more easily disciplined work force in the South, so have American firms taken advantage of transportation and communications improvements to move their production facilities abroad in recent times. Fields in Mexico, once used to raise corn, beans, and squash for the local campesinos, now raise out-of-season asparagus to be flown by jet to North American markets. Assembly processes especially, which are labor intensive, have been sent to southeast Asia, the Caribbean, or Mexico.

Cheaper labor (which no longer needs to have the old skills), tax incentives, and an absence of environmental regulations make the moves attractive. The picture is complicated, however, by the fact that while Fords may now be made in Mexico and Dodges in Japan, some Hondas are made in Ohio. Not only does this seeming anomaly confuse the patriotic urge to "buy American" (does one support American jobs [Honda] or American capital [Chrysler]?), but it suggests that the old method of cutting the wages of labor is not the only, or perhaps the best, way to remain competitive today.

The dream has died hard. At least since Karel Capek's play *R.U.R.* introduced the word *robot* in 1923, the notion of using such units to perform the difficult, dirty work of life has been a persistent ideal. They would be more easily controlled than flesh-and-blood workers and less subject to fatigue, would work unceasingly with predictable accuracy, and would never go on strike or sue for injury. Ira Levin's novel *The Stepford Wives* (1972) suggested that the domestic and sexual work of women might also be performed by robots, although of course ones that looked like the real thing. The literature of science fiction kept the dream alive, and by 1988 some 15,000 robots were actually at work in American industry. Although they lacked the recognizable features of the familiar android, they performed tasks such as spot welding automatically, replacing human workers in a manner by now expected and accepted. Robots, however, were as much a sign of the old ways of mass production as an indication of new departures. Computers made it possible to quickly reprogram general-purpose machine tools, thus avoiding the use of dedicated tools that could make only one part and therefore reintroducing a measure of flexibility into the productive process. In most cases, however, these existed as "islands of automation" rather than as integrated parts of a new and larger scheme.

Lean production, on the other hand, with its related techniques of just-in-time inventory control and the subcontracting of such tasks as secretarial work and delivery of finished product, has begun to seriously erode the old sureties of Fordism and Taylorism. Developed by the Japanese automobile industry, particularly at Toyota by Taiichi Ohno and Eiji Toyoda beginning in the 1950s, this new process came to embody elements of both mass production and the even older methods of craft (batch) production.

Under the craft method of making cars, parts were bought from suppliers who worked independently, and then skilled craftspeople fitted them together with a great deal of filing. Gradually the car was built up, piece by piece, each fitted to the last, and because of "dimensional creep" no two turned out to match each other, let alone the original plans. Such a system was expensive, of course, and the resulting machines were difficult and expensive to repair.

The system of mass production pioneered by Henry Ford insisted that all parts supplied be designed and manufactured to a single gauge. They were then assembled by unskilled or semiskilled workers who were able to simply put them together without further fitting. Special machine tools were dedicated to making only a single part, thus sacrificing flexibility for accuracy and speed of production. Finally, assembly workers were placed along a moving line of vehicles to further speed up the work.

A line of specialized workers extended from both ends of this assembly process. On one end, a host of engineers and managers designed the components and the machines to manufacture them. On the other, an army of tool-repair specialists, quality inspectors, and "rework" people took over specialized tasks that before had been the responsibility of the craft workers themselves. Especially important were the reworkers, who corrected the mistakes on cars before they left the factory. A single manager had the power to stop the line, but because mistakes were more easily corrected than avoided, the incentive to "move the metal" was strong. Indeed, when Toyoda visited Ford's River Rouge plant in 1950, it was producing 7,000 cars a day. In Japan his family's firm, Toyota, had made only 2,685 cars in 13 years. By a judicial combination of borrowing and innovation, Toyoda and Ohno were able not only to vastly increase their production, but to do so with half the investment in human resources, manufacturing space, tools, and engineering hours spent in model design: hence the term *lean*.

One key to the process was having assembly workers operate in teams, giving them the work elsewhere assigned to the repair, inspection, and rework people, with the assumption that mistakes could be prevented, not just repaired. Each worker had the ability and responsibility to stop the line if mistakes appeared. Since lifetime employment meant that workers were a fixed cost, the company invested in upgrading and using the full skill of their employees. Thus the actual assemblers added even more value to the product, and the army of reworkers, which in American plants did not add value at all, was virtually eliminated.

Auto assembly, however, accounts for only 15 percent of the manufacturing process of a car; the rest consists of making the ten thousand component parts. Both Fordism and the corporate structure developed at General Motors by Alfred P. Sloan dealt with a small army of parts and subcomponent suppliers, some in-house and many outside the organization. These suppliers (typically 1,000 to 2,500 of them for American cars) were challenged to make the lowest possible bid for each part and were kept independent of both the automobile company and each other. Under the Japanese system, the auto company dealt with many fewer suppliers; they were arranged in tiers, from suppliers of single parts on up to suppliers of complex subcomponent systems, and they were urged to trade information and cooperate with both the purchaser and each other. Finally, the suppliers delivered their parts to the assembly plant often, on a hourly basis; new parts were not made until the last box of them had been returned empty (hence the notion of just-in-time inventory). The result of all these changes is a greater flexibility of production, a combination of the economies of mass production, without the scale. The monolithic imperative—best expressed in Henry Ford's remark that customers could have cars in any color so long as it was black—promises to be replaced by some of the variety and choice of the handicraft-production era.

Lean production, then, is a thorough and systemic reordering of the method and assumptions of the mass production that was emblematic of American production for nearly the entire twentieth century. By 1990 it had begun to make some American automobile plants nearly as efficient as the best Japanese plants, but only with a wrenching effect on the social context of manufacturing in this country. Like every technology, automobile production is embedded in a complex environment of geography, class, economics, gender, and politics. A society painfully adjusted to mass production over many decades is not easily accommodated to another system.

Lean production, however effective in beginning to make American manufacturing more profitable and however suggestive of a large-scale change in the way technology is structured in American society, addresses only a part of the problem of modernist logic. The entire question of American "competitiveness," as it is called, must address the great questions of public policy as well as private gain, since the actions and options of governments at all levels, especially those of the federal government, have always played a decisive role in shaping the nation's technology.

Despite the fact that much of what has passed for "science policy" in the years since World War II has, in reality, dealt with technology, the United

States has no explicit policy for technology itself. From the 1950s on, American policy has been to invest heavily in military technology, with the idea that civilian applications would "spin off" and prove adequate to keep the nation in the forefront of industrial progress. As one veteran science policy observer noted in 1991, "In the era immediately following World War II, the United States had a virtual monopoly on new technology. This was fostered by spin-offs from defense R & D. . . . However defense R & D gradually ceased to be a stimulus to the civilian economy. Global competition in high technology emerged."[7] Critics such as Seymor Melman had for years argued that "profit without production," that is, the wholesale investment of the nation's technical resources in research and development for the military rather than the market, was having a devastatingly distorting effect on the American economy.

Much of this cold war effort was funneled through the Defense Advanced Research Projects Agency (DARPA). Over the years, such technologies as computers and artificial intelligence were forced ahead strongly by this agency. Still, in 1991, of the $1.3 billion the federal government spent on "advanced manufacturing and materials," over 40 percent came from the defense budget, including $100 million a year to improve the manufacture of computer chips and $311.5 million for a program known as Manufacturing Technologies (Mantech), which looks particularly at machine tools and automatic machinery.

During the 1980s, however, DARPA began to concentrate more on strictly military hardware. One Pentagon agency, the Strategic Computing Project, which worked on such devices as a robot tank, alone had a budget of more than $100 million a year beginning in 1983. Congressional demands for a technology policy were resisted by the administrations of Ronald Reagan and George H. W. Bush as calls for an "Industrial Policy," which would make the criteria of federal technology spending more open, explicit, and, presumably, democratic.

As Representative George Brown, Jr. (Democrat of California), commented, "We have been imprinted with a national security mentality in interpreting the federal role for technology," but "commercial technology development can no longer be a stepchild of the defense structure."[8] By the 1990s, the U.S. government spent only 0.2 percent of its research and development budget for industrial development, compared with 4.8 in Japan and 14.5 in West Germany. For defense, however, the United States spent

65.6 percent, compared with 4.8 for Japan and 12.5 for Germany. For energy, the United States spent 3.9 percent of its R & D budget, Japan 22.8 percent, and Germany 7.8 percent.

Faced with White House refusal to alter its priorities, in 1987 the Congress reconstituted the old (1901) National Bureau of Standards into the new National Institute of Standards and Technology to focus on civilian industrial problems and opportunities. The Reagan and G. H. W. Bush administrations, however, failed to put any significant resources into it. In the early 1990s a Critical Technologies Institute was created under the Office of Science and Technology Policy to pursue civilian initiatives, but its funds, too, were provided by the Pentagon. The difficulty in shifting from cold war patterns of technological support to something different, despite mounting evidence of its doing more harm than good as the century closed, is clear evidence of the fact that the political matrix of technology, like the hardware itself, is designed to privilege certain power relationships and deny others. The military-industrial complex's monopoly on the public definition and support of technology is deeply rooted in the nation's historical experience.

Since World War II the technological efforts and priorities of the U.S. government have been warped by science as well as by the military. The notion championed by Vannevar Bush, that new technology flows from basic scientific research, was embedded in the common phrase "research and development" and implied a system of priorities as well as an explicit theory of change. The phrase reversed—development and research—seems not only awkward to pronounce but wrongheaded in meaning. To suggest that technological change should be aimed at a purpose (development), and that research should follow only as a way around problems as they come up, flies against the priority built into "R & D," that one should do research for its own sake and then develop whatever possibilities it might present. The historic conflation of technology with science policy continues to obscure such political advantage.

In his book *The Condition of Postmodernity* (1989), the geographer David Harvey argues that "there has been a sea-change in cultural as well as in political-economic practices since around 1972." He writes of "the rise of postmodernist cultural forms" and "shifting dimensions of time and space," as well as "the emergence of more flexible modes of capital accumulation."[9] During the postwar boom, the technologies (and techniques) of the interwar years were perfected and formed the basis of both American prosperity and American worldwide hegemony: mass-produced consumer goods, petrochemicals, radio and television, steel and construction. During the 1970s and 1980s, wartime and postwar technologies, aerospace, and especially the computer and other aspects of electronics came to the fore, along with newer forms of production, consumption, labor control, and finance.

Postmodernity is most easily seen, of course, in its artistic manifestations, especially architecture, for which, in fact, the term was coined. But the notion is also helpful in making sense of a host of other phenomena, from cultural pluralism and feminism to the spread of nuclear, chemical, and biological weapons to emerging nations; from the Rust Belt and factories along the Mexican border to the success of Japanese consumer durables and the frenzy of corporate mergers financed by junk bonds. The demise of the "metanarrative" and the shift in power from the center to the peripheries is true both within and outside the field of technology. Neither capitalism nor industrial technology is likely to disappear any time soon, but it is even more certain that they will not remain "modern" in the sense that we have come to know them over the past century and a half.

Mumford's plea to form a clear conception of "what constitutes a valid human life" moves the argument far beyond the balance of trade and even the number of manufacturing jobs created. The industrialization of American agriculture in the twentieth century, for example, broadly laid out along both Taylorist and Fordist lines, made it the most productive in terms of produce per person-hour in the world. But here, as elsewhere in technology, the process has created a tightly organized agroindustrial complex, at the cost of fragmenting other, perhaps more important, relationships, with people driven from the land and small towns dying; soils and groundwater poisoned, some perhaps forever; crops bred, grown, and processed to maximize value-added in economic terms but stripped of nutritional value and taste; petroleum-based systems of cultivation that are energy sinks (requiring more British thermal units to grow than they produce in crops); the loss of a way of life that had its own values, not least of all a realistic sense of scale, cause, and effect. Whether one looks at industrial production, biological reproduction, domestic workplaces, the mechanism of war, or the machinery of government, the same restructuring of meaning through modernist technology can be seen and appears problematic.

Noam Chomsky has asked how it came to be that the freest, most democratic nation in the world has such a narrow and impoverished range of political discourse, how ideas not acceptable to the dominant authorities are so easily and effectively marginalized and trivialized. It can also be asked how modern American technology, so ripe with the promise of variety and empowerment, leads instead to monolithic systems that constrict rather than liberate, that often undermine rather than reinforce our professed values as a nation.

Whether Mumford's jeremiad is heard or ignored as the United States moves into the twenty-first century, one part of the NAS's insight into technology assessment is correct: human behavior and purpose are the true and proper subjects of our concern. The academy erred, however, in thinking that human behavior and purpose are separate from technology. The three are, in fact, inseparable; technology can be understood and used to best advantage only when seen as the very embodiment of human behavior and purpose.

V

GLOBALIZATION

14
OUR (UN)WIRED WORLD

THE HISTORIAN ROSALIND WILLIAMS perceptively notes that "the outstanding feature of modern cultural landscapes is the dominance of pathways over settlements." She goes on to identify the "powerful cultural assumption that it is the destiny of the West to promote circulation through technological systems-building" and the "constant tendency . . . to incorporate more and more of the world's surface into these networks," resulting in a global system of networks. "A second tendency," she notes, "is for the rate of circulation to become increasingly rapid and easy." Williams points out that, as has always been true, "the pathways of modern life are also corridors of power, with power being understood in both its technological and political senses." Finally, she notes that "the project of conquering space inherently and inevitably entails the devaluation of place," both a consequence and a goal of modernity.[1] A striking example was provided by Secretary of Defense Donald Rumsfeld, who early in 2006 argued that Al Qaeda, presumably operating from some cave on the Afghanistan-Pakistan border, was mounting a propaganda campaign superior to that of the U.S. government. Today's weapons of war, he claimed, included e-mail, BlackBerries, instant messaging, digital cameras, and blogs. As a Pentagon study had recently proclaimed, "Victory in the long war [against terrorism] ultimately depends on strategic communication."[2] The realization that the new electronic environment was a critical element in warfare should have come as no surprise, because its roots were deep in the research and development programs carried on by the Pentagon since World War II.

Harvard University's MARK I became the nation's first large electromechanical computer in 1944, designed to make fast calculations related to defense work. The next year the University of Pennsylvania's ENIAC, paid for by the army and also designed to do war work, became the first numerical computer to use electronic rather than mechanical switches, and by 1951 there were perhaps half a dozen other computers at work in the country. That year companies in England and the United States began to make them for general sale. Then in 1953 IBM brought out its Model 650, of which the company thought it might sell 250. In fact they sold thousands, making it what historian Paul Ceruzzi has called "the Model T of computers."[3]

Howard Aiken, who had designed MARK I, had predicted that four or five such machines would satisfy the nation's needs for computers. Along with IBM's conservative estimate of sales for the Model 650, Aiken's wildly erroneous prediction was the result of three factors, in Ceruzzi's estimation. First, the machines were considered fragile and unreliable: ENIAC, for example, had 18,000 vacuum tubes that burned out with distressing regularity. Second, the computers were developed by scientists and engineers who wanted only to deal with very large amounts of data, and they assumed that any other users would have the same esoteric needs. Third, even the computer designers misunderstood the nature of computers, how they worked, and what they could do. Only gradually did they realize, for example, that the machines could store their own programs internally and make them accessible to users as a "menu" of simple commands.

Several more technical developments proved important before the computer could become a truly user-friendly appliance. In the technical hothouse that characterized cold war America, the electronic advances made during the war were pushed even further by an eager military. Bell Laboratories was the research arm of AT&T (the Bell telephone system), and its director, Frank B. Jewett, had been a member of the Office of Scientific Research and Development (OSRD). During the war Western Electric, AT&T's manufacturing arm, had been the largest industrial contractor of the OSRD, and, taken together, these companies represented an enormous reservoir of both technical knowledge and political influence. In 1947 a team at Bell's Murray Hill, New Jersey, laboratory learned how to make a piece of semiconductor crystal act like a vacuum tube. It held the potential to become the basis for a solid-state electronic switch that would avoid the many negative characteristics of vacuum tubes. A radio and a television with these new transistors rather than vacuum tubes were publicly demonstrated the following year. By 1953 the Pentagon was supplying half the transistor research costs of Bell, and paying for their production facilities as well.

Civilian uses for transistors began to develop that same year. In December 1952 the Sonotone Corporation began to sell a hearing aid that used two vacuum tubes but also a transistor, and by 1954 at least 97 percent of such devices used transistors only. The first all-transistor Regency radio (with four transistors) appeared in time for the Christmas market in 1954 and proved popular. The technology for the Regency had been produced by Texas Instruments, where in 1958 Jack Kirby was one of the pioneers in the development of integrated circuits (microchips), most of which were used in military applications. The idea of the microchip was to put an entire electronic circuit on a piece of semiconductor material—Kirby used germanium. Texas Instruments wanted a product that would introduce the integrated circuit into the civilian market, and in 1965 chose the "slide-rule computer" as its vehicle.

The design problem was to produce a handheld device into which data could be entered, then manipulated, and finally read out, all relying on its own power source and providing the user with a "a keyboard that was thin, simple, cheap, reliable—[and would] work a million times," according to one member of the team.[4] A patent for the device was issued to Kirby and his colleagues in 1967; the Japanese firm Canon licensed the technology and had a calculator on the market in 1970. The Pocketronic weighed a pound and a half and cost $395, but a decade later truly pocket-sized four-function calculators were being sold for $10 or less. Kirby went on to win the Nobel prize in physics in 2000.

If design was a challenge, the problem of manufacturing microchips was even more so. In 1958 Jean Hoerni, of the new Fairchild Semiconductor company, invented a planar technique for layering up silicon and silicon dioxide on a thin wafer of silicon crystal. It proved to be an excellent way of producing a number of small, flat transistors at one time. Robert Noyce, the head of Fairchild, then demonstrated that the planar process could be used to produce circuits on a single piece of silicon, with transistors reduced in size to dots under a microscope. IBM had begun to use transistors in its 7090 machine in 1955, but even so, computers with 200,000 components were becoming hopelessly complex, and ones with millions were already being planned.

By this time the locus of activity in electronics innovation had begun to move west, particularly to the former fruit-growing region locally called the Valley of Heart's Delight, just south of San Francisco. Soon after the establishment of Stanford University in 1885, electrical engineers there had begun to work closely with the large electrical utilities and radio interests based in San Francisco, giving the school a strong tradition of practical research aimed at improving current technologies and coming up with new ones. It was a tradition built on by Frederick E. Terman, whose father was a psychologist on the faculty. The younger Terman graduated in chemical engineering from Stanford, then went to the Massachusetts Institute of Technology to earn his doctorate in electrical engineering under Vannevar Bush. He returned to join the Stanford faculty with the realization that his school could, in a real sense, become the MIT of the West in terms of leadership in electrical engineering research. In 1934 two of his students, William Hewlett and David Packard, graduated, and two years later Terman encouraged them to form a company to commercialize an audio oscillator that Hewlett had developed while a student. It was World War II, however, that gave Terman the impetus he needed to build a research powerhouse at Stanford. In 1942 Bush contacted Terman and asked him to move to Boston to take over Harvard's secret Radio Research Laboratory. Bush's experience taught Terman that after the war "there would be, for the first time, real money available to support engineering research and graduate students. This new ballgame would be called sponsored research."[5]

After the war, with Terman as dean of engineering, Stanford blossomed as a research center. Protégés Hewlett and Packard were already located nearby, as was another Stanford graduate, Charles Litton, who had formed Litton Engineering Laboratories. Also before the war, the Varian brothers, one of whom was a Stanford graduate, moved their laboratory to the university, where they developed a new electron tube called the klystron. After the war the Varians moved off campus to nearby San Carlos and opened a business specializing in the development of ideas discovered in research that had been done at the university. In 1950 Varian Associates asked to lease land owned by Stanford and was the first of many high-technology firms to occupy the new Stanford Industrial Park. Hewlett-Packard moved there, as did Ampex and Lockheed's Space and Missile Division. By 1960 over forty companies were located in the university's development. As one historian has written, "Cashing in on the vacuum tube, radar, microwave technology, the cold war, and space exploration, giant national firms of the tube era of electronics as well as young new entrepreneurs set up shop from Palo Alto [the site of Stanford] south to San Jose."[6] Then in 1956 William B. Shockley, coinventor of the transistor at Bell Telephone, returned to Palo Alto, where he had grown up, and recruited bright young researchers for what was the first semiconductor company in the valley. A year later eight of the best broke away and, with funding from Fairchild Camera and Instrument Corporation of New Jersey, formed Fairchild Semiconductor in nearby Mountain View.

Over the next few years engineers drifted away from Fairchild, too, and by the early 1970s forty-one new semiconductor companies had been formed by former Fairchild employees, many staying in the valley. Indeed, it is said that of the four hundred semiconductor engineers attending a 1969 conference in Sunnyvale, all but two dozen had once worked at Fairchild. Robert Noyce had been one of the original Shockley engineers who had gone on to Fairchild, and he left that company as well to start his own firm, the Intel Corporation. As historian Jim Williams tells the story, Noyce then hired Marcian "Ted" Hoff, who had recently earned a Ph.D. in electrical engineering at Stanford. Hoff, in turn, further developed in 1969 the microprocessor that in 1971 became the first to be widely marketed. This "computer on a chip," Intel's 4004, sparked a renewed cycle of growth in the area now called Silicon Valley; its 8080 chip provided the basis for the Altair computer, the first that was really affordable, and the improved 8088 chip was used in the first IBM personal computer.

Altair's 8800 computer, which was sold unassembled for $400, was first advertised in the January 1975 issue of *Popular Electronics,* an event that led to the formation of Silicon Valley's Homebrew Computer Club. Foremost among the enthusiasts belonging were Stephen Wozniak and Steven Jobs, who in 1977 began to manufacture their new Apple computer in Jobs's garage. Eight years later their company had sales of $1.5 billion, creating the paradigmatic Silicon Valley example of explosive growth and financial success. A similar though cautionary tale was that of Adam Osborn, who in June 1981 introduced his 24-pound "portable" computer at the West Coast Computer Faire. By the end of the year he had orders for 8,000, and in 1982 he took orders for 110,000. But by 1983 his company was bankrupt, beset by manufacturing difficulties and a large inventory that could not be sold because Osborn had unwisely already announced that a new and improved computer would be available in the near future.

William H. "Bill" Gates and Paul G. Allen were also inspired by the *Popular Electronics* introduction of the Altair 8800. The manufacturer had assumed that the buyer who could put the computer together could also program it, so initially no program was provided. Gates and Allen started with a programming language devised for mainframe computers and adapted it for the Altair, calling it Altair Basic, a name later changed to Microsoft Basic. According to industry lore, a paper-tape copy of Altair Basic was stolen from a Silicon Valley motel in 1975 and fifty newly made copies were distributed at a Homebrew Club meeting within a few days. Software was only just emerging as a business, and a kind of countercultural attitude of freely sharing new software was still widely held among hobbyists. Ironically now, in the early years of the twenty-first century, there appears to be a movement away from programs that are resident in individual computers and mainframes, which have been replaced with "software as a service" or (as it might be called at other sites of lean production) "software on demand." In this model, software is not sold in a box and then installed on the purchaser's computer, but rather remains on servers belonging to the vendor, from which it is delivered, when needed, by way of the Internet.

The iconic figure of "Cap'n Crunch" (the pseudonym adopted by John T. Draper) personified the somewhat anarchistic tendency of the early computer hobbyists.[7] In 1969 Draper pulled his Volkswagen microbus up to two isolated telephone booths off a freeway ramp in northern California and, using a blue box (a special tone-generating device) he had built, was able, without putting in any coins, to use one phone to place a call through Tokyo, India, Greece, South Africa, South America, and New York to himself in the second phone booth. In the mid-nineteenth century, when the American magnate Cyrus W. Field planned to lay a telegraph cable under the Atlantic to Europe, it was said that the president of the United States could then send a message directly to Queen Victoria in England. Allegedly a contemporary cynic (or snob) asked what the American president could possibly say that the queen would want to hear. In eerie confirmation of that canard, Cap'n Crunch, having phoned himself free around the world, could only say: "Hello, test one, test two, test three." Significantly, Wozniak and Jobs earned a few dollars, while students at the University of California, by assembling and selling these same blue boxes.

Despite the rhetoric of freedom and individual empowerment, the future of the personal computer owed as much to its Pentagon roots as to its anarchist idealism. According to anthropologist Bryan Pfaffenberger, "The personal computer revolution was no revolution" at all, at least in terms of the excited expectations surrounding its introduction.[8] By then marginal young men had begun to work on the cheap, user-friendly "home" computer, the typical large mainframes used by business and government had become symbols of hierarchy, repression, dehumanization, and the centralization of power. In trying to reconstruct the ideology of computing around "decentralization, democratic autonomy, and the restoration of nature,"[9] these young pioneers succeeded largely in appealing to and reinforcing the very values they professed to reject. Trying to reform rather than destroy "the system," they created a machine that was more likely to be found on the work desks of secretaries and white-collar workers, and when it made its way to the home, often had the effect of merely lengthening the workday for those same workers. Significantly, they created a hypermasculine culture in which one's individual self-worth was measured in terms of technical abilities.

In the 1970s, many of the components of the electronic environment came together in a vast network named the Internet. One reaction to the orbiting of *Sputnik I* by the Soviet Union in 1957 had been the establishment within the Defense Department of ARPA (the Advanced Research Projects Agency), charged with providing funds for high-tech research. Of critical concern to the Pentagon at the time were problems involving "command, control, and communication," called C3. This concern was motivated in part by a desire to somehow use computers under battlefield conditions, but also in part by a desire to ensure continued C3 after a full-blown nuclear exchange with the Soviets. The military, like the rest of the country, essentially relied on the Bell telephone network, which appeared to be much too fragile to survive an attack since any segment of the line broken had to be fixed before communication could resume.

It was a happy coincidence that at this point the interests of the military, for secure C3, and those of the nation's growing cadre of computer engineers and scientists began to converge. As Joseph Carl Licklider, the head of ARPA's Information Processing Techniques Office from 1962 to 1964 put it, "What the military needs is what the businessman needs is what the scientist needs."[10] One early and critical contribution was made by Paul Baran, a researcher at the RAND (Research and Development) Corporation, a South-

The plan to build an "information highway" across the nation was an infrastructural project that deliberately recalled the interstate highway network of the 1950s. Given a boost by the election of Bill Clinton to the White House in 1992, the "highway" was a direct descendent of a computer network developed by the Pentagon's Advanced Research Projects Agency (ARPA) in the early 1970s. Note that the "users" are all universities, think tanks, and government agencies. *Science,* 175 (March 10, 1972), iv.

ern California think tank supported by the air force. Baran and his colleagues advocated "packet switching," by which messages to be transmitted were broken down into a number of separate bundles that were then sent along any free and operating communication lines, switching from one to another until they were finally reassembled at their destination. In this way messages could pick their own way through the maze of surviving lines, so no single break into the system could stop them. This concept of a nonhierarchical "distributed adaptive message block network" was first proposed by Baran in 1962 and published in 1964.

A desire parallel to ARPA's for secure C3 networks was that for a network to link all the various supercomputer projects that ARPA was supporting across the country. In theory at least, they believed, it should be possible to move from a model of one person at one computer to a model of a network of computers, each time-shared by any number of researchers. The system, to be called ARPANET, was handed to the electrical engineering company Bolt Beranek and Newman (BBN), which had close ties to MIT, for development. Beginning in 1969, BBN worked with other ARPA contractors to put together the system using Baran's block network idea. A team of researchers headed by BBN's Frank Heart tackled the project with many of the attitudes and prejudices of the counterculture of the 1960s, of which some had been a part. Much like the young engineers and scientists of the Silicon Valley in later years, they preferred consensus to hierarchy, wore sneakers to high-level meetings, and tried hard to ignore the fact that they were being lavishly funded by the Pentagon.

The team was trying to design both the hardware and the software for what some called "packet switches" and others "interface message processors," or IMPs. Information or communications theory had little to do with it: as one member recalled, "It becomes an engineering problem as opposed to a theory problem. . . . We had to send these bits down the wire: how do you put a header on the front; how do you put a trailer on the back?"[11] In September of 1969, BBN delivered the first IMP to the University of California, Los Angeles. Within a short period they had also provided IMPs to the other four sites that were to be networked: the University of California, Santa Barbara; Stanford Research Institute; and the University of Utah. A network working group had developed the basic ARPANET protocols in 1972. That same year the first International Conference on Computer Communication was held at a Hilton Hotel in Washington, D.C., and the builders of ARPANET chose it as the venue for the first public demonstration of their work. Over a thousand people watched as computers set up in a display room at the hotel networked with others all over the country. One of the team said that the demonstration converted the doubters: "It was almost like the train industry disbelieving that airplanes could really fly until they actually saw one in flight."[12]

ARPANET had been intended for the exchange of work between Pentagon researchers, but within a few months e-mail accounted for three-quarters of its traffic. Soon mailing lists of like-minded users appeared, the first for science fiction fans, called SF-LOVERS. Since this use was unauthorized, and indeed unintended, ARPA shut it down, but its members successfully argued that they were really testing the network's mail capacity. Other lists followed quickly, taking ARPANET in directions undreamed of by those in the Pentagon who were worried about C3. One science fiction writer was quoted as saying it was "as if some grim fallout shelter had burst open and a full-scale Mardi Gras parade had come out."[13]

In 1974 the Transmission Control Protocol / Internet Protocol was defined as the protocol for interfacing networks, and when in 1983 the Pentagon made that a requirement for all host computers on ARPANET, they enabled what is now called the Internet. That same year the Defense Department split off from ARPANET a network called MILNET to serve an exclusively military purpose, and in 1986 the National Science Foundation set up NSFNET to connect five supercomputers that they had funded. Between 1988 and 1992, the number of computers linked to the Internet doubled each year. As the Internet grew, no longer dominated by researchers and government contractors, ARPANET as such was shut down in 1990 and NSFNET five years later, leaving the Internet as a flourishing resource of many facets, uses, and communities of users.

In 1990 Tim Berners-Lee, working at CERN (the European nuclear research organization) developed a hypertext system called the World Wide Web, and in the next few years browsers such as Netscape and Internet Explorer were developed. By 2006 more than a billion people around the world were using the Internet, and a survey in the United Kingdom by Google discovered that the average Briton spent more time each day online than in front of the television: 164 minutes of surfing as compared with only 148 minutes of TV watching. As is often the case, the great power that these two innovations placed in the hands of average users was enormous, but the result was something less than revolutionary. Lists made by search firms in 2005 showed the overwhelming influence of popular culture. When sexual references were ignored, America Online reported that lottery queries were most common, followed by horoscopes and tattoos. The top ten search terms on Yahoo! were, first, Britney Spears (Lycos listed Paris Hilton first), followed by the rapper 50 Cent, the Cartoon Network, Mariah Carey, and so on, through Paris Hilton (seventh) and Eminem (eighth).

Since its inception, the Internet had been controlled by the United States, first directly by the Department of Commerce and then, as it became a significant international presence, through a private body created by the government called the Internet Corporation for Assigned Names and Numbers. By 2005 American claims to global leadership (or, some would say, domination) were being widely challenged and U.S. control of the Internet had become a hotly debated issue. The United Nations was one candidate to be the new supervisor, but the European Union also sought that role. This challenge has been a threat to American dominance that the U.S. government has stubbornly refused to compromise, but the issue is not likely to go away.

By the turn of the twenty-first century the participle "wired" had come to mean connected, aware, and "in-the-know." And the term "hard-wired," used to denote connection by fiber optics rather than simply by telephone landline, had evolved to also mean genetically (and therefore presumably unalterably) programmed for certain patterns of behavior. At just that time, however, positive, even progressive, connotations of these terms were overshadowed by one borrowed from early twentieth-century radio—"wireless." The federal government auctioned off parts of the nation's radio band in 1994 to companies planning to provide wireless communication services. Indeed, at least three systems were already in use in a number of cities, using incompatible, proprietary technologies. Then in 2000 the Institute of Electrical and Electronic Engineers (IEEE) established a standard known as IEEE 802.11b using the 2.4 gigahertz radio spectrum. Following this breakthrough approximately two hundred firms in the industry formed the WiFi Alliance, which certified the "interoperability" of some 802 products.

Wireless quickly became available not only in offices and homes, but in so-called hot spots such as airports, coffee shops, hotels, and university campuses. By the end of 2005 there were an estimated 100,000 hot spots around the world, of which 37,000 were in the United States, followed by 12,668 in the United Kingdom and 9,415 in South Korea. Among the ten most-wired cities, however, only two were in the United States (San Francisco and New York) and two were in Europe (London and Paris), while six were in Asia, including two in South Korea. Several American cities, hoping to provide an amenity but also no doubt to give them a competitive edge in attracting high-tech industries, were also moving into establishing municipal, citywide systems. Philadelphia was the acknowledged leader, but New Orleans, Chicago, and San Francisco were also investing in networks.

At least one agricultural region was also providing wireless services. In eastern Oregon, a hot spot described as the world's largest covered 700 square miles. Echoing the reluctance of privately owned electric utilities to provide rural service in the 1920s and 1930s, local telephone companies had not found the potential profits attractive; in this case an individual spent $5 million of his own funds to create a wireless "cloud" and hoped to recoup costs from state and local government agencies and from large corporate farms. Meanwhile, all farmers in the area could check the moisture in their crops, turn irrigation sprinklers on and off, or just play computer games as their tractors moved about the fields. The interconnectedness that the WiFi Alliance sought for appliances had become characteristic of people. The lines between work and play, between accessibility and privacy, even between civility and rudeness had been severely compromised.

Helping to blur those distinctions further were developments in personal entertainment. Tape recorders had been developed in Germany during World War II, and in 1965 Philips Electronics had patented the cassette tape. Building on this technology, the Sony Corporation of Japan brought out its famous monaural Pressman in 1977, followed by its iconic stereo Walkman in 1979. Sony founder Akio Morita predicted, "This is the product that will satisfy those young people who want to listen to music all day. They'll take it everywhere with them."[14] Morita proved correct: by 1999 the cassette Walkman had sold 186 million units, and the CD Walkman, introduced in 1984, had sold 46 million. One element in the success of the Walkman was its innovative earphones, which weighed less than 50 grams, compared with the standard of 300 to 400 grams.

In 2001 Apple released the iPod, a device that became the widespread successor to the Walkman. The critical technical difference was that the iPod stored music on a hard drive, but even this was not an Apple innovation, since other devices with hard drives were already on the market. Using a hard drive made by Toshiba, an operating system from Pixo, and the MP3 from PortalPlayer, Apple added a long-life battery, a high-speed FireWire interface, and an aesthetically exciting exterior design, including white earpieces, then shrank it all to a quarter of the size of competing products. The iPod had a capacity of one thousand songs, but was seen as expensive at $400. Updated versions expanded the song capacity to ten thousand, and by late 2003 about 1.4 million had been sold.

Like similar devices, the iPod typically played songs downloaded from computers, short-circuiting the actual physical records and, frequently, the copyrights of artists and record companies as well. Just as the compact disc (CD), introduced in 1984, eventually drove vinyl record technology out of the market, two decades later it was itself marked for extinction. One owner of three large record store chains in Australia announced in 2006 that each store would soon be fitted with "Fast Track Kiosks" at which customers could download any of 100,000 songs onto iPods or other MP3 players. In many ways this rapid introduction of new technologies used to access music, and the struggles of companies and consumers with heavy investments in older technologies, was reminiscent of the changes originally wrought in the film industry by television, then video and DVD, and now also by the new technologies of computers, iPods, and cell phones.

The "wristwatch radio" had been a futuristic device familiar to all Americans since the World War II years, when the famous cartoon detective Dick Tracy had used one to fight crime. The idea for Tracy's two-way radio has been attributed to Al Gross, an electrical engineer who held a large number of patents for inventions anticipating such devices as cell phones, paging systems, garage-door openers, and the walkie-talkie, a portable device he developed in 1938 that allowed him to communicate with other ham operators while he was moving about. During World War II Gross developed a ground-to-air two-way battery-operated radio for the military.

The ubiquitous cell phone, or mobile as it is called in other parts of the world, is attributed to Motorola, particularly to its project manager Martin Cooper. As Cooper tells it, Bell Laboratories was working on cellular technology and planning to "take over" the entire business of personal communications. Motorola had already become interested in personal portable phones, and in 1972 a team headed by Cooper built one over several months. It weighed two pounds, but it worked, and standing on the street outside a Hilton hotel in midtown Manhattan, Cooper placed the first public call on a hand-held cellular phone—to the man in charge of the project at Bell Laboratories. Thirty years later, 144 million Americans had such phones.

The United States was by no means the leader in cell phone use. In 2005, some 100 million Africans (one out of every nine persons) had cell phones, ranging from the very rich to the relatively poor. The use of telephones on that continent had long been limited by inadequate technological infrastructure in the communication field, with expensive telephones, months-long waiting lists for installation, and frequent loss of service because of equipment failure, bad weather, and the theft of copper wires. Cheap cell phones and prepaid phone cards (the latter a $2 billion a year business employing thousands of small-time vendors) cut through the problems and created a $25 billion industry. In Australia "mobile" phones were in wide use by 2005, and 20 percent of phone users said that they would abandon landlines altogether the next time they moved.

As the market approached apparent saturation by technologies such as cell phones (it was estimated in 2005 that 80 percent of Americans between 18 and 65 owned such phones), new markets were developed. Only 10 percent of preteens, for example, had phones, so toy and educational game makers moved in to provide colorful phones for children that have built-in or programmable parental control mechanisms.

For many years the "perpetual next big thing" was the picture phone, which would allow people to see each other as they talked. AT&T began testing such a device as early as 1931, but a lack of consumer interest led them to drop the device until the 1950s. In 1964 they unveiled the PicturePhone, but by 1972 it, too, was abandoned, called by one AT&T executive "the most famous failure in the history of the Bell system."[15] Now, in the early twenty-first century, however, cell phones that take pictures are common, violating the privacy of people on beaches and toilets, but also providing pictures from within prisons and from inside the London underground system during the July 7, 2005, terrorist attacks. Newer-generation cell phones are a combination of MP3 music player, still camera, video camera, web browser, and news service, with motion pictures and television programs available as well. Although it is estimated that 90 percent of cell phone users actually use fewer than 10 percent of the features already available, manufacturers continue to hope that the perpetual next big thing will continue to find a market.

Another area of electronic success has been games. Electronic games have become so successful that they are displacing the more traditional "toys" with which children played for generations. Gaming began in the years immediately after World War II, first on TV screens (video games) and later on computer screens. In 1958 Willy Higginbotham, an employee of the Brookhaven National Laboratories, came up with a *Tennis for Two* game. Then in 1962 a group of students at MIT came up with *Spacewar,* a simple object-in-motion program. Within months it had evolved into a game with two spaceships, each with a supply of rockets and fuel and controlled by a player with a joystick or keyboard controls and lightning reflexes. The program was carried on a punched paper tape and circulated to other laboratories and universities. Nolan Bushnell, then a graduate student at the University of Utah, was inspired by the game and went on to found the company Atari. He developed *Pong,* another tennis game, which in 1972 became the first commercially successful video game, played in bars and other sites around the country. In Japan Toru Iwatani came up with *Pac-Man,* which became a rage in the United States in 1980.

The Japanese influence was overwhelming. Nintendo had been a manufacturer of playing cards, established in 1889. In 1969 it expanded into toys and arcade equipment, then in 1977 reached an agreement with Mitsubishi Electronics to produce video games. *Donkey Kong* and *Mario Brothers* (followed by *Super Mario Brothers*) became bestsellers. Then in 1989 Nintendo launched their new console Game Boy, which by 2005 had sold 180 million copies worldwide. The games developed to be played on Game Boy and other devices, such as Sony's PlayStation Portable (PSP), have continued to proliferate, usually immersing players in violent "first-person shooter" scenarios set in deep space, the Middle Ages, urban streets, or (at the turn of the twenty-first century) the Middle East. Despite the production of games aimed specifically at them, girls continue to be an audience not yet captured by gaming. Racial minorities also are underrepresented among gamers. Gaming remains predominantly an enthusiasm of white boys and young men.

Perhaps not surprisingly, gaming formed a close relationship with the military. The latter had long used simulators for training purposes, but by the 1990s had discovered the power of computer games with realistic graphics to influence the behavior of soldiers. By the end of 2005, over 29 million free copies had been distributed of *America's Army,* a first-person shooter game developed by the army to help recruit players into the military. Once there, soldiers were likely to be trained, in part, by other games, and the military was supplying games to troops in Iraq that would allow them to hone their skills during free time spent with a Game Boy or PSP. The link between even commercial games and actual warfare in Iraq was sobering. One marine, after a bloody patrol through an Iraqi town, was quoted as saying: "I was just thinking one thing when we drove into that ambush. *Grand Theft Auto: Vice City.* I felt like I was living it when I seen the flames coming out of the windows, the blown-up car in the street, guys crawling around shooting at us. It was f___ing cool."[16]

Today, at the beginning of the twenty-first century, computers have worked their way into many of the things with which Americans surround themselves. A digital camera can have two hundred functions, the uses of which are explained in a 205-page instruction manual. Automobiles carry as many as fifty computers to monitor and control everything from airbags and cruise control to transmission and the "power distribution box module." In 2002 the luxury 745i BMW came with a system called iDrive, which had some seven hundred functions. In many dealerships and garages, a single-purpose computer is used to diagnose engine problems no longer accessible to mechanics' skills alone. Airplanes are controlled by three overall computers: one to fly the plane, a second to monitor the first and make sure it is working properly, and a third to settle disputes in case the first two do not agree. Using these computers is called "flying by wire."

One area of interest to those who advocate "pervasive computing" is the home, where "smart" appliances can be developed to "think" about what they do and what they need to do it. Enthusiasm is expressed (mainly by academic and corporate engineers) for refrigerators that will sense when the milk is low and automatically order a new carton. Or for appliances that will contact their manufacturers when they are in need of repair. Appliances, toasters, and refrigerators that will "talk" to each other are often mentioned. The enthusiasm of the engineers has been met by skepticism, both as to how all this might work (what if a child replaced the carton of milk in the "wrong" spot on the refrigerator shelf?) and why anyone would want appliances that talked to each other in the first place. Such domestic appliances would fit well into that other dream technology, the "smart" house. In such complex and expensive environments, heating, cooling, lighting, and security would adjust automatically to whoever was home at the time. Although all of these ideas have been advertised as making life better for people, it seems obvious that this would be true only for the world's wealthy, who could afford the initial cost and that of upkeep.

Japan, not surprisingly, has led the way in this futuristic world. In early 2002 the Japan Electronic Industries Association and the Ministry of Economy, Trade, and Industry organized a demonstration home in a Tokyo suburb equipped with smart appliances by twenty of the country's high-tech consumer electronics companies. The innovations seemed endless. By sending a message from anywhere using a cell phone, the owners could take delivery of goods, run their bath water, feed the cat, turn on the washing machine, or roast a chicken. When the owners arrived at home, a sensor would recognize their fingerprints and open the door while a robot dog did a customized dance of welcome. In a culture where elderly relatives often live with younger family members, a smart bed would monitor the occupant's vital signs and an electronic thermos, filled with hot tea and left for their refreshment, would send an e-mail if it had not been used for a certain period of time. A brave new world, indeed.

Electronics are also increasingly in evidence in schools, if not turning lights on and off then at least enabling (and perhaps disabling) students. In Australia it was estimated in 2005 that a student's backpack might well contain A$4,500 worth of electronic devices, including a laptop computer, iPod, USB flash drive, mobile phone, PSP, and personal organizer. A survey of teenagers in the United States at the same time found that 90 percent of children 12 through 17 had Internet access (compared with 66 percent of adults). Three quarters of the "wired" teens said they used IM (instant messaging) to communicate with their friends, although they used e-mail with parents, teachers, and other adults. Only 45 percent of those surveyed said they had cell phones, which was probably a lower percentage than in most industrialized countries.

The smart house, filled with pervasive computing appliances, would clearly be a luxury in a country, and in a world, where so many people lack access to even the most rudimentary of shelters, and not every teenager could be expected to carry several thousands of dollars' worth of electronic equipment to school and back every day. As the American "digital divide" had shown, race and class are powerful predictors of access to the information age. While cell phones and iPods may be common even among the poor, personal computers are not, and, even after accounting for class, people of color in particular live in what are called "cyberghettos." A survey reported in 1998 showed that income, educational level, and race were all factors in determining who owned and used computers, both at home and at work. For households with an income of under $40,000 per year, for example, whites were twice as likely to have a computer at home. For those with an income of over $40,000, African-Americans were slightly more likely to have a home computer than whites, and a much higher proportion had access to one at work. The explanation offered was that, in that higher-income bracket, African-

Americans were more likely than comparable white people to have college educations, were younger, and were more likely to work in computer-related occupations. The largest difference in home computer ownership occurred among students: 73 percent of white students had access to a computer at home, while only 32 percent of African-American students did.

The actual situation may well have been even worse. One African-American web designer noted that the survey on which these conclusions were based may have been biased. The data were collected by a telephone survey, and 18 percent of African-American households did not have even that piece of nineteenth-century communications technology. The same was true, of course, for Native Americans living on reservations and perhaps for Hispanics living in barrios or workers housed on corporate farms. Deploring this divide between the tech-rich and the techno-poor, he urged, "We need to make sure that the glass ceiling isn't replaced by a silicon ceiling."[17]

15

AMERICA'S GLOBAL REACH

THE UNITED STATES EMERGED into the twenty-first century as the world hegemon. The end of the cold war in 1989, and the breakup of the Soviet Union, Europe's last real empire, two years later, had left America as the surviving superpower, based on its military, economic, and cultural strength and its worldwide presence. Over the past half century America had developed a menu of technologies, both commercial and military, that helped bring about and then supported this global dominance.

In 1988 America's Council of Academies of Engineering and Technological Sciences announced what had already become obvious: "The effects of technological change on the global economic structure are creating immense transformations in the way companies and nations organize production, trade goods, invest capital, and develop new products and processes."[1] Eight years later the Department of Commerce's undersecretary for technology noted the market for technologies which—"in the form of products, know-how, intellectual property, people and companies—are being traded, transferred, hired, bought and sold on a global basis." She went on to point out that "global competition is taking place on two levels. First is the competition between companies. Second is the competition between nations to attract and retain the engines of wealth creation that increasingly skip around the globe looking for the best opportunities."[2]

Writing in 2003, the geographer David Harvey underscored the basic fact that "much of the world's research and development is done in the U.S. This gives it a sustained technological advantage, and biases the global paths of technological change towards its own interests (particularly those centered in the military-industrial complex). It generates a flow of technological rents from the rest of the world." He warned, however, that "while the US lead in technological innovation still remains substantial (thanks in large part to its research universities), there are many signs that it is declining."[3] It was a warning that has often been sounded during the past half century, but one that has been given new urgency under the conditions of globalization.

No better example of the globalization of the effects of technology can be found than the problem of global warming. It had long been suspected that human activity affected the climate, but in 1896 a Swedish scientist offered the specific hypothesis that burning coal and other fossil fuels would release sufficient carbon dioxide into the atmosphere to actually cause it to warm—the so-called greenhouse effect. In 1961, as a result of years of government-funded research, measurements showed that buildup of the gas was indeed occurring. Spurred on by the rise of the environment movement in the 1970s, increasing numbers of scientists became alarmed that the buildup was happening fast enough, and was sufficiently severe, to cause real dangers to both the planet and human society. Much of this increasing certainty was driven by the use of computers to model the greenhouse effects. By the early twenty-first century, climate change was already dramatically reducing the polar ice caps, disrupting the world's fisheries, and threatening to raise ocean levels to a degree sufficient to threaten coastal cities.

Growing alarm led to an international conference on the subject in Kyoto, Japan, in 1997. Participants in the conference drew up a protocol that committed signatory nations to the very modest goal of cutting greenhouse gases 5.2 percent by the year 2010, taking 1990 as a base. Developing nations were exempted from the quotas on the grounds of economic need. Although thirty-eight industrialized countries signed on to the protocol, the United States did not. President Bill Clinton supported it but was unable to get it ratified by Congress. President George W. Bush announced in March 2001 that the United States would not sign on to the agreement, claiming that, especially in light of the exemptions for such countries as China and India, the American economy would be adversely effected. Ironically, by mid-2005 over 150 American cities had voluntarily committed to meet the Kyoto standards.

Although the machines of the U.S. industrial economy and its accompanying petroleum-based transportation systems have been identified as the major culprits in global warming, it is still possible to invoke the familiar argument that the proper remedy for poor technologies is more and better technologies. While many American industries rely on the political shield for their activities provided by President Bush rather than attempting to invent new "green" technologies, Japan is forging ahead along the path of innovation. Antipollution directives in the countries of the European Union are creating an attractive market for Japanese green technologies, including solar energy and biodegradable plastics made of sugar beets. Domestically, the electronics firm Fujitsu set itself the goal of cutting back on its use of electricity, gas, and oil by 25 percent in the year to March 2004, and exceeded that goal with a 28.6 percent reduction. Toyota and Honda both took the lead in producing hybrid gasoline-electric automobiles, a technology later licensed to American makers, and in 2004 Japanese industry was producing about half of the world's supply of solar photovoltaic cells. For American technology, the Organization of Petroleum Exporting Countries oil embargo and the hopeful appropriate technology movement of the mid-1970s appeared, a generation later, to have been an opportunity lost. At the 2005 international meeting in Montreal to plan for post-Kyoto emission cuts, the United States again refused to agree to limits, but pointed proudly to what it said was an expenditure of $5 billion a year on science and technology. Development of an environmental program akin to that oft-invoked standard of scale and urgency, the Apollo program to land an American on the moon, seemed to many the appropriate response for the United States but was not on the political agenda.

The critical role of technology was emphasized in 2005 by a report from the British Institute of Physics. The British government had committed itself two years earlier to generating 10 percent of its electricity from renewable source by 2010, but skeptics doubted its ability to do so. The institute's chief executive expressed doubt that there were sufficient scientists and engineers at work on the problem and claimed that "a simple, clearer system of funding linked to more post-graduate and research opportunities in renewable energy research is urgently needed to kick-start this process."[4] One British government response was to propose a new round of building nuclear power plants.

Early in the new millennium, a series of natural and human-caused disasters, some of which could plausibly be linked to global climate change, were being met with a combination of institutional and engineering infrastructure planning that extended, at the same time, the Enlightenment dream of development and the capitalist need for new venues of investment. From Hurricane Mitch, which ravaged parts of Central American in 1998, through the devastation resulting from American military action in Afghanistan and Iraq and the December 2004 tsunami in south and southeast Asia to Hurricanes Katrina and Rita along the American Gulf Coast in 2005 and the earthquake in Kashmir that same year, the post–World War II American dream of global engineering was played out. As President Bush remarked with respect to New Orleans, lands cleared by disaster represented "opportunity zones."

During the nineteenth century, when large American cities like Chicago lost their older centers to dramatic fires, local developers and politicians seized the opportunity to rebuild on a grander scale, one more exemplary of the modern urban ideal and, of course, better suited to the needs of commerce and manufacturing. So now, at the beginning of the twenty-first century, American global interests, working through such international agencies as the World Bank and International Monetary Fund and with the instruments of large American engineering firms, are at the forefront in the task of "reconstruction."

The same month she became secretary of state, Condoleezza Rice characterized the southeast Asian tsunami as "a wonderful opportunity" for the United States, since the peasant culture of small fishing villages spread along the coastlines had been destroyed and could be replaced with more "modern" development. Because upscale tourist resorts, hotels and casinos, and large industrial fish farms were much more valuable, they were to be built along the beaches stripped clean by the tsunami, made accessible by new highways and airports. After the 2005 hurricanes in Louisiana, large engineering firms like Bechtel, Halliburton, and Fluor, which were already handling enormous contracts to build American bases and rebuild destroyed infrastructure in Iraq, were called on to start the cleanup and reconstruction.

Agriculture, perhaps the most basic of all technologies, is deeply enmeshed in the process of globalization in at least two important ways. First, food crops are increasingly delinked from local production and consumption so that even perishable crops are grown and then flown halfway around the world to consumers. Second, crops of genetically modified organisms (GMOs), most often developed in the United States, are sold to producers in other countries, where they threaten local seed stock and small farmers alike.

The global circulation of foodstuffs is an old activity, in America having begun with the New England export of salt cod and import of molasses in colonial times. Entire regimes were installed, such as the rice culture in the Carolinas that was brought over by enslaved workers from West Africa. By the end of the nineteenth century, American beef was being exported in refrigerated ships and wheat from the American Midwest, Argentina, Australia, and Russia crisscrossed the globe in search of markets. A hundred years later, however, that traffic has expanded to include perishable fruits and vegetables, which are available on American tables at all seasons of the year. In 2002 one visiting a local grocery in the Midwest could find tangerines from South Africa, apples from New Zealand, bananas from Costa Rica, and asparagus from Mexico. In 1988 trucks brought produce to Chicago over an average distance of 1,518, miles and by the end of the century refrigerated jumbo jets were flying in an increasing proportion of food originating in other parts of the world.

Likewise, the selective improvement of crop and animal species by farmers was nothing new, but the corporate manipulation of the genetic makeup of plants and animals was a controversial new departure from that traditional activity. By 2002, something like 34 percent of the corn, 75 percent of the soybeans, 70 percent of the cotton, and 15 percent of the canola grown in the United States were genetically engineered, compared with virtually none six years before. The case of corn was perhaps the best known. One type, known as Bt, took six years to move from the laboratory to one out of every five acres planted in corn in the country. It had had its genetic material altered to allow it to produce its own toxins to ward off the European corn borer. Other crops were designed to resist the herbicides with which fields were drenched to prevent weeds. But even more exotic combinations have been made: flounder genes have been used in strawberries, mice genes in potatoes, cow genes in sugarcane and soy, and chicken genes in corn.

Each new genetically engineered plant or animal is, of course, a new invention, and as such is patented by the firm that produces it. The American chemical giant Monsanto owned or licensed in 2002 an astonishing 90 percent of all the GMO seed planted around the world. Like other GMO producers, it has bought up many of the competing seed firms. The global reach of GMOs has large and largely unstudied ramifications, both for the biodiversity of the planet and for the lives and welfare of indigenous farmers. GM corn is licensed to farmers for use for only one growing season, so they cannot save any for planting in the future. In 1998 the U.S. Department of Agriculture (USDA) and the Delta and Pine Land Company announced that they had developed and patented a new biotechnology called control of plant gene expression. Known also as the "terminator technology," it allows its owners to create seeds that are programmed to kill their own embryos so that all the seed sold is sterile and must be replaced by the firm for the next planting cycle. One USDA scientist was quoted as saying that "the need was there to come up with a system that allowed you to self-police your technology, other than trying to put on laws and legal barriers to farmers saving seed, and to try and stop foreign interests from stealing the technology."[5]

Farmers in Oaxaca, Mexico, plant sixty different varieties of local corn, but the GM seed has already been found there as well. Coupled with the high government subsidies provided to farmers in the industrialized countries (35 percent of farm income in Europe, 20 percent in the United States), GMOs—designed to fit into large-scale, chemical-intensive, mechanized, and therefore capital-intensive agricultural regimes—make corn growing in Oaxaca, even for subsistence, a losing proposition. As one critic charged, "Genetic-corporate agriculture is in fact a system of feeding on the world rather than for feeding the world."

A powerful early marker of technology's role in globalization was the way in which electronic communications had created, if not a "global village," at least a "network society." Beginning during World War II and continuing during the years of the cold war, the American government, including the military and American corporations, established a communications network designed to promote the hegemonic American ideals of free trade, free enterprise, the free flow of information, and a consumer-based society. The government's desire to shape the world through such agencies as the Voice of America, the desire of business to penetrate foreign markets for television programs as well as a host of consumer goods from Coca-Cola to running shoes, and the desire of the military for instant and secure communication with far-flung "trouble spots" around the world all pointed to the importance of a comprehensive, American-dominated global communications system.

The launching of *Sputnik I* by the Soviet Union in 1957, the sending of Yuri Gagarin into space in 1961, and President John F. Kennedy's commitment to go to the moon overshadowed the passage of the Communications Satellite Act of 1962. This act created Comsat, which helped set up and manage Intelsat, the first international system of global communications based on orbiting satellites. In 1945 the British science fiction writer Arthur C. Clark had suggested that a system of geosynchronous satellites would provide global communications. In the 1950s AT&T had begun to consider such a system and eventually orbited a medium-altitude satellite called *Telstar.* The Kennedy administration, however, elected to construct a system in which the government played a decisive role, and technically, one based on geosynchronous satellites. In 1965 *Intelsat I,* also known as *Early Bird,* became the first commercial satellite, providing a single path for signals between the United States and Europe. By 1977 the INTELSAT system comprised eight larger, more powerful satellites serving 150 earth stations.

The building of such a system, essentially controlled by the United States, had both commercial and diplomatic consequences. The chair of the House Committee on Science and Astronautics urged in 1961 that the system be used for the global distribution of television programs because, as he believed, "the nation that controls world-wide communications and television will ultimately have that nation's language become the universal tongue."[6] The extension of American influence could be reinforced by a disruption of old imperial ties. Newly emerging African nations, for example, could communicate directly with each other and with the United States without having to rely on fixed cables that had been laid to connect the former colonies with the old imperial seats of London and Paris. The integrated nature of the system was illustrated by the 1969 Apollo moon landing, which was watched by an estimated 500 million people in forty-nine countries relying on the INTELSAT system. This decisive triumph of American technological vanguardism over the Soviet Union due to the Soviets' "failure" (or perhaps only irrelevance) was revealed to the world directly by an American-built and controlled communications system based on much of the same technology used by Apollo itself.

Also working in the context of the space race with the Soviet Union, scientists at Johns Hopkins University worked out a way to use radio signals from satellites to give naval vessels accurate determinations of their positions at sea. In the 1960s the U.S. Air Force undertook a similar program, and in 1973 the navy and air force programs were merged to form the Navigation Technology Program, which itself evolved into the Navstar Global Positioning System (GPS), which began to come on line with the launch of the first Block IIA satellite in early 1989. The GPS "constellation" consists of at least twenty-four satellites that orbit the earth in six distinct planes at an altitude of 10,900 nautical miles, the last of which was put into operation in 1993. Each satellite orbits the earth twice daily, sending down radio beams. By the end of the century the system was providing precise space-based positioning for such activities as mapping, geodetic surveying, and search and rescue missions.

Although the system was developed by and primarily for the use of the military, since 1997 approximately 1.4 million civilian receivers have been produced each year, and in 2000 the entire GPS market was estimated to be $6.2 billion. In agriculture, for example, GPS units were used to map fields in such a way that soil samples could be taken and fertilizer and pesticide application tailored to the particular needs of the site. Rice growers were using the system to map field elevations and place dikes for flooding the crops. An even more common civilian application has been use in automobiles to allow drivers to follow a map displayed on the dashboard and receive directions via car phones. Some systems automatically notify an operator of a collision; if the driver does not respond to a call, emergency services are notified and directed to the exact location of the accident. In 1994 the car rental firm Avis installed GPS equipment in thirty cars on an experimental basis, making a driver, as one newspaper commented, "feel like a Magellan of the interstates." In the following years a variety of systems became available as optional equipment on a range of new cars sold.

In a somewhat similar reaction to the perceived "technology gap" of the 1960s, the European Union launched its own first GPS satellite at the end of 2005. When all the thirty planned satellites are in place, sometime in 2008, the system, called Galileo, is expected to have cost $4.5 billion. Designed to break the American monopoly on space-based navigation systems, Galileo, in the words of the French foreign minister, would represent "the independence of the European Union." Ironically, the first satellite was launched from

Kazakhstan atop a Russian rocket even though the Russians themselves were moving forward with their own system. Called GLONASS, it is scheduled to be in operation in 2010.

Satellite imaging has its own implications for the networked world. Now, in the early years of the twenty-first century, highly magnified and detailed pictures of the earth are already available commercially. One can, for a price, buy a picture of one's own neighborhood and have a clear image of one's own house and yard, including perhaps a tool shed and swimming pool. In a putative age of terrorism, such precise information is alarming to some. In 2005 the president of India complained that a commercial Internet site provided clear and detailed images of India's parliament building, surrounding government offices, and the president's house. Government officials in South Korea and Thailand made similar complaints, as did legislators in the Netherlands.

The GPS developed between the 1950s and 1990s for the U.S. Air Force and Navy, although widely available for civilian use, had its intended application in the tracking of military activities and the guidance of new ("smart") missiles. Smart bombs, that is, those that could be individually guided to their targets, were first developed in 1966 for use in Vietnam; cheap "strap-on" kits were attached to conventional weapons, then guided to targets by laser beams. The Gulf War against Iraq in 1991 was the first in which smart weapons became an important part of the arsenal. Used operationally for the first time, the Tomahawk cruise missile was preprogrammed to hit certain targets identified by cameras that compared the terrain with satellite photos stored in on-board computers, and its flight path could be modified using GPS guidance. Tomahawks were launched from ships far from their targets and, with their claimed accuracy, became a weapon of choice for American forces. Subsequently Tomahawks were used in Bosnia, Yugoslavia, and Afghanistan and formed an important component of the "shock and awe" terror bombing of Baghdad that initiated the war in Iraq in 2003. The more "conventional" 21,500-pound MOAB (Massive Ordnance Air Blast) superbomb, designed both for shock and awe and to penetrate buried bunkers, was shipped to Iraq but arrived too late to be used during the bombardment of Baghdad. Its smaller predecessor, the GBU-28 "bunker buster," had been used in the Gulf War in 1991.

A less successful new weapon used in the Gulf War, the Patriot ground-to-air missile, was highly praised for its accuracy and effectiveness in bringing down Iraqi Scud ground-to-ground missiles fired at Israel and Saudi Arabia. The claims of 100 percent success early in the war, however, had more realistically been reduced to 0 percent by the time of congressional hearings on the matter in 1992. Basically, there was little agreement on what constituted success in the circumstances in which the missile was used, and great difficulties in establishing the Patriot's actual performance within each of the possible scenarios. Twenty-eight American troops were killed in their barracks when a Scud slipped through a hole opened up by a small error in the Patriot computer program, an error that, after hours of replication, became a very great danger. Israeli authorities flatly claimed that no incoming Scuds had been destroyed by Patriots. Neither weapon played a part in the second Gulf War of 2003.

Research and development efforts since the Vietnam War had led, by the mid-1990s, to the deployment of a UCAV (uninhabited combat air vehicle) called the Predator drone. A contract was let for the aircraft in 1994, and the next year, again in the skies over Bosnia, the plane was being tested as an intelligence-gathering vehicle. In 2001 the Predator was for the first time fitted out with Hellfire laser-guided missiles. In both Afghanistan and Iraq, as well as in "informal" situations, the Predator was used to find, track, and attack targets of interest to the military. The Predator B, introduced in 2005, could fly at 50,000 feet, carry seven times the munitions of its predecessor, and stay aloft for thirty hours.

One putative "success" of the controversial Patriot missile in the first Gulf War was the revival of President Reagan's Strategic Defense Initiative, better known as Star Wars. Largely abandoned after Reagan left office because of its great expense and its still unproven technologies, Star Wars gained renewed support from the purported ability of the Patriot to knock Scuds out of the sky. During his presidential campaign in 2000, candidate George W. Bush called for a renewal of the full Reagan program, and as president in 2002 he ordered the installation of sixteen antimissile missiles in Alaska and another twenty at sea. Like other weapons systems before it, Star Wars seemed to lead a charmed life.

Early in 2006 the Pentagon released its new "20-year defense strategy," the first since the attack on America on September 11, 2001 (an event known now simply as 9/11). Whereas the previous plan had rather conventionally called for the development of military ability to conduct operations in four regions—Europe, the Middle East, the Asian coast, and northeast Asia—the new plan anticipated a "long war," with American troops fighting in dozens of countries, often clandestinely. At the same time China was singled out as the future's most important military competitor.

This vision of a long war against "Islamic extremists," with no boundaries of time or place, rests on two interrelated initiatives. First, highly mobile troops, trained to be "culturally sensitive" and fluent in Arabic, Chinese, or Farsi, will maintain "a long-term, low-visibility presence in many areas of the world where US forces do not traditionally operate." Second, "the US will work to ensure that all major and emerging powers are integrated as constructive actors and stakeholders into the international system. It will also seek to ensure that no foreign power can dictate the terms of regional or global security."[7]

The last half of the twentieth century was shaped in large part by the cold war and America's military, diplomatic, and cultural response to the perceived threat from the Soviet Union. Now the long war seems set to replace that with a new paradigm of peril to the nation (now renamed the "homeland"). In 2006 the Pentagon released its new Quadrennial Defense Review, which acknowledged that "we have been adjusting the US global force posture, making long overdue adjustments to US basing by moving away from a static defense in obsolete cold war garrisons, and placing emphasis on the ability to surge quickly to trouble spots across the globe."[8]

In a document titled *The National Defense Strategy of the United States of America,* the Defense Department declared in 2005 that "our role in the world depends on effectively projecting and sustaining our forces in distant environments where adversaries may seek to deny US access." Bases, many of which were set up in Germany, Japan, and South Korea during the cold war to "contain" the Soviet Union were to be shut down and others opened in areas such as Eastern Europe, Central Asia and the Caucasus, Africa, Asia and the Pacific, and the Persian Gulf, where fourteen "enduring bases" were already under construction in Iraq. This redeployment from older-style garrisons to what were now called "forward operating sites" and "cooperative security locations," marked a clear shift from containing the Soviet empire to advancing the American.[9]

A different mission dictated a different mix of weapons. In terms of military technology, the long war against terrorism was said to require a large increase in such reconnaissance weapons as the Predator and the expansion of a secure global information grid. At the same time, the worry over China was said to call for greater capabilities in the long-range air strike force and submarines. The annual budget that accompanied the "Quadrennial Defense Review" called for $72.2 billion for research, development, testing, and evaluation, targeted primarily at space-based weapons. Meanwhile, $84.2 billion was earmarked for procurement, including $10.4 billion for the "missile defense shield"; $10.4 billion for the "Joint Strike Fighter"; $3.3 billion for a DDX destroyer; $2.6 billion for another nuclear submarine; and $3 billion for transport planes. Two older technologies were to see some cuts: Minuteman III land-based nuclear missiles were to be cut back from 500 to 450, and a small number of Trident submarine missiles were to be converted from nuclear to conventional warheads. All in all, however, no existing weapons systems were to be sacrificed to the new conditions, though the disappointing results of the shock and awe campaign against Baghdad and the conventional forces on the ground in Iraq signaled a shift away from conventional warfare.

Indeed there was continuing disagreement even within the military over the role of ever more elaborate and sophisticated technologies. On the one hand, echoing the concerns within the navy over the widespread adoption of radios by the fleet in the early twentieth century, critics worried that communications advances that made battlefield data instantly available to senior officers far behind the lines (and perhaps even in the United States) would increase the ability of those officers to micromanage tactical operations from afar. On the other hand, there was general concern that "the debate over visions of empire" inflated the role of technology in what was, after all, a complex task of introducing ideas, keeping order, and, in the end, "nation building." As one former Special Forces officer charged, "In the glow of the apparent effectiveness of 'precision' munitions during Operation Desert Storm in 1991, the [defense] department adopted strategies that promise success through creation of massive technologically-oriented support structures that would make smaller field forces much more effective." The Pentagon hope, he said, was that "a revolution in military affairs (RMA) will dramatically, if miraculously, improve its capabilities, primarily through achievement of information superiority, which it defines as 'the capability to collect, process, and disseminate an uninterrupted flow of information.'" The record showed, in his opinion, that "technology contributes virtually nothing to complex civil-military operations, like recent ones in Haiti, Somalia,

Bosnia, and Kosovo, in which the US military has not performed particularly well."[10]

Arguably, one of the most striking manifestations of the postmodern shift of power to the periphery and away from the center can be seen in the increasing technological evolution of Asia. The West, particularly America, had long seen itself as the very epitome of the modern, and technology as the measure of that modernity. Pre-modern peoples, which included peasant societies in all parts of the world, were demonstrably "backward" in their failure to have what the British writer C. P. Snow called "jam." Snow optimistically wrote in 1959: "It is technically possible to carry out the scientific revolution in India, Africa, South-east Asia, Latin America, the Middle East, within fifty years."[11] Whether or not that was possible, or even desirable, the point was made that these parts of the world had not participated in the Scientific Revolution, which in the seventeenth and eighteenth centuries had done much to define the West's transition to modernity.

Moreover, such areas were necessary in order to define and measure the great transformations of the West. And in terms of the standards of hierarchy and racism that marked Enlightenment thinking, the Orient, which spread from the Ottoman Empire and biblical lands of the Near East through southern Asia to the "Far East," was certainly exotic, for many erotic, and above all inferior to the modernized West. It was a great shock, therefore, when in the 1990s Japan seemed suddenly to be the site of the best modern technology, from automobiles to cameras and from flat-screen televisions to solar voltaics. By 1989 polls showed that Japan had passed Russia as the country Americans feared most. It was not just that in many areas of technology, such as the automobile, "theirs" was demonstrably superior to "ours," but what that might mean for America's vanguardist self-identification as the most modern of nations by virtue of its unassailed role as the font of the newest and the best of technologies.

Worst of all, Japan seemed in many ways to be the first postmodern nation, and in this new world it, not America, was the model and measure. "In the future," wrote Jean Baudrillard, "power will belong to those peoples with no origins and no authenticity, who know how to exploit that situation to the full."[12] Japanese cartoons (Anime) seemed to capture the new culture, and the so-called *otaku* generation (kids "lost to everyday life" through their saturation with computer reality) were seen to be, like the *noir* oriental streets of Los Angeles in the film *Blade Runner,* a frightening preview of the post-modern world. To alarmed Americans, the old racist canard that the Japanese were "cold, impersonal and machine-like, [with] an authoritarian culture lacking emotional connection to the rest of the world,"[13] represented a merging of the oriental (pre-modern) and the technological (modern).

By the end of the twentieth century Japan was not alone in representing a threat to American technological leadership. The rise of both China and India as economic giants had come to provide another set of problems. Their exemption from the Kyoto protocol as "developing" nations had, for example, provided a major excuse for the United States not to sign on either. China's vast hydroelectric projects, space program, nuclear capability, increasing appetite for petroleum, and heavily polluted rivers and air all marked that country as a technological powerhouse. The tremendous Three Gorges Dam, planned to be the world's largest, required the displacement of 1.13 million people and was expected to impound 265 billion gallons of raw sewage and industrial waste every year. In 1956 China opened its first institute of missile and rocket research, headed by a Chinese scientist who had returned from the United States. In 1992 the country announced its intention to send people into space, and in 2003 it successfully launched its first astronaut into orbit.

Like that of China, the technological development of India seems more like modernization than a leap into the postmodern. India, too, developed successful nuclear and missile programs, the mark of a modern state in the late twentieth century. In addition, foreign investment and "brain drain" both play dominant roles on that subcontinent. The writer Arundhati Roy vividly captured the unevenness of development when she wrote: "In the lane behind my house, every night I walk past road gangs of emaciated laborers digging a trench to lay fiber-optic cables to speed up our digital revolution. In the bitter winter cold, they work by the light of a few candles." Is globalization, she asks, "about 'eradication of world poverty,' or is it a mutant variety of colonialism, remote controlled and digitally operated?"[14] Roy charges that India's agricultural peasants, cultivating their own small plots often with hand tools, are being forced to compete with American corporate farms with their GMO crops and capital-intensive technologies. Big dams built over the past half century in India have displaced 30 million people, often the poorest and most despised of the population, as part of a $32 to $46 billion international dam construction industry. In 1993 Enron signed an agreement whereby it could build the first private power project in India. In 1996 they renegotiated the contract, estimated to be worth $30 billion—the largest in the history of India. In all of these projects the technology (and the profits) have been American, the labor and payments Indian.

In 1952 the Walter-McCarran Act introduced a preference for highly educated and talented people within national quotas for immigration to the United States. In 1957 some 5,373 scientists, engineers, and physicians were admitted under this rule. Five years later such immigrants were allowed to enter without regard to national quotas, and by 1966 their number had risen to 9,534 per year. These large numbers helped fill America's burgeoning need for trained technical workers and led many countries of the world, including those in Europe, to fear that they were suffering a "brain drain" to the United States. As late as 2003, European nations were alarmed that 73 percent of their young people who earned doctorates in the United States chose to stay here to build their careers. It was one reason that only 5.36 per thousand European workers had Ph.D.s in science or engineering, while in America it was 8.66 per thousand. Significantly, Japan reported 9.72 per thousand.

By this time, however, there was beginning to be a significant new element to this global circulation of technical talent. In 2005 it was reported that Intel Corporation was preparing to invest $800 million in expanding its Indian operations, including the building of a new research and development center in Bangalore, designed to employ 2,800 workers. Additionally, the company's venture capital arm was planning to invest another $250 billion in local Indian companies. This was all on top of $700 million already invested since the mid-1990s. For years American firms had been outsourcing computer and software development work to engineers in India, where, it was estimated in 2003, they earned about $3,000 a year compared with the $50,000 to $60,000 earned by an American engineer. Software development was the work most commonly exported, but the amount of customer service "call center" work and "back-office accounting" work outsourced was rapidly expanding. An estimated half of America's Fortune 500 firms were making use of these and other technical services on the subcontinent.

The internationally famous call centers are large collections of trained Indian workers who were hired to answer calls from customers of American (and some British) companies who had questions or complaints about purchases, equipment failure, insurance claims, airline reservations, or other such business. They are to use American- or English-sounding names, and training begins with intensive lessons meant to give them an American or British accent so as to disguise, as much as possible, the fact that although "this call is important to us," it is being farmed out halfway around the world. A study released in 2005 called the workers "cyber coolies" who, despite education and intelligence, are exploited in almost Victorian workhouse conditions of control, surveillance, and impossible hours and conditions. Calls made from the United States during normal working hours, for example, have to be answered by workers in India in the middle of their night.

The efficiency and convenience of sending electronic office work overseas has also, of course, raised the specter of fraud and invasion of privacy. Kaiser Permanente, the large West Coast health care provider, in 2003 was sending to India U.S. patients' medical data; members' personal information, including financial records; and payroll information for its 135,000 employees and 11,000 physicians. Included in the latter was information on salaries and benefits and also Social Security numbers. Such information transfers created a situation ripe for "identity theft," since American regulatory bodies had no authority over the contracting firms in India. The "brains" of Indian workers may be drained to America, but their bodies remain in India.

Journalist Thomas L. Friedman argued in 2004 that all this was not simply a matter of cheap wages in the East; rather America, which had once skimmed off the best technical people in the world and had both attracted and kept the best science and engineering students from around the world, was letting that global attraction atrophy. A host of problems, from tightened post-9/11 procedures for obtaining visas to tax holidays offered by foreign countries, were conspiring to dry up the tribute in minds that the United States had exacted of the rest of the world for half a century. Indeed, there was evidence that the brain drain was being reversed, as places like Singapore were offering unrestricted research on stem cells and large corporations were moving significant research and development facilities to India and China.

The brain drain was not a new concern, for even while Europeans had discovered it in the 1960s, some American observers had been warning of an eroding American "competitiveness." In 1994 the journal *Science* proclaimed that "from the end of World War II until the 1970s, U.S. research and development, technologies, and related industries were dominant internationally. But during the past 15 years, strong global competition has emerged." It pointed out that thirty large American corporations had established 114 laboratories overseas, and that while in 1990 six Asian nations (including Japan) had graduated a quarter million "first degree" engineers, the United States had graduated only 65,000.[15] That same year an economist warned that it was time for the United States to get better at borrowing new technologies from abroad. Japan was widely thought to have built its success on borrowing, and as more innovations appeared abroad, America should not be too proud to learn the same lesson. It was not a proposal that fit well with the nation's proud sense of its own vanguard role in the world.

The smashing of two giant commercial jet airplanes into the World Trade Center towers on September 11, 2001, drew a cruel line across post–World War II American history. In part this was because of what the nation chose to do about it: the administration of George W. Bush committed the nation to the long war for American empire later spelled out in the Defense Review of 2006 when it retaliated against the Taliban in Afghanistan and, more controversially, chose to launch a preemptive war against Iraq. That decision cost hundreds of billions of dollars in arms, the goodwill of much of the world, and a portion of the very civil liberties for the protection of which Americans were presumably fighting.

But regardless of the reaction, the event itself—relentlessly shown on television in vivid detail—had a powerful effect on the nation. One group of commentators called the attack "September's terror" and noted that "it made no demands, it offered no explanations. It was premised on the belief (learned from the culture it wishes to annihilate) that a picture is worth a thousand words—that a picture, in the present condition of politics, is itself, if sufficiently well executed, a specific and effective piece of statecraft."[16] What these writers refer to as "spectacle" is very close to what the historian David Nye calls the "American Technological Sublime": "The sublime underlies . . . {Americans'} enthusiasm for technology," he writes. "One of the most powerful human emotions, when experienced by large groups the sublime can weld society together." Writing at an earlier time, the English philosopher Edmund Burke wrote that "the passion caused by the great and sublime in nature, when those causes operate most powerfully, is Astonishment; and astonishment is that state of the soul, in which all motions are suspended, with some degree of horror. In this case the mind is so entirely filled with its object, that it cannot entertain any other, nor by consequence reason on that object which employs it. Hence arises the great power of the sublime, that far from being produced by them, it anticipates our reasonings, and hurries us on by an irresistible force."[17] Little wonder, then, that Americans, weaned on a century of the technological sublime and specifically schooled in the sublime spectacle of imploded buildings collapsing into their own base, felt the horror devoid of understanding.

But behind the spectacle was its meaning. Historian Michael Adas has written: "Long before the rise of the United States to global hegemony, the nation's dependence on advanced technological systems made it increasingly vulnerable to weapons designed to inflict mass devastation on civilian society."[18] On August 6, 1945, the United States had dropped the first atomic bomb on Hiroshima, where 66,000 people died immediately. Three days later a second bomb was dropped in the city of Nagasaki, and another 40,000 were killed instantly. But the terror had not started there and then. In the summer of 1943, the bombing of Hamburg had created a firestorm that killed 50,000. Then on February 13, 1945, the Anglo-American bombing of Dresden created a firestorm that killed somewhere between 35,000 and 100,000 people. A month after that raid, British Prime Minister Winston Churchill wrote to the air marshal in charge of Bomber Command ordering him to stop raids such as that on Dresden, which were "simply for the sake of increasing terror." During the years of the cold war both the United States and the Soviet Union simply wrote off their civilian populations as undefendable and therefore expendable should the "balance of terror" give way and a full nuclear exchange begin.

In 1987 Carol Cohn, describing what she referred to as "Sex and Death in the Rational World of Defense Intellectuals," identified what she called the "technostrategic" language of that world. "In learning the language," she wrote, "one goes from being the passive powerless victim to the competent, wily, powerful purveyor of nuclear threats and nuclear explosive power. The enormous destructive effects of nuclear weapons systems become extensions of the self, rather than threats to it."[19] It was a reversal of place many Americans have experienced in visiting the Smithsonian Institution's National Air and Space Museum in Washington, D.C. Standing among the cold and beautiful missiles, one is comforted by the assumption that, standing on their fins pointing skyward, these weapons are poised to defend the American "homeland." How different would be the experience if instead they were hung "upside down" from the ceiling, as though within a split second of wiping the nation's capital off the map.

Over half a century since Hiroshima, Americans had learned to identify with the nation's technological vanguardism, locating in themselves the power to research, develop, manufacture, deploy, and use complex technologies as diverse as airliners and skyscrapers for their own purposes and on their own behalf. More specifically, a half century of ever-elaborating and ever-more-powerful military technology used against dozens of "enemy" states across the globe had taught the false lesson that, since such interventions were technologically possible and were always justified in terms of America's civilizing mission, they were without consequences for the country's civilian population. The fallacy of that delusion was exposed by 9/11. In this globalized world, which is also postmodern, significant power has shifted to the periphery, putting the metropolis at new risk. The "medieval" Middle Eastern terrorists have replaced the Japanese challenge in the American nightmare and given "techno-orientalism" a renewed currency.

NOTES

INTRODUCTION

1. Lynn White, Jr., *Medieval Technology and Social Change* (Oxford, 1962), 134.

2. Quoted ibid.

3. Carlo M. Cipolla, *Guns, Sails and Empires: Technological Innovation and the Early Phases of European Expansion, 1400–1700* (New York, 1965), 136.

CHAPTER 1: THE TOOLS BROUGHT OVER

1. Quoted in Patrick M. Malone, "Changing Military Technology among the Indians of Southern New England, 1600–1677," *American Quarterly,* 25 (March 1973), 53–54.

2. William Cronon, *Changes in the Land: Indians, Colonists, and the Ecology of New England* (New York, 1983), 160–61.

3. Quoted ibid., 162.

4. Testimony of William Vickers, before the British Parliament, *Second Report . . . Exportation of Machinery* (1841), 73.

5. Daniel Drake, *Pioneer Life in Kentucky, 1785–1800,* ed. Emmet Horine (New York, 1948), 44–45.

6. Quoted in Leroy L. Thwing, "Lighting in Early Colonial Massachusetts," *New England Quarterly,* 11 (March 1938), 169.

7. Benjamin Franklin, *An Account of the Newly Invented Pennsylvanian Fire-Place* (Philadelphia, 1744), 1.

8. *Pennsylvania Gazette,* February 28, 1765.

9. Thomas Ellicott, "The Practical Mill-Wright," in Oliver Evans, *The Young Mill-Wright and Miller's Guide* (Philadelphia, 1795), 285–86.

10. Quoted in Greville and Dorothy Bathe, *Oliver Evans: A Chronicle of Early American Engineering* (Philadelphia, 1935), 12.

11. Quoted in Floyd L. Vaughan, *The United States Patent System: Legal and Economic Conflicts in American Patent History* (Norman, 1956), 14–15.

12. Quoted in Introduction by Thomas R. Adams to Franklin's *Account of the Newly Invented,* v.

CHAPTER 2: IMPORTING THE INDUSTRIAL REVOLUTION

1. Quoted in Carroll W. Pursell, Jr., "Thomas Digges and William Pearce: An Example of The Transit of Technology," *William and Mary Quarterly,* 21 (October 1964), 551.

2. Quoted ibid.

3. Quoted ibid., 553.

4. Henry Wansey, *An Excursion to the United States of North America, in the Summer of 1794* (London, 1798), 68–69.

5. *Gazette of the United States,* June 6, 1792.

6. Tench Coxe, *View of the United States . . .* (Philadelphia, 1794), 443.

7. (Philadelphia) *Aurora,* October 1, 1816.

8. (Wilmington) *American Watchman,* January 11, 1822.

9. Quoted in *Hunt's Merchants' Magazine,* 15 (October 1846), 371–72.

10. *The Diaries of George Washington* (Charlottesville, 1979), 5:479–80, entry for October 28, 1789.

11. "A Topographical and Historical Description of Boston, 1794," *Collections of the Massachusetts Historical Society,* ser. 1, 3 (Boston, 1810), 279.

12. J. C. Dyer, "Notes on the Origins of several Mechanical Inventions, and their subsequent application to different purposes," *Proceedings, Manchester Literary and Philosophical Society,* 5 (1865–66), 6–7.

13. Ibid., 7.

14. Testimony of G. Withers, in Great Britain, Parliament, *First Report . . . Exportation of Machinery* (1841), 72.

15. "Growth, Trade, and Manufacture of Cotton," *DeBow's Review,* 16 (January 1854), 4.

16. Quoted in "History of the Steam Engine in America," *Journal of the Franklin Institute,* 102 (October 1876), 259–60.

17. *Pennsylvania Gazette,* March 14, 1799.

18. *Agricultural Museum,* 2 (1811), 104.

19. Thomas Cooper, "Iron," *Emporium of the Arts and Science,* n.s. 1 (1813), 232.

20. *Port Folio,* 3 (1817), 198.

CHAPTER 3: IMPROVING TRANSPORTATION

1. T. S. Ashton, *The Industrial Revolution, 1760–1830* (London, 1948), 81.

2. David Stevenson, *Sketch of the Civil Engineering of North America* (2d ed., London, 1859), 185.

3. Quoted in Thomas H. MacDonald, "The History and Development of Road Building in the United States," *Transactions of the American Society of Civil Engineers,* 92 (1928), 1183, 1184.

4. Quoted in Seymour Dunbar, *A History of Travel in America* (Indianapolis, 1915), 3:746.

5. [Joseph] Whitworth and [George] Wallis, *Industry of the United States in Machinery, Manufactures, and Useful and Ornamental Arts* (London, 1854), 20–21.

6. G. W. Smith, *Internal Improvement. Rail Roads, Canals, Bridges, &c.* (Philadelphia, 1825), 4.

7. Quoted in Robert F. Hunter, "Turnpike Construction in Antebellum Virginia," *Technology and Culture,* 4 (Spring 1963), 198.

8. B. Franklin to Samuel Rhoads, August 22, 1772, in *The Papers of Benjamin Franklin,* 19 (New Haven, 1975), 279.

9. Quoted in Charles B. Stuart, *Lives and Works of Civil and Military Engineers of America* (New York, 1871), 59.

10. Cadwallader D. Colden, *Memoir . . . at the Celebration of the Completion of the New York Canals* (New York, 1825), 12.

11. Stevenson, *Sketch,* 119.

12. Quoted in Samuel Smiles, *The Life of George Stephenson, Railway Engineer* (Boston, 1863), 49–50.

13. Quoted in Robert E. Carlson, "British Railroads and Engineers and the Beginnings of American Railroad Development," *Business History Review,* 34 (Summer 1960), 140.

14. Stevenson, *Sketch,* 258.

15. Ibid., 258.

16. Testimony of Alexander Jones, M.D., in *Parliamentary Papers: First Report . . . Exportation of Machinery* (1841), 204.

17. Stevenson, *Sketch,* 253.

18. Minus Ward, *Remarks, Propositions and Calculations, Relative to A Rail-Road and Locomotive Engines to be used upon the same, from Baltimore to the Ohio River* (Baltimore, 1827), iv.

19. Stevenson, *Sketch,* 239–40.

CHAPTER 4: THE EXPANSION OF AMERICAN MANUFACTURES

1. W. F. Durfee, "An Account of the Experimental Steel Works at Wyandotte, Michigan," *Transactions of the American Society of Mechanical Engineers,* 6 (1884), 41.

2. *Thirteenth Annual Report of the Commissioner of Labor, 1898. Hand and Machine Labor,* 1 (Washington, 1899).

3. Ibid.

4. Ibid.

5. Quoted in Gustavus A. Weber, *The Patent Office: Its History, Activities, and Organization* (Baltimore, 1924), 3.

6. Joseph Barnes, *Treatise on the Justice, Policy, and Utility of Establishing an Effectual System for Promoting the Progress of Useful Arts, by Assuring Property in the Products of Genius* (Philadelphia, 1792), 9–10.

7. Frances Trollope, *Domestic Manners of the Americans* (New York, 1949), 219.

CHAPTER 5: THE MECHANIZATION OF FARMING

1. Whitworth and Wallis, *Industry of the United States,* 19–20.

2. *Niles' Weekly Register,* 44 (July 27, 1833), 355.

3. Frank Corry, "California Harvesting Machinery," *Scientific American,* 79 (October 8, 1898), 235.

4. Quoted in Peter H. and Jo Ann E. Argersinger, "The Machine Breakers: Farmworkers and Social Change in the Rural Midwest of the 1870s," *Agricultural History,* 58 (July 1984), 401.

5. Ibid., 402.

6. Ibid., 407, 406.

7. Whitworth and Wallis, *Industry of the United States,* 20.

8. James F. W. Johnston, *Notes on North America: Agricultural, Economical, and Social,* 1 (Edinburgh, 1851), 160–61.

9. Benjamin Butterworth, *The Growth of Industrial Art* (Washington, D.C., 1892), 1.

10. Johnston, *Notes on North America,* 2:319.

CHAPTER 6: CREATING AN URBAN ENVIRONMENT

1. Robert Ridgway, "The Modern City and the Engineer's Relation to It," *Transactions of the American Society of Civil Engineers,* 88 (1925), 1245.

2. Quoted in Constance McLaughlin Green, *The Rise of Urban America* (New York, 1965), 42.

3. Stevenson, *Sketch,* 220.

4. Charles F. Brush, "Development of Electric Street Lighting," *Journal of the Cleveland Engineering Society,* 9 (September 1916), 55.

5. Quoted in Mel Gorman, "Charles F. Brush and the First Public Electric Street Lighting System in America," *Ohio Historical Quarterly,* 70 (April 1961), 139–40.

6. Quoted in Frank J. Sprague, "The Electric Railway," Century, 70 (July 1905), 520.

7. *Scientific Press,* 24 (February 17, 1872), 99.

8. Herbert Croly, "The New York Rapid Transit Subway: How It Will Affect the City's Life and Business," *American Monthly Review of Reviews,* 30 (September 1904), 306.

9. *Niles' Weekly Register,* 23 (September 7, 1822), 1.

10. J. Leander Bishop, *A History of American Manufactures from 1608 to 1860,* 2 (Philadelphia, 1861), 404.

11. "Establishment of Manufactures at New Orleans," *DeBow's Review,* 8 (January 1850), 13.

12. Charles F. McKenna, "The Centering of Great Industries in the New York Metropolitan District," *Transactions of the American Institute of Chemical Engineers,* 2 (1909), 81.

13. W J McGee, "Fifty Years of American Science," *Atlantic Monthly,* 82 (September 1898), 311–12.

14. John W. Reps, *The Forgotten Frontier: Urban Planning in the American West before 1890* (Columbia, 1981), 5, 3.

15. Quoted in James B. Allen, *The Company Town in the American West* (Norman, 1966), 86.

16. Sam Bass Warner, *Streetcar Suburbs: The Process of Growth in Boston, 1870–1900* (Cambridge, 1962).

CHAPTER 7: WESTWARD THE COURSE OF INDUSTRY

1. *Scientific American,* 11 (June 14, 1856), 313.

2. Remarks of James Douglas in *Transactions of the American Institute of Mining Engineers,* 29 (1899), lii.

3. Rodman Wilson Paul, "Colorado as a Pioneer of Science in the Mining West," *Mississippi Valley Historical Review,* 47 (June 1960), 47.

4. Samuel B. Christy, "The Growth of American Mining-Schools and Their Relation to the Mining Industry," *Transactions of the American Institute of Mining Engineers,* 23 (1893), 452.

5. Ibid., 457.

6. B. L. Thane, "Stoping with Machine Drills," *Transactions of the American Institute of Mining Engineers,* 29 (1899), 770.

7. Quoted in Mark Wyman, "Industrial Revolution in the West: Hard-Rock Miners and the New Technology," *Western Historical Quarterly,* 5 (January 1974), 42.

8. Walt Whitman, "Passage to India" (1868), in *Leaves of Grass and Selected Prose,* ed. John Kouwenhoven (New York, 1950), 322.

9. Ibid.

CHAPTER 8: EXPORT, EXPLOITATION, AND EMPIRE

1. *Dictionary of National Biography,* 6 (London, 1917), 288.

2. J. C. Dyer, "Notes on the Origin of several Mechanical Inventions, and their subsequent application to different purposes," *Proceedings, Manchester Literary and Philosophical Society,* 5 (1865–66), 8.

3. Oliver Evans, *The Abortion of the Young Steam Engineer's Guide* (Philadelphia, 1805), vi.

4. Quoted in *Niles' Weekly Register,* 64 (July 29, 1843), 341.

5. Testimony of Matthew Curtis, before British Parliament, in House of Commons, *Sessional Papers* (1841), 7, question 1544, 111.

6. Quoted in K. R. Gilbert, "The Ames Recessing Machine: A Survivor of the Original Enfield Rifle Machinery," *Technology and Culture,* 4 (Spring 1963), 210.

7. Quoted in Robert F. Dalzell, Jr., *American Participation in the Great Exhibition of 1851* (Amherst, 1960), 29.

8. Quoted ibid., 48.

9. Quoted ibid., 51.

10. Quoted ibid., 52.

11. *Scientific American,* 7 (November 8, 1851), 59.

12. Ibid., 6 (November 23, 1850), 75.

13. Ibid., 79 (September 24, 1898), 194.

14. *Lehigh University: Ten Year Book of the Class of 1896* (n.p., 1906), 24.

15. Ibid., 132.

16. J. A. L. Waddell, *Memoirs and Addresses of Two Decades,* Frank W. Skinner, ed. (Easton, Pa., 1928), 11.

17. "Some Observations on the Regeneration of China and the Engineering Work Involved Therein," in Waddell, *Memoirs,* 1046.

18. Octave Chanute, "The Effect of Invention upon the Railroad and Other Means of Intercommunication," *Proceedings of the Celebration of the American Patent System* (Washington, D.C., 1891), 169.

19. "Truman's Point Four Program, June 24, 1949," in Henry Steele Commager, *Documents of American History,* 7th ed. (New York, 1963), 558–59.

CHAPTER 9: THE COMING OF SCIENCE AND SYSTEMS

1. Johnston, *Notes on North America,* 1:162–63.

2. Newton C. Blanchard et al., eds., *Proceedings of a Conference of Governors in the White House, Washington, D.C., May 13–15, 1908* (Washington, D.C., 1909), 1.

3. Ibid., 3.

4. Ibid., 405.

5. Charles S. Howe, "The Function of the Engineer in the Conservation of the Natural Resources of the Country," *Science,* 28 (October 23, 1908), 539, 547–48.

6. James Douglas, "Conservation of Natural Resources," *Transactions of the American Institute of Mining Engineers,* 40 (1909), 419, 426.

7. Samuel P. Stadtler, "Conservation and the Chemical Engineer," *Transactions of the American Institute of Chemical Engineers,* 2 (1909), 107.

8. *Engineering News,* 76 (December 14, 1916), 1144–45.

9. "Taylor's Testimony before the Special House Committee," in Frederick Winslow Taylor, *Scientific Management* (New York, 1947), 29–30.

10. Frederick Winslow Taylor, *The Principles of Scientific Management* (New York, 1911), 7.

11. Ibid., 25.

12. Ibid., 36.

13. *Engineering News,* 28 (August 25, 1892), 181.

14. Morris Llewellyn Cooke, "Some Factors in Municipal Engineering," *Mechanical Engineering,* 37 (February 1915), 82.

15. Leonard D. White, *The City Manager* (Chicago, 1927), ix.

16. Ibid., 295.

17. Quoted ibid., 94.

18. Ibid., 43.

19. F. W. Ballard, "President's Address," *Journal of the Cleveland Engineering Society,* 10 (July 1917), 46.

20. Ridgway, "The Modern City," 1254.

21. "Dr. Jackson's Address before the American Institute," *Scientific American,* 7 (November 1, 1851), 51. Alonzo Potter, *The Principles of Science Applied to the Domestic and Mechanic Arts* (Boston, 1841), 266–67.

22. Oliver Evans, *The Abortion of the Young Steam Engineer's Guide* (Philadelphia, 1805), 139.

23. "The Potentialities of Chemical Research," *Scientific American,* 70 (January 20, 1894), 34.

CHAPTER 10: THE DECADE OF PROSPERITY AND CONSUMPTION

1. Edwin E. Slosson, "Back to Nature? Never! Forward to the Machine," *The Independent,* 101 (January 3, 1920), 37.

2. Ibid., 37–39.

3. *Recent Social Trends in the United States,* 1 (New York, 1933), xxv.

4. W. F. Ogburn, with the assistance of S. C. Gilfillan, "The Influence of Innovation and Discovery," *Recent Social Trends,* 1:122.

5. *Recent Social Trends,* 1:xxviii.

6. A. Hunter Dupree, *Science in the Federal Government: A History of Policies and Activities to 1940* (Cambridge, 1957), 288.

7. National Resources Committee, *Research—A National Resource. I.—Relation of the Federal Government to Research* (Washington, D.C., 1938), 5.

8. National Resources Planning Board, *Research—A National Resource. II.—Industrial Research* (Washington, D.C., 1940), 1.

9. Fred Wilbur Powell, *The Bureau of Mines: Its History, Activities and Organization* (New York, 1922), 29.

10. J. C. Hunsaker, "Forty Years of Aeronautical Research," *Annual Report . . . of the Smithsonian Institution, 1955* (Washington, D.C., 1956), 252.

11. Gustavus A. Weber, *The Bureau of Standards: Its History, Activities, and Organization* (Baltimore, 1925), 212.

12. Ibid., 211.

13. Carroll H. Wooddy, *The Growth of the Federal Government, 1915–1932* (New York, 1934), 549.

14. Laurence F. Schmeckebier, *The Federal Radio Commission: Its History, Activities and Organization* (Washington, D.C., 1932), 31.

15. *The Memoirs of Herbert Hoover,* II: *The Cabinet and the Presidency, 1920–1933* (New York, 1952), 148.

16. James J. Flink, "Three Stages of American Automobile Consciousness," *American Quarterly,* 24 (October 1972), 457.

17. Henry Ford, "Mass Production," *Encyclopaedia Britannica,* 13th ed., Suppl. vol. 2 (1926), 821–23.

18. Ibid.

19. "Henry Ford, the Road Promoter," *Engineering News-Record,* 79 (September 6, 1917), 434.

20. Fred Levis, "Highways as Elements of Transportation," *Transactions of the American Society of Civil Engineers,* 95 (1931), 1021.

21. Ibid., 1023.

22. Ruth Schwartz Cowan, "Ellen Swallow Richards: Technology and Women," in *Technology in America: A History of Individuals and Ideas,* ed. Carroll W. Pursell, Jr. (2d ed., Cambridge, 1990), 149.

23. George Basalla, "Keaton and Chaplin: The Silent Film's Response to Technology," in *Technology in America: A History of Individuals and Ideas,* ed. Carroll W. Pursell, Jr. (2d ed., Cambridge, 1990), 228.

24. Thordis Simonsen, ed., *You May Plow Here: The Narrative of Sara Brooks* (New York, 1986).

CHAPTER 11: DEPRESSION: STUDY AND SUBSIDY

1. Charles A. Beard, ed., *Whither Mankind? A Panorama of Modern Civilization* (New York, 1928), 24.

2. Charles A. Beard, ed., *Toward Civilization* (London, 1930), 304.

3. Howard Scott, "A Rendezvous with Destiny," *American Engineer,* 6 (October 1936), 10.

4. President Hoover's Research Committee on Social Trends, *Recent Social Trends* (New York, 1933), 1:122, 153–56.

5. Harry Jerome, *Mechanization in Industry* (New York, 1934), 387.

6. "Obsolete Men," *Fortune,* 6 (December 1932), 27.

7. *Engineering News-Record,* 108 (May 5, 1932), 639.

8. "In the Ditch!", *New Outlook,* 163 (February 1934), 35–36.

9. D. B. Steinman, "Engineers and Unemployment," *American Engineer,* 1 (February 1935), 23.

10. U.S. Department of Agriculture (USDA), *Technology on the Farm* (Washington, D.C., 1940), v.

11. Quoted in Carroll W. Pursell, Jr., "Government and Technology in the Great Depression," *Technology and Culture,* 20 (January 1979), 168.

12. Ibid., 168.

13. *Time,* 27 (March 23, 1936), 60.

14. USDA, *Technology on the Farm,* 10.

15. John Steinbeck, *The Grapes of Wrath* (New York, 1939), 52–53.

16. Quoted in Carroll W. Pursell, Jr., "A Preface to Government Support of Research and Development: Research Legislation and the National Bureau of Standards, 1935–41," *Technology and Culture,* 9 (April 1968), 145.

17. Thomas K. McCraw, *TVA and the Power Fight, 1933–1939* (Philadelphia, 1971), 62.

18. James W. Carey and John J. Quirk, "The Mythos of the Electronic Revolution," *American Scholar,* 39 (Spring 1970), 222.

19. Stuart Chase, *Men and Machines* (New York, 1929), 347.

20. USDA, *Technology on the Farm,* xi.

CHAPTER 12: WARS AND THE "AMERICAN CENTURY"

1. Amasa S. Bishop, *Project Sherwood: The U.S. Program in Controlled Fusion* (Reading, Mass., 1958), vii.

2. *Public Papers of the President of the United States, Dwight D. Eisenhower, 1960–61* (Washington, D.C., 1961), 1035–40.

3. Henry Steele Commager, ed., *Documents of American History,* 7th ed. (New York, 1963), 558–59.

4. George Terborgh, *The Automation Hysteria* (New York, 1966), 15–16.

5. The Ad Hoc Committee for the Triple Revolution, *The Triple Revolution* (Santa Barbara, 1964), 5.

6. *Technology and the American Economy,* Report of the National Commission on Technology, Automation, and Economic Progress, 1 (February 1966), 110–11.

7. *Wall Street Journal,* June 17, 1968.

8. Henry R. Luce, *The American Century* (New York, 1941), 30, 37.

9. Raymond Williams, *Television: Technology and Cultural Form* (New York, 1975), 20.

10. Ibid., 26.

CHAPTER 13: CHALLENGE AND CHANGE IN A POSTMODERN WORLD

1. *New York Review of Books,* 6 (April 28, 1966), 3, 5.

2. Ibid., 5.

3. National Academy of Sciences, *Technology: Processes of Assessment and Choice* (Washington, D.C., 1969), 3, 11, 15, 78–79.

4. Quoted in Bathe, *Oliver Evans,* 97.

5. *Science,* 197 (July 15, 1977), 241.

6. Society of Women Engineers, *Achievement Award, 1952–1974* (n.p., n.d.).

7. Philip H. Abelson in *Science,* 252 (May 10, 1991), 757.

8. *Science,* 252 (April 5, 1991), 21.

9. David Harvey, *The Condition of Postmodernity* (Oxford, 1989), vii.

CHAPTER 14: OUR (UN)WIRED WORLD

1. Rosalind Williams, "Cultural Origins and Environmental Implications of Large Technological Systems," *Science in Context,* 6, 2 (1993), 381, 382, 389, 395.

2. Quoted in the *Sydney Morning Herald,* February 19, 2006.

3. Paul Ceruzzi, "An Unforeseen Revolution: Computers and Expectations, 1935–1985," in *Imagining Tomorrow: History, Technology, and the American Future,* ed. Joseph J. Corn (Cambridge, 1986), 191.

4. Quoted in Mike May, "How the Computer Got Into Your Pocket," *Invention and Technology,* 15 (Spring 2000), 49.

5. Quoted in James C. Williams, "Frederick E. Terman and the Rise of Silicon Valley," in *Technology in America: A History of Individuals and Ideas,* ed. Carroll W. Pursell, Jr. (2d ed., Cambridge, 1990), 284.

6. Williams, "Frederick E. Terman," 288.

7. The following account is taken from Bryan Pfaffenberger, "The Social Meaning of the Personal Computer; or, Why the Personal Computer Revolution Was No Revolution," *Anthropological Quarterly,* 61 (1988), 39. For Draper's version see his Web site at www.webcrunchers.com/crunch/.

8. Ibid.

9. Ibid., 44.

10. Quoted in Thomas P. Hughes, *Rescuing Prometheus* (New York, 1998), 265.

11. Quoted in ibid., 280.

12. Quoted in ibid., 292.

13. Quoted in Edwin Diamond and Stephen Bates, "The Ancient History of the Internet," *American Heritage,* 46 (October 1995), 45.

14. Quoted in Mary Bellis, "The History of the Sony Walkman," http://inventors.about.com/od/wstartinventions/a/Walkman_p.htm.

15. Quoted in the *New York Times,* April 13, 2000.

16. Quoted in the *Sydney Morning Herald,* November 17, 2005.

17. Frederick L. McKissack, Jr., "Cyberghetto: Blacks Are Falling through the Net," *The Progressive,* 62 (June 1998), 22.

CHAPTER 15: AMERICA'S GLOBAL REACH

1. H. Guyford Stever and Janet H. Muroyama, "Overview," in *Globalization of Technology: International Perspectives: Proceedings of the Sixth Convocation of The Council of Academies of Engineering and Sciences* (Washington, D.C., 1988), 1.

2. Mary Lowe Good, "Globalization of Technology Poses Challenges for Policymakers," *APS News,* July 1996 edition, www.aps.org/apsnews/0796/11568.html.

3. David Harvey, *The New Imperialism* (Oxford, 2003 [2005]), 221, 222.

4. Quoted in Mark Tran, "Green Energy Targets in Jeopardy, Report Warns," *The Guardian,* October 18, 2005.

5. Quoted in Vandana Shiva, *Stolen Harvest: The Hijacking of the Global Food Supply* (Cambridge, Mass., 2000), 82.

6. Quoted in Hugh R. Slotten, "Satellite Communications, Globalization, and the Cold War," *Technology and Culture,* 43 (April 2002), 336.

7. Simon Tisdall and Ewen MacAskill, "America's Long War," *The Guardian Digital Edition,* February 15, 2006, www.guardian.co.uk/print/0,,5399823–110878,00.html.

8. Tisdall and MacAskill, "America's Long War."

9. Michael T. Klare, "Imperial Reach: The Pentagon's New Basing Strategy," *The Nation,* 280 (April 25, 2005), 17.

10. John Gentry, "Doomed to Fail: America's Blind Faith in Military Technology," *Parameters* (Winter 2002–3), 88, 91.

11. C. P. Snow, *The Two Cultures and the Scientific Revolution* (New York, [1959] 1961), 48.

12. Quoted in David Morley and Kevin Robins, "Techno-Orientalism: Futures, Foreigners and Phobias," *New Formation,* 16 (Spring 1992), 141.

13. Ibid., 154.

14. Arundhati Roy, *Power Politics* (Cambridge, Mass., 2001), 2, 14.

15. *Science,* 266 (December 9, 1994), 1623.

16. Iain Boal et al., *Afflicted Powers: Capital and Spectacle in a New Age of War* (London, 2005), 26.

17. David E. Nye, *American Technological Sublime* (Cambridge, Mass., 1994), xiii; Burke quoted on 9.

18. Michael Adas, *Dominance by Design: Technological Imperatives and America's Civilizing Mission* (Cambridge, Mass., 2006), 387.

19. Carol Cohn, "Sex and Death in the Rational World of Defense Intellectuals," *Signs,* 12 (Summer 1987), 707.

FURTHER READING

THE BOOKS LISTED BELOW are the ones that I found most useful in writing this history of American technology. I have limited myself to published material, almost all of it monographic. The list is, of course, far from exhaustive. The most complete record of publications dealing with the history of technology appears annually in the journal *Technology and Culture.* An index to the first twenty-five volumes of that indispensable journal was published in 1991. An older but still useful listing is found in Eugene S. Ferguson, *Bibliography of the History of Technology* (1968). Brooke Hindle, *Technology in Early America: Needs and Opportunities for Study* (1966), has not been superseded for its bibliography and fine introductory essay. Carroll Pursell, ed., *A Companion to American Technology* (2005), contains twenty-two authoritative essays on aspects of the subject: the first two are on the colonial period and the nineteenth century, while the others concentrate on the twentieth century. Judy Wajcman, *Feminism Confronts Technology* (1991), while not dealing specifically with American technology, is a useful introduction to the way gender and technology interact generally.

GENERAL THEMES

Brooke Hindle and Steven Lubar, *Engines of Change: The American Industrial Revolution, 1790–1860* (1986), is the best survey available of the early years of technology in America and can be supplemented with Thomas P. Hughes, *American Genesis: A Century of Invention and Technological Enthusiasm* (1989). John F. Kasson, *Civilizing the Machine: Technology and Republican Values in America, 1776–1900* (1976), is a classic account from an American studies perspective. The giant in this category, of course, is still Leo Marx, *The Machine in the Garden: Technology and the Pastoral Ideal in America* (1946). Howard P. Segal, *Technological Utopianism in American Culture* (1985), is also standard. David E. Nye, *American Technological Sublime* (1994), is stimulating reading, and Merritt Roe Smith and Leo Marx, eds., *Does Technology Drive History? The Dilemma of Technological Determinism* (1994), is a good introduction to the subject.

Anne L. Macdonald, *Feminine Ingenuity: Women and Invention in America* (1992), is a not very satisfactory start on that important subject. Judy Wajcman, *Feminism Confronts Technology* (1991), although it does not concentrate on America, is useful for pointing out the issues. The best guides remain two brief pieces by Judith A. McGaw, "Women and the History of American Technology," *Signs*, 7 (Summer 1982), 798–828, and "No Passive Victims, No Separate Spheres: A Feminist Perspective on Technology's History," in *In Context: History and the History of Technology* (1989), ed. Stephen H. Cutcliffe and Robert C. Post, 172–91. Two books that survey the subject of technology and African-Americans are Bruce Sinclair, ed., *Technology and the African-American Experience: Needs and Opportunities for Study* (2004), and Carroll Pursell, ed., *A Hammer in Their Hands: A Documentary History of Technology and the African-American Experience* (2005). A brief survey is Portia R. James, *The Real McCoy: African-American Invention and Innovation, 1619–1930* (1989).

CHAPTER 1: THE TOOLS BROUGHT OVER

The best introductions to (British) colonial American technology remain Brooke Hindle, *The Pursuit of Science in Revolutionary America, 1735–1789* (1956), and Silvio A. Bedini, *Thinkers and Tinkers: Early American Men of Science* (1975). James Deetz, *In Small Things Forgotten: The Archeology of Early American*

Life (1977), is a classic study of material culture in the colonial era, and Brooke Hindle, ed., *America's Wooden Age: Aspects of Its Early Technology* (1975), is a collection of fine essays on various aspects of wood technology. Three books give us a good introduction to the metal industries in early America: Arthur Cecil Binning, *Pennsylvania Iron Manufacture in the Eighteenth Century* (1938), is still a standard; James A. Mulholland, *A History of Metals in Colonial America* (1981), covers copper as well as iron; and Joseph E. Walker, *Hopewell Village: A Social and Economic History of an Iron-Making Community* (1966), covers a late colonial Pennsylvania iron furnace that lasted well into the nineteenth century. Agriculture has been neglected, but William Cronon, *Changes in the Land: Indians, Colonists, and the Ecology of New England* (1983), is excellent for the time and place it covers. An important contribution to the field is Judith Ann Carney, *Black Rice: The African Origins of Rice Cultivation in the Americas* (2001).

Domestic technology has also been neglected, although Ruth Schwartz Cowan has made a start in the first pages of her classic *More Work for Mother: The Ironies of Household Technology from the Open Hearth to the Microwave* (1983). John McPhee, *The Survival of the Bark Canoe* (1975), while not a formal history, is a powerful and detailed evocation of that precious gift of Native American technology to the early European settlers. Patrick M. Malone, *The Skulking Way of War: Technology and Tactics among the New England Indians* (1991), explains both the genius of Native American warfare and why it proved inadequate. The end of the colonial era is the period discussed by Neil Longley York in *Mechanical Metamorphosis: Technological Change in Revolutionary America* (1985). Carl Bridenbaugh, *The Colonial Craftsmen* (1961), though more social than technological, is still useful. John Francis Bannon, *The Spanish Borderlands Frontier, 1513–1821* (1970), is a standard work. An excellent collection that goes beyond this period is Judith A. McGaw, ed., *Early American Technology: Making and Doing Things from the Colonial Era to 1850* (1994).

CHAPTER 2: IMPORTING THE INDUSTRIAL REVOLUTION

Darwin H. Stapleton, *The Transfer of Early Industrial Technologies to America* (1987), covers several of the important industries and skills brought to

America in the early nineteenth century. For steam engines see Carroll W. Pursell, *Early Stationary Steam Engines in America* (1969), and for textiles David J. Jeremy, *Transatlantic Industrial Revolution: The Diffusion of Textile Technologies between Britain and America, 1790–1830s* (1981). A classic account of one small early textile community in Pennsylvania is given in Anthony F. C. Wallace, *Rockdale: The Growth of an American Village in the Early Industrial Revolution* (1978).

CHAPTER 3: IMPROVING TRANSPORTATION

In part because of a long-term interest in the subject by economic historians, a good bit is known about early transportation in America. George Rogers Taylor, *The Transportation Revolution, 1815–1860* (1951), and Forest G. Hill, *Roads, Rails, and Waterways: The Army Engineers and Early Transportation* (1957), are still good starting points. W. Turrentine Jackson, *Wagon Roads West: A Study of Federal Road Surveys and Construction in the Trans-Mississippi West, 1846–1869* (1964), is primarily a study of policy. Louis C. Hunter, *Steamboats on the Western Rivers* (1949), is a classic that can be supplemented by James Thomas Flexner, *Steamboats Come True* (1944). John F. Stover, *American Railroads* (1961), is brief and general, but still a good introduction to the subject. A more literary approach is made in John Seelye, *Beautiful Machine: Rivers and the Republican Plan* (1991).

The subject seems particularly rich in biographical studies. F. Daniel Larkin, *John B. Jervis: An American Engineering Pioneer* (1990), deals with both canals and railroads, while Robert F. Hunter and Edwin L. Dooley, Jr., *Claudius Crozet: French Engineer in America, 1780–1864* (1989), also covers turnpike construction. Edith McCall, *Conquering the Rivers: Henry Miller Shreve and the Navigation of America's Inland Waterways* (1984), tells the story of the inventor of the snag boat. Thomas A. Kinney, *The Carriage Trade: Making Horse-Drawn Vehicles in America* (2004), carries this story to the coming of the automobile.

CHAPTER 4: THE EXPANSION OF AMERICAN MANUFACTURES

Merritt Roe Smith, *Harpers Ferry Armory and the New Technology: The Challenge of Change* (1977), is the key book for understanding the rise of the American system of manufacturing. Otto Mayr and Robert C. Post, eds., *Yankee Enterprise: The Rise of the American System of Manufactures* (1981), covers some aspects of the subject not addressed in Smith's monograph. Carolyn C. Cooper, *Shaping Invention: Thomas Blanchard's Machinery and Patent Management in Nineteenth-Century America* (1991), covers one key machine used in armory practice as well as illuminating the entire question of how patents were shaped and manipulated. Two of the classic contemporary descriptions of American manufactures at this time are introduced and reprinted in Nathan Rosenberg, ed., *The American System of Manufactures: The Report of the Committee on the Machinery of the United States 1855 and the Special Report of George Willis and Joseph Whitworth 1854* (1969). David A. Hounshell, *From the American System to Mass Production: The Development of Manufacturing Technology in the United States* (1964), is the standard treatment of that important subject. Further discussion can be found in Donald R. Hoke, *Ingenious Yankees: The Rise of the American System of Manufactures in the Private Sector* (1990).

Louis C. Hunter produced a massive three-volume study, *A History of Industrial Power in the United States, 1780–1930.* The first volume is subtitled *Waterpower* (1979), the second *Steam Power* (1985), and the third, written with Lynwood Bryant, *The Transmission of Power* (1991). A book that brings together the history of technology and the environment is Theodore Steinberg, *Nature Incorporated: Industrialization and the Water of New England* (1991).

Jeanne McHugh, *Alexander Holley and the Makers of Steel* (1980), covers the industry from the vantage point of one of its major players. Philip Scranton covers the Philadelphia textile industry in two volumes, *Proprietary Capitalism: The Textile Manufacture at Philadelphia, 1800–1885* (1983) and *Figured Tapestry: Production, Markets, and Power in Philadelphia Textiles, 1885–1941* (1989). Scranton's *Endless Novelty: Specialty Production and American Industrialization, 1865–1925* (1997), is an important corrective to the notion that mass production was inevitable and appropriate in all cases.

The literature on the engineering profession is impressive. An early contribution was Daniel H. Calhoun, *The American Civil Engineer: Origins and Conflicts* (1960). Two excellent books cover the mechanical engineer:

Monte A. Calvert, *The Mechanical Engineer in America, 1830–1910: Professional Cultures in Conflict* (1967), and Bruce Sinclair, *A Centennial History of the American Society of Mechanical Engineers, 1880–1980* (1980). Clark C. Spence, *Mining Engineers and the American West: The Lace-Boot Brigade, 1849–1933* (1970), and A. Michal McMahon, *The Making of a Profession: A Century of Electrical Engineering in America* (1984), cover those branches of the profession. Raymond H. Merritt, *Engineering in American Society, 1850–1875* (1969), includes a useful chapter on urban engineers, and Edwin T. Layton, *The Revolt of the Engineers: Social Responsibility and the American Engineering Profession* (1971), concentrates on the early twentieth century.

CHAPTER 5: THE MECHANIZATION OF FARMING

Agricultural history is a long-established field, and agricultural technology has been extensively covered. Now out of date, Vivian B. Whitehead, *A List of References for the History of Agricultural Technology* (1979), is still useful. For an excellent overview see Deborah Fitzgerald, "Beyond Tractors: The History of Technology in American Agriculture," *Technology and Culture,* 32 (1991), 114–26. William Cronon, *Changes in the Land: Indians, Colonists, and the Ecology of New England* (1983), provides a beginning for the colonial period, and Clarence H. Danhof, *Change in Agriculture: The Northern United States, 1820–1870* (1969), has information on technology and techniques as well.

Margaret W. Rossiter, *The Emergence of Agricultural Science: Justis Liebig and the Americans, 1840–1880* (1975), chronicles the bringing of a scientific ideal to this country from Germany and Britain. For more recent times, some contributions of science can be followed in Deborah Fitzgerald, *The Business of Breeding: Hybrid Corn in Illinois, 1890–1940* (1990). An informed attack on how scientific agriculture has worked itself out in the twentieth century can be found in Jim Hightower, *Hard Tomatoes Hard Times: A Report of the Agribusiness Accountability Project on the Failure of America's Land Grant College Complex* (1973).

Reynold Wik's classic *Steam Power on the American Farm* (1955) looks particularly at traction engines and threshers. A more recent study is J. Sanford Rikoon, *Threshing in the Midwest, 1820–1940: A Study of Traditional Culture and Technological Change* (1988). Barbed wire is covered in Henry F.

McCallum and Frances T. McCallum, *The Wire That Fences the West* (1965), and T. Lindsay Baker, *A Field Guide to American Windmills* (1985), is definitive. Earl W. Hayter, *The Troubled Farmer, 1850–1900: Rural Adjustment to Industrialism* (1968), described the ways in which farmers struggled to adapt to new technologies.

The tractor was the great universal machine on American farms, and Robert C. Williams has addressed this subject in his *Fordson, Farmall, and Poppin' Johnny: A History of the Farm Tractor and Its Impact on America* (1987). Reynold M. Wik, *Henry Ford and Grass-Roots America* (1972), covers the tractor, but much more as well. James H. Street, *The New Revolution in the Cotton Economy: Mechanization and Its Consequences* (1957), looks at the cotton-picking machine, and a provocative analysis of the impact of the lettuce-picking machines is given in William H. Friedland, Amy E. Barton, and Robert J. Thomas, *Manufacturing Green Gold: Capital, Labor, and Technology in the Lettuce Industry* (1981).

Deborah Fitzgerald, *Every Farm a Factory: The Industrial Ideal in American Agriculture* (2003), explains the ideological justification for the rise of agribusiness.

CHAPTER 6: CREATING AN URBAN ENVIRONMENT

The literature on technology and the city is scattered. Transportation, particularly the streetcar and automobile, has attracted some attention. Especially useful is Paul Barrett, *The Automobile and Urban Transit: The Formation of Public Policy in Chicago, 1900–1930* (1983). Clay McShane, *Technology and Reform: Street Railways and the Growth of Milwaukee, 1887–1900* (1974), is also useful, as is Scott L. Bottles, *Los Angeles and the Automobile* (1987). Howard L. Preston, *Automobile Age Atlanta: The Making of a Southern Metropolis, 1900–1935* (1979), is one of the few books to take race seriously in terms of technology. John R. Stilgoe, *The Metropolitan Corridor: Railroads and the American Scene* (1983), deals with the urban landscape, and Alan Trachtenberg, *Brooklyn Bridge: Fact and Symbol* (1965), is a classic. Arwen P. Mohun, *Steam Laundry: Gender, Technology, and Work in the United States and Great Britain, 1880–1940* (1999), is an excellent account of that urban amenity.

Nelson Blake, *Water for the Cities* (1956) is a pioneer study, and Fern L. Nesson, *Great Waters: A History of Boston's Water Supply* (1983), deals with one city in a similar vein. Harold L. Platt, *The Electric City: Energy and the Growth of the Chicago Area, 1880–1930* (1991), is the best study of a single city, while Sarah Pressey Noreen, *Public Street Illumination in Washington, D.C.: An Illustrated History* (1975), is perhaps unique. Harold L. Platt, *City Building in the New South: The Growth of Public Services in Houston, Texas, 1830–1915* (1983), is useful, while *Pollution and Reform in American Cities, 1870–1930* (1980), edited by Martin V. Melosi, contains essays on a number of problems associated with particular technologies. Joel A. Tarr and Gabriel Dupuy, eds., *Technology and the Rise of the Networked City in Europe and America* (1988), provides a comparative approach. S. J. Kleinberg, *The Shadow of the Mills: Working-Class Families in Pittsburgh, 1870–1907* (1989), has excellent material on the impact of providing (or withholding) utilities to mills and working-class neighborhoods. John W. Reps, *The Forgotten Frontier: Urban Planning in the American West before 1890* (1981), and James B. Allen, *The Company Town in the American West* (1966), are helpful. Dolores Hayden, *Building Suburbia: Green Fields and Urban Growth, 1820–2000* (2003), is an excellent overview.

CHAPTER 7: WESTWARD THE COURSE OF INDUSTRY

The pioneer attempt to assess the role of technology in the settling of the West is the classic by Walter Prescott Webb, *The Great Plains* (1931), but my own thinking has been most influenced by Alan Trachtenberg, *The Incorporation of America: Culture and Society in the Gilded Age* (1982). William Cronon, *Nature's Metropolis: Chicago and the Great West* (1991), is an innovative study of Chicago and its regional influence on the West, told in part in terms of technology.

The best survey of mining is still Rodman Wilson Paul, *Mining Frontiers of the Far West, 1848–1880* (1963). Three books that look at labor and technology in the industry are Otis E. Young, Jr., *Black Powder and Hard Steel: Miners and Machines on the Old Western Frontier* (1976); Mark Wyman, *Hard Rock Epic: Western Miners and the Industrial Revolution, 1860–1910* (1979); and Richard E. Lingenfelter, *The Hardrock Miners: A History of the Mining Labor Movement in the American West, 1863–1893* (1974). The Comstock district is the major focus of Robert E. Stewart, Jr., *Adolph Sutro: A Biography* (1962), and of particular use is Clark C. Spence, *Mining Engineers and the American West: The Lace-Boot Brigade, 1849–1933* (1970). A unique view of the subject is given by the wife of a mining engineer in Rodman W. Paul, ed., *A Victorian Gentlewoman in the Far West: The Reminiscences of Mary Hallock Foote* (1972).

Water problems in general, and irrigation in particular, are covered in Donald Worster, *Rivers of Empire: Water, Aridity and the Growth of the American West* (1985). Thomas F. Glick, *The Old World Background of the Irrigation System of San Antonio, Texas* (1972), covers one of the oldest systems in the country. William L. Kahrl, *Water and Power: The Conflict over Los Angeles' Water Supply in the Owens Valley* (1982), describes the search for water by a great city, while Donald J. Pisani, *From Family Farm to Agribusiness: The Irrigation Crusade in California and the West, 1850–1931* (1984), looks at the needs of farming interests. Michael C. Robinson, *Water for the West: The Bureau of Reclamation, 1902–1977* (1979), is a brief history of a key federal agency. Donald C. Jackson, *Building the Ultimate Dam: John S. Eastwood and the Control of Water in the West* (1995), follows the career and work of one of the most important engineers to work with western waters.

For logging and sawmilling see Thomas R. Cox, *Mills and Markets: A History of the Pacific Coast Lumber Industry to 1900* (1974). The best basic primer on land surveys is Marion Clawson, *The Land System of the United States: An Introduction to the History and Practice of Land Use and Land Tenure* (1968). A. Hunter Dupree, *Science in the Federal Government: A History of Policies and Activities to 1940* (1957), is the standard source for the involvement of the federal government in the exploration and surveying of the West, as in many other areas. Joan Didion, *Where I Was From* (2003), is a powerful exploration of the role of federal subsidies to technology in the development of California. James C. Williams, *Energy and the Making of Modern California* (1997), covers the subject from pioneer days to the present.

CHAPTER 8: EXPORT, EXPLOITATION, AND EMPIRE

This important topic has not received anything like the attention it deserves, although Michael Adas, *Dominance by Design: Technological Imperatives and America's Civilizing Mission* (2006), is a first-rate beginning. Joseph

Bradley, *Guns for the Tsar: American Technology and the Small Arms Industry in Nineteenth-Century Russia* (1990), is a pioneer effort to trace the role of the American system abroad. David McCullough still provides the best description of the Panama Canal in *The Path Between the Seas: The Creation of the Panama Canal, 1870–1914* (1977).

Gerald H. Nash, *The Life of Herbert Hoover: The Engineer, 1874–1914* (1983), is a good account of the future president's early years, many of them spent abroad. Merle Curti and Kendall Birr, *Prelude to Point Four: American Technical Missions Overseas, 1838–1938* (1954), is a neglected book on a neglected subject. Michael Gelb, ed., *An American Engineer in Stalin's Russia: The Memoirs of Zara Witkin, 1932–1934* (1991), is a window into that pivotal era.

CHAPTER 9: THE COMING OF SCIENCE AND SYSTEMS

The rise of industrial research is covered in a number of fine monographs. Paul Israel, *From Machine Shop to Industrial Laboratory: Telegraphy and the Changing Context of American Invention, 1830–1920* (1992), provides the necessary background on the shift to industrial research. George Wise, *Willis R. Whitney, General Electric, and the Origins of U.S. Industrial Research* (1985), deals with the leading laboratory, and a comparative view is provided by Leonard S. Reich, *The Making of American Industrial Research: Science and Business at GE and Bell, 1876–1926* (1985). The biography of one key person at GE is presented in Ronald R. Kline, *Steinmetz: Engineer and Socialist* (1992). David A. Hounshell and John Kenly Smith, Jr., *Science and Corporate Strategy: Du Pont R&D, 1902–1980* (1988), covers one of the giants. The biography of the research director at General Motors is told in Stuart W. Leslie, *Boss Kettering: Wizard of General Motors* (1983). An excellent account of events in the radio field is provided in Hugh G. J. Aitken, *The Continuous Wave: Technology and American Radio, 1900–1932* (1985). JoAnne Yates, *Control through Communication: The Rise of System in American Management* (1989), looks at the office rather than the laboratory.

Any study of scientific management should begin with a reading of Frederick Winslow Taylor's wonderful tract *The Principles of Scientific Management* (1911). The classic denunciation of the labor process under Taylorism is Harry Braverman, *Labor and Monopoly Capital: The Degradation of Work in the*

Twentieth Century (1974). Daniel Nelson, *Frederick W. Taylor and the Rise of Scientific Management* (1980), is a good introduction to both the man and his method. Hugh G. J. Aitken, *Scientific Management in Action: Taylorism at Watertown Arsenal, 1908–1915* (1960), covers a significant attempt to impose Taylorism at one site. The spread of Taylor's dream of "efficiency" beyond the shop is told in Samuel Haber, *Efficiency and Uplift: Scientific Management in the Progressive Era, 1890–1920* (1964). Martha Banta, *Taylored Lives: Narrative Productions in the Age of Taylor, Veblen, and Ford* (1993), sets the whole story in a rich cultural context. Samuel P. Hays, *Conservation and the Gospel of Efficiency: The Progressive Conservation Movement, 1890–1920* (1969), is still the best book on the subject.

CHAPTER 10: THE DECADE OF PROSPERITY AND CONSUMPTION

Two key government agencies of the 1920s are covered in Rexmond C. Cochrane, *Measure for Progress: A History of the National Bureau of Standards* (1966), and Alex Roland, *Model Research: The National Advisory Committee for Aeronautics, 1915–1958* (2 vols., 1985). Bruce E. Seely, *Building the American Highway System: Engineers as Policy Makers* (1987), traces federal involvement through the 1950s. James J. Flink, *The Automobile Age* (1988), is the best single volume on the car, and Scott L. Bottles, *Los Angeles and the Automobile: The Making of the Modern City* (1987), treats the most automobilized of all American cities. Virginia Scharff, *Taking the Wheel: Women and the Coming of the Motor Age* (1991), covers a neglected subject.

Jeffrey L. Meikle, *Twentieth Century Limited: Industrial Design in America, 1925–1939* (1979), traces the development of industrial design as a separate field. Terry Smith, *Making the Modern: Industry, Art, and Design in America* (1993), is also excellent. Joseph J. Corn, *The Winged Gospel: America's Romance with Aviation, 1900–1950* (1983), has much material on the 1920s. The telephone has been ably covered in Claude S. Fischer, *America Calling: A Social History of the Telephone to 1940* (1992). David E. Nye provides a fine account in his *Electrifying America: Social Meanings of a New Technology* (1990). Katherine Jellison, *Entitled to Power: Farm Women and Technology, 1913–1963* (1993), is splendid. Robert Friedel, *Zipper: An Exploration in Novelty* (1994), is a richly rewarding cultural history of this modest appliance.

Cecelia Tichi, *Shifting Gears: Technology, Literature, Culture in Modernist America* (1987), is the standard work on that subject, and Karen Lucic, *Charles Sheeler and the Cult of the Machine* (1991), deals with that great painter and photographer. The radio is well described in Susan Smulyan, *Selling Radio: The Commercialization of American Broadcasting, 1920–1934* (1994). John M. Jordan, *Machine-Age Ideology: Social Engineering and American Liberalism, 1911–1939* (1994), tracks the social scientists between the wars.

CHAPTER 11: DEPRESSION: STUDY AND SUBSIDY

Harry Jerome, *Mechanization in Industry* (1934), is still a good starting place for the period. Joseph E. Stevens, *Hoover Dam: An American Adventure* (1988), covers the history of what was surely the signature engineering project of the decade. *Rural Lines—USA: The Story of the Rural Electrification Administration's First Twenty-Five Years, 1935–1960* (1960) is a good starting place for the REA. Subcommittee on Technology to the National Resources Committee, *Technological Trends and National Policy, Including the Social Implications of New Inventions, June 1937* (1937), is a basic document of the time. Amy Sue Bix, *Inventing Ourselves Out of Jobs? America's Debate over Technological Unemployment, 1929–1981* (2000), follows a key controversy.

William E. Aitken, *Technocracy and the American Dream: The Technocracy Movement, 1900–1941* (1977), is standard. Thomas K. McCraw, *TVA and the Power Fight, 1933–1939* (1971), is one of the best books on the controversial project. Donald Worster, *Dust Bowl: The Southern Plains in the 1930s* (1979), while not purporting to be a history of technology, is the best introduction to that disaster. The chapters "Manly Work" and "Masculine Expertise" in Barbara Melosh, *Engendered Culture: Manhood and Womanhood in New Deal Public Art and Theater* (1991), address this neglected subject.

CHAPTER 12: WARS AND THE "AMERICAN CENTURY"

Herbert F. York, *Race to Oblivion: A Participant's View of the Arms Race* (1970), is an excellent perspective on the cold war arms race. Paul Boyer, *By the Bomb's Early Light: American Thought and Culture at the Dawn of the Atomic Age* (1985), covers topics from the Bikini tests to the "peaceful atom." Walter McDougall, . . . *The Heavens and the Earth: A Political History of the Space Age* (1985), is still excellent.

Ann Markusen et al., *The Rise of the Gun Belt: The Military Remapping of Industrial America* (1991), is a first-rate guide to the geographic (and of course human) consequences of the rise of some industries and regions. Stuart W. Leslie, *The Cold War and American Science: The Military-Industrial-Academic Complex at MIT and Stanford* (1993), helps explain how some regions and industries became privileged over others, and Jon C. Teaford, *Cities of the Heartland: The Rise and Fall of the Industrial Midwest* (1993), shows how some paid the price for that success. Carroll W. Pursell, Jr., *The Military-Industrial Complex* (1972), has both commentary and key documents on this subject.

The best historical study of automation in industry is David F. Noble, *Forces of Production: A Social History of Industrial Automation* (1984). Max Holland, *When the Machine Stopped: A Cautionary Tale from Industrial America* (1989), chronicles the rise and fall of one machine tool maker caught in shifting markets and leveraged buyouts.

Herbert I. Schiller, *Mass Communications and the American Empire* (1971), traces the international implications of America's communications reach, and Lynn Spigel, *Make Room for TV: Television and the Family Ideal in Postwar America* (1992), suggests some of the effects of that powerful technology. Dolores Hayden, *Redesigning the American Dream: The Future of Housing, Work, and Family Life* (1984), ties the design of postwar domestic spaces to related themes. Gail Cooper, *Air-conditioning America: Engineers and the Controlled Environment, 1900–1960* (1998), is authoritative. Adam Rome, *The Bulldozer in the Countryside: Suburban Sprawl and the Rise of American Environmentalism* (2001), is an important look at the technologies of suburban sprawl and their consequences. David Brodsly, *L.A. Freeway: An Appreciative Essay* (1981), is a short interpretive study. Christopher Finch, *Highways to Heaven: The Auto Biography of America* (1992), places the car in a rich social and cultural context. Robert C. Post, *High Performance: The Culture and Technology of Drag Racing, 1950–1990* (revised edition, 2001), is splendid.

CHAPTER 13: CHALLENGE AND CHANGE IN A POSTMODERN WORLD

An extremely helpful introduction to the concept of postmodernity is David Harvey, *The Condition of Postmodernity* (1989), and I have been strongly influenced by it. The book gives, among other things, a useful context in which to place James P. Womack et al., *The Machine That Changed the World* (1990), a good summary of mass production in the automobile industry and its evolution into "lean production" at Toyota and other Japanese firms. Otis L. Graham, Jr., *Losing Time: The Industrial Policy Debate* (1992), provides a policy context for the failure of the federal government to provide a coherent program for technological development. Langdon Winner, *The Whale and the Reactor: A Search for Limits in an Age of High Technology* (1986), is a welcome corrective to the great deal of nonsense that is written about contemporary technology. Janine Marie Morgall, *Technology Assessment: A Feminist Perspective* (1993), provides another view of that neglected subject. Andrew Ross, *Strange Weather: Culture, Science and Technology in the Age of Limits* (1991), is a provocative cultural critique of postmodern technology. Joseph J. Corn, ed., *Imagining Tomorrow: History, Technology, and the American Future* (1986), is the best introduction to why futurology seems to work so badly.

Michael Sorkin, ed., *Variations on a Theme Park: The New American City and the End of Public Space* (1992), is a provocative assessment of the urban environment. Steven Lubar, *InfoCulture: The Smithsonian Book of Information Age Inventions* (1993), is an excellent examination to the much-hailed subject. Sanford Lakoff and Hebert F. York provide a good introduction to Star Wars in *A Shield in Space? Technology, Politics, and the Strategic Defense Initiative* (1989), as does Frances FitzGerald in *Way Out There in the Blue: Reagan, Star Wars and the End of the Cold War* (2000). Star Wars is put into a larger cultural perspective in H. Bruce Franklin, *War Stars: The Superweapon and the American Imagination* (1988).

CHAPTER 14: OUR (UN)WIRED WORLD

Steven Lubar, *InfoCulture: The Smithsonian Book of Information Age Inventions* (1993), is a well-illustrated narrative covering the days of the telegraph through the computer. Michael Riordan and Lillian Hoddeson, *Crystal Fire:*

The Birth of the Information Age (1997), is a deft telling of the development of the transistor, and John Markoff, *What the Dormouse Said: How the 60s Counterculture Shaped the Personal Computer Industry* (2005), adds a fascinating dimension to that story. Paul N. Edwards, *The Closed World: Computers and the Politics of Discourse in Cold War America* (1996), shows how the political, military, and technical worlds of the cold war period converged. Thomas P. Hughes, *Rescuing Prometheus* (1998), contains an excellent chapter on networking. Ellen Ullman, *Close to the Machine: Technophilia and Its Discontents* (1997), is a wonderful memoir by a software programmer. The best detailed overview of the subject is Jeffrey R. Yost, "Computers and the Internet: Braiding Irony, Paradox, and Possibility," in *A Companion to American Technology,* ed. Carroll Pursell (2005), 340–60.

CHAPTER 15: AMERICA'S GLOBAL REACH

The closer historians work to the present, the fewer historical monographs they have to depend on and the more they have to rely on what are, in effect, journalistic or social scientific accounts written in the swirl of events. An excellent overview is provided by Michael Adas, *Dominance by Design: Technological Imperatives and America's Civilizing Mission* (2006).

David Harvey, *The New Imperialism* (2003), explains how America's power grew through the late twentieth century and how it has been deployed around the world, and Arundhati Roy, *Power Politics* (2001), describes how it works out in one country, India. Anthony Giddens, *Runaway World: How Globalization Is Reshaping Our Lives* (2000), is an assessment by an influential sociologist. An attempt to understand 9/11 and the new imperialism can be found in Iain Boal et al., *Afflicted Powers: Capital and Spectacle in a New Age of War* (2005). David Morley and Kevin Robins, "Techno-Orientalism: Futures, Foreigners, and Phobias," *New Formations,* 16 (Spring 1992), 136–56, presents a provocative argument. For an excellent discussion of the effectiveness of the Patriot missile in the Gulf War, see Harry Collins and Trevor Pinch, *The Golem at Large: What You Should Know about Technology* (1998).

Two critical looks at the impact of globalization on agriculture are Brewster Kneen, *Farmageddon: Food and the Culture of Biotechnology* (1999), and Vandana Shiva, *Stolen Harvest: The Hijacking of the Global Food Supply* (2000). Spencer Weart, *The Discovery of Global Warming* (2004), provides a fine history of the science behind this critical problem.

INDEX